THE WHOLE STORY BEHIND BLIND ADAPTIVE EQUALIZERS/ BLIND DECONVOLUTION

Dr. Monika Pinchas

Department of Electrical and Electronic Engineering, Ariel University Center of Samaria, Ariel 40700, ISRAEL

DEDICATION

I dedicate this book to my lovely husband Michael and to our daughters Maya, Merav and Moran, who are my inspiration in everything I do and every choice I make.

CONTENTS

FOREWORD

The phenomena of inter-symbol interference in band limited channels, pose severe limitations on the rate of data that can be transferred in digital communications. Equalizers are often being used in order to reduce inter-symbol interference and allow recovery of the transmitted symbols. Blind deconvolution techniques found to be efficient in carrying an online estimation of the equalizer filter, while relying on the transmitted signal statistics only. As a result, the recovered signal is obtained by convolving the impulse response of the estimated filter with the distorted received signal.

Over the years, many algorithms for the solution of the blind equalization problem have been developed. However, not much was written on their different features, explaining the advantages and drawbacks. This book not only explains the whole story behind blind adaptive equalization in such a way, that also a beginner in this field will feel comfortable with but also supplies new developments, new ideas where simulation results are supplied to support the theory. This book can be used as a teaching book and may be of particular interest to advanced undergraduate students, graduate ones, university instructors and research scientists in related disciplines.

Yosef Pinhasi
Dean, Faculty of Engineering
Ariel University Center of Samaria

PREFACE

The problem of blind deconvolution arises comprehensively in various applications such as digital communications, seismic signal processing, speech modeling and synthesis, ultrasonic nondestructive evaluation and image restoration. Many papers are written on blind adaptive equalizers using the cost function approach or Bayes rules. But, up to now there is no book available that makes order in the various techniques, namely, explains the differences, advantages, disadvantages and relationship between the cost function and Bayesian approach.

It is well known that the equalizer's tap length, step-size parameter, channel power and source signal have a great affect on the convergence speed as well as on the residual Intersymbol Interference (ISI). Thus, it is important to understand the relationship between the equalization performance and the various parameters involved in the system in order to obtain optimal equalization performance from the residual ISI point of view and convergence speed. In this book, a whole chapter deals with the relationship between the equalizer's parameters and the equalization performance. Closed-form approximated expressions are given for the mean square error (MSE) as a function of time (for the real valued and two independent quadrature carrier case and low ISI) and residual ISI for type of serially blind adaptive equalizers where the error that is fed into the adaptive mechanism which updates the equalizer's taps can be expressed as a polynomial function of order three of the equalized output.

The book explains why higher order statistics (HOS) are used in the blind equalization field and why HOS based blind equalization methods can not be applied to Gaussian sources. In addition, it explains the two different classes of HOS-based equalization methods used in the literature and supplies examples for each class.

The book describes the single input single output (SISO) and single input multiple output (SIMO) system where the condition for perfect equalization performance from the residual ISI point of view is given for each case.

In the literature, the convolutional noise is often assumed to be a Gaussian process. In this book we propose a new model for the convolutional noise probability density function (pdf). The new model is based on the Edgeworth expansion which is close related to the Gaussian model. Based on this new model for the convolutional noise pdf, a new closed-form approximated expression for the conditional expectation is obtained. It should be pointed out that the new derived expression for the conditional expectation, is also based on the Maximum Entropy approach, Laplace integral method and is valid for the noiseless, real valued and two independent quadrature carrier case. According to simulation results carried out for signal to noise ratio (SNR) $SNR = 30$ [dB], the new derived expression leads to improved equalization performance for the 16QAM input constellation case.

In summary, this book not only explains the whole story behind blind adaptive equalizers/blind deconvolution in such a way, that also a beginner in this field will feel comfortable with the book but also supplies new developments, new ideas where simulation results are supplied to support the theory and a simple Matlab program for a blind adaptive equalizer (Godard's algorithm) for the 16QAM constellation input which is sent via an easy channel (FIR channel). This book can be used as a teaching book and may be of particular interest to advanced undergraduate students, graduate students, university instructors and research scientists in related disciplines.

MONIKA PINCHAS

Department of Electrical and Electronic Engineering, Ariel University Center of Samaria, Ariel 40700, ISRAEL

E-mail: monika.pinchas@gmail.com

Chapter 1

INTRODUCTION

MONIKA PINCHAS

Department of Electrical and Electronic Engineering, Ariel University Center of Samaria, Ariel 40700, ISRAEL

ABSTRACT

In this chapter the blind deconvolution problem is defined, the different applications are given in which the blind deconvolution arises and the difference between the Polyspectral , Bussgang-type and Probabilistic algorithms is explained.

KEYWORDS

Blind deconvolution, Polyspectral algorithm, Bussgang-type algorithm, Probabilistic algorithm, FIR filter, Intersymbol Interference (ISI), non-blind equalizer, semi-blind equalizer, adaptive filtering algorithms, nonlinearity function

INTRODUCTION

We consider a blind deconvolution problem in which we observe the output of an unknown, possibly nonminimum phase, linear system from which we want to recover its input using an adjustable linear filter (equalizer) [1]. The problem of blind deconvolution arises comprehensively in various applications such as digital communications, seismic signal processing, speech modeling and synthesis, ultrasonic nondestructive evaluation, and image restoration [2].

It is well known that ISI (Intersymbol Interference) is a limiting factor in many communication environments where it causes an irreducible degradation of the bit error rate (BER) thus imposing an upper limit on the data symbol rate [3]. In order to overcome the ISI problem , an equalizer is implemented in those systems. Among the three types of equalizers: non-blind , semiblind and blind , the blind equalizer has the benefit of bandwidth saving and no need of going through a training phase. More about advantages as well as disadvantages of blind adaptive equalizers compared with its non-blind version may be found in chapter 10. It should be pointed out that training and pilot symbols (used in non-blind equalizers) have a significant benefit, and often performance for a block type (e.g., mobile wireless) system is substantially better than that achievable with a blind system. In some cases, a semi-blind equalizer can provide a good tradeoff and come close to performance that relies solely on training.

Blind equalization algorithms are essentially adaptive filtering algorithms designed such that they do not require the external supply of a desired response to generate the error signal in the output of the adaptive equalization filter [4]. The algorithm itself generates an estimate of the desired response by applying a nonlinear transformation to sequences involved in the adaptation process [4]. Very often, blind equalization algorithms are classified according to the location of their nonlinearity in the algorithm chain [3]. Proakis and Nikias [5], for example, distinguish between three different types:

1. Polyspectral algorithms
2. Bussgang-type algorithms
3. Probabilistic algorithms

In the first method the nonlinearity is located at the output of the channel, right before the equalizer filter. The nonlinearity has thus the function of estimating the channel and feeding that information to the equalizer for adapting the filter taps [3]. For further literature see for example [6] or [7].

In the second type the nonlinearity is situated at the output of the equalizer filter. Nonlinearities are often memoryless. Since Bussgang-type algorithms often have shorter convergence times than polyspectral methods , which need larger amounts of data for an equivalent estimation variance, they are more popular [3]. Among the traditional Bussgang-type algorithms we find Sato's [8] , Godard's [9], Benveniste et al. [10], Benveniste-Goursat's [11] and the Stop-and-Go [12] algorithm.

In the third class of algorithms, directly locating the nonlinearity is more problematic compared to the first two groups since the nonlinearity is combined with the data detection process [3]. While these algorithms can extract considerable information from relatively little data, this is often accomplished at a huge computational cost [3]. An excellent tutorial about such techniques and other blind equalization methods can be found in [13].

The nonlinearity is designed to minimize a cost function that is implicity based on higher order statistics (HOS) according to one approach [8, 9] or calculated directly according to the Bayes rules [1, 14, 15, 16]. The main difference between the Bussgang type algorithms lies in the choice of the memoryless nonlinearity [4]. Obviously, the performance of such kind of blind equalizer depends strongly on the memoryless nonlinearity .

In this book, we focus on serially blind adaptive equalizers (belonging to the Bussgang-type algorithms), which are more attractive in broadcast systems where data is streamed over a slowly varying channel although also block adaptive schemes are of strong interest in many scenarios, such as mobile wireless (e.g., cellular) where transmission is not continuous for lengthy data sizes.

The reader may refer to [17], [18], [19], [20], [21], [22] and [23] in order to consider how the early works on blind equalization have led to a number of more sophisticated algorithms for equalization and multiuser detection with applications to various systems.

In this book, the reader may not only find explanations but also new developments, new ideas where simulation results are supplied to support the theory.

ABBREVIATIONS

ISI= Intersymbol Interference

BER=   bit error rate

HOS=Higher Order Statistics

DISCLOSURE

Part of the information included in this chapter has been previously published in Signal Processing Volume 90, Issue 6, June 2010, Pages 1940-1962 and is reused here with permission from Elsevier.

REFERENCES

[1] M. Pinchas and B. Z. Bobrovsky, "A maximum entropy approach for blind deconvolution," *Signal Processing Journal (Eurasip)*, vol. 86, pp. 2913–2931, 2006.

[2] C. Feng and C. Chi, "Performance of cumulant based inverse filters for blind deconvolution," *IEEE Transaction on Signal Processing*, vol. 47, pp. 1922–1935, 1999.

[3] M. Pinchas, "A closed approximated formed expression for the achievable residual intersymbol interference obtained by blind equalizers," *Signal Processing Journal (Eurasip)*, vol. 90, pp. 1940–1962, 2010.

[4] C. L. Nikias and A. P. Petropulu, Eds., *Higher-Order Spectra Analysis A Nonlinear Signal Processing Framework*. Prentice-Hall, 1993.

[5] J. G. Proakis and C. L. Nikias, "Blind equalization (overview paper)," in *SPIE The International Society for Optical Engineering*, 1991, pp. 1565:76–87.

[6] S. Bellini and F. Rocca, Eds., *Blind deconvolution: Polyspectra or Bussgang techniques?, in Digital Communications*. Elsevier Science Publishers B.V., 1986.

[7] D. Hatzinakos and C. L. Nikias, Eds., *Blind equalization based on higher-order statistics (H.O.S.), in Blind Deconvolution (S. Haykin,ed.)*. Prentice Hall, 1994.

[8] Y. Sato, "A method of self-recovering equalization for multilevel amplitude-modulation systems," *IEEE Transaction on Communications*, vol. COM-23, pp. 679–682, 1975.

[9] D. N. Godard, "Self recovering equalization and carrier tracking in two-dimensional data communication systems," *IEEE Transaction on Communications*, vol. COM-28, pp. 1867–1875, 1980.

[10] M. G. A. Benveniste and G. Ruget, "Robust identification of a nonminimum phase system: Blind adjustment of a linear equalizer in data communications," *IEEE Transactions on Automatic Control*, vol. 25, pp. 385–399, 1980.

[11] A. Benveniste and M. Goursat, "Blind equalizers," *IEEE Transaction on Communications*, vol. COM-32, pp. 871–883, 1984.

[12] G. Picchi and G. Prati, "Blind equalization and carrier recovery using a 'stop-and-go' decision-directed algorithm," *IEEE Transaction on Communications*, vol. COM-35, pp. 877–887, 1987.

[13] L. T. J. K. Tugnait and Z. Ding, "Single-user channel estimation and equalization," *IEEE Signal Processing Magazine*, vol. 17, pp. 16–28, 2000.

[14] S. Bellini, "Bussgang techniques for blind equalization," in *IEEE Global Telecommunication Conference Records*, 1986, pp. 1634–1640.

[15] ——, "Blind equalization," *Alta Freq.*, vol. 57, pp. 445–450, 1988.

[16] S. Haykin, Ed., *Adaptive Filter Theory*. Prentice-Hall, Englewood cliffs,NJ, 1991.

[17] Z. Xu and M. Tsatsanis, "Adaptive minimum variance methods for direct blind multichannel equalization," *Signal Processing Journal (Eurasip)*, vol. 73, no. 1-2, pp. 125–138, Feb. 1999.

[18] Z. Xu and M. K. Tsatsanis, "Blind adaptive algorithms for minimum variance CDMA receivers," *IEEE Transaction on Communications*, vol. 49, no. 1, Jan. 2001.

[19] C. Xu, G. Feng and K. S. Kwak, "A Modified Constrained Constant Modulus Approach to Blind Adaptive Multiuser Detection," *IEEE Transaction on Communications*, vol. 49, no. 9, 2001.

[20] J. K. Tugnait and T. Li, "Blind detection of asynchronous CDMA signals in multipath channels using code-constrained inverse filter criterion," *IEEE Transaction on Signal Processing*, vol. 49, pp. 1300–1309, July 2001.

[21] Z. Xu and P. Liu, "Code-Constrained Blind Detection of CDMA Signals in Multipath Channels," *IEEE Signal Processing Letters*, vol. 9, no. 12, Dec. 2002.

[22] R. C. de Lamare and R. Sampaio-Neto, "Blind Adaptive Code-Constrained Modulus Algorithms for CDMA Interference Suppression in Multipath," *IEEE Communications Letters*, vol. 9, no. 4, pp. 334–336, Apr. 2005.

[23] R. C. de Lamare and R. Sampaio-Neto, "Blind adaptive MIMO receivers for space-time block-coded DS-CDMA systems in multipath channels using the constant modulus criterion," *IEEE Transaction on Communications*, vol. 58, no. 1, pp. 21–27, Jan. 2010.

Chapter 2

SYSTEM DESCRIPTION

MONIKA PINCHAS

Department of Electrical and Electronic Engineering, Ariel University Center of Samaria, Ariel 40700, ISRAEL

ABSTRACT

In this chapter, the Single Input Single Output (SISO) and Single Input Multiple Output (SIMO) system are described. For each system the condition for perfect equalization is given. In addition, the meaning of convolutional noise and Intersymbol Interference (ISI) often used as a measure of performance in equalizers' applications are explained.

KEYWORDS

Blind deconvolution, Single Input Single Output (SISO) system, Single Input Multiple Output (SIMO) system, perfect equalization, convolutional noise, Intersymbol Interference (ISI), equalizers's adaptation mechanism, nonlinear function, conditional expectation, FIR filter

FIR BASED CHANNEL AND EQUALIZER

In this chapter two different systems are going to be described. Namely, we will consider the Single Input Single Output (SISO) and Single Input Multiple Output (SIMO) case. For both cases, we wish to recover the sent sequence by using the serially adaptive blind equalizer's scheme. We start with the SISO case.

The SISO CASE

Let us consider the system illustrated in Fig. (**2.1**), where we make the following assumptions:

1. The input sequence $x[n]$ is a real or two independent quadrature carrier case constellation input with variance σ_x^2 where $x_1[n]$ and $x_2[n]$ are the real and imaginary parts of $x[n]$ respectively.

2. The unknown channel $h[n]$ is a possibly nonminimum phase linear time-invariant filter (FIR filter) in which the transfer function has no "deep zeros", namely, the zeros lie sufficiently far from the unit circle.

3. The equalizer $c[n]$ is a FIR filter.

4. The noise $w[n]$ is an additive Gaussian white noise with zero mean and variance $\sigma_w^2 = E[w[n]w^*[n]]$ ($E[\cdot]$ is the expectation operator and $()^*$ is the conjugate operation).

5. The function $T[\cdot]$ is a memoryless nonlinear function. It satisfies the following condition: $T[z[n] = T[z_1[n]] + jT[z_2[n]]$ where $z_1[n]$ and $z_2[n]$ are the real and imaginary parts of $z[n]$ respectively.

The sequence $x[n]$ is sent via the channel $h[n]$ and is corrupted with noise $w[n]$. The received sequence $y[n]$ at the input of the equalizer is defined as:

$$y[n] = x[n] * h[n] + w[n] \tag{2.1}$$

where "*" denotes the convolution operation. According to [1], we obtain for the ideal case:

$$z[n] = x[n - \tau]e^{j\phi} \tag{2.2}$$

where τ is a constant delay and ϕ is a constant phase shift. The expression in (2.2) implies that for the ideal case we have:

$$c[n] * h[n] = \delta[n - \tau]e^{j\phi} \tag{2.3}$$

where δ is the Kronecker delta function. In this book we assume, as is usually done in the literature ([2], [3], [4], [5]) that $\tau = 0$ and $\phi = 0$, since τ does not affect the reconstruction of the original input sequence $x[n]$ and ϕ can be removed by a decision device [1]. At the beginning of the deconvolution process, the ideal coefficients of the equalizer $c[n]$ are unknown. Usually, we initiate the equalizer by setting the center tap equal to one and all others to zero. Let us denote $\tilde{c}[n]$ as the equalizer with non-ideal coefficients. Thus we may write:

$$\tilde{s}[n] = \tilde{c}[n] * h[n] - \delta[n] + \zeta[n] \tag{2.4}$$

where $\xi[n]$ stands for the difference (error) between the ideal and non-ideal coefficients of $c[n]$ and $\tilde{c}[n]$ respectively. Convolving $\tilde{c}[n]$ with the received sequence $y[n]$ and using (2.1), we obtain:

$$z[n] = y[n] * \tilde{c}[n] =$$
$$x[n] * h[n] * \tilde{c}[n] + w[n] * \tilde{c}[n] \tag{2.5}$$

substituting (2.4) into (2.5) yields:

$$z[n] = x[n] * (\delta[n] + \xi[n]) + w[n] * \tilde{c}[n] =$$
$$x[n] + p[n] + \tilde{w}[n] \tag{2.6}$$

where $p[n]$ is the convolutional noise ($p[n] = x[n] * \xi[n]$), namely, the residual intersymbol interference (ISI) arising from the difference between $\tilde{c}[n]$ and $c[n]$ and $\tilde{w}[n] = w[n] * \tilde{c}[n]$. The ISI is often used as a measure of performance in equalizers' applications, defined by:

$$ISI[n] = \frac{\sum_{\tilde{m}} |\tilde{s}[\tilde{m}]|^2 - |\tilde{s}|^2_{max}}{|\tilde{s}|^2_{max}} \tag{2.7}$$

where $|\tilde{s}|_{max}$ is the component of $\tilde{s}$, given in (2.4), having the maximal absolute value. In order to understand (2.7), let us assume for a moment that the length of the channel $h[n]$ and equalizer with non ideal coefficients $\tilde{c}[n]$ is N_1 and N_2 respectively. Thus the length of $h[n] * \tilde{c}[n]$ is $N_1 + N_2 - 1$ which is described in (2.7) as $\tilde{m}$. Let us take the channel described in [6]:

$h_i[n] = 0$ for $i < 0$; -0.4 for $i = 0$ $0.84 \cdot 0.4^{i-1}$ for $i > 0$ and $\tilde{c}[n]$ initialized with the following coefficients:

$[0 \quad 0 \quad 0 \quad 0 \quad 0 \quad 0 \quad 1 \quad 0 \quad 0 \quad 0 \quad 0 \quad 0 \quad 0]$. The outcome of convolving $h[n]$ with $\tilde{c}[n]$ defined by $\tilde{s}[\tilde{m}]$ may be described in Fig. (**2.2**) where we named for convenience $\tilde{s}$ as "out". The ideal case of convolving $h[n]$ with $\tilde{c}[n]$ is when $\tilde{c}[n] = c[n]$, satisfying (2.3) ($\tau = 0$ and $\phi = 0$) is described in Fig. (**2.3**). Please note that if $|out|_{max} = 1$ or equivalently $|\tilde{s}|_{max} = 1$ it implies that there is no gain between the equalized output and input signal. Namely, the gain between the equalized output and input signal is equal to one. Let us go back to Fig. (**2.2**). Obviously, this is not the ideal case where we have a single tap with non-zero value while the other taps are set to zero . According to Fig. (**2.2**), we have several taps with non-zero values which implies having ISI . The amount of ISI is determined by (2.7). For the ideal case where we have a single tap with non-zero value while the other taps are set to zero, the ISI given by (2.7) is zero since the single tap is also the tap of $\tilde{s}$ having the maximal absolute value.

Next we turn to the adaptation mechanism of the equalizer. According to [4], [3], there are two approaches to derive this adaptation scheme . According to the first approach, we define some estimator for the input sequence $x[n]$, which is produced by the function $T[z[n]]$. Thus the error signal is:

$$e\,[n] = T\,[z\,[n]] - z\,[n] \qquad (2.8)$$

This error is fed into the adaptive mechanism which updates the equalizer's taps. In a compact filter vector notation, the updated equation for the taps of the equalizer looks quite similar to the LMS algorithm for the non-blind case [7]:

$$\underline{c}_{eq}[n+1] = \underline{c}_{eq}[n] + \mu \cdot e\,[n]\,\underline{y}^*\,[n] \qquad (2.9)$$

where $()^*$ is the conjugate operation, μ is the step-size parameter and $\underline{c}_{eq}[n]$ is the equalizer vector where the input vector is $\underline{y}[n] = [y[n] \ldots y[n-N+1]]^T$. The operator $()^T$ denotes for transpose of the function $()$ and N is the equalizer's tap length. The main concern in this approach is to develop a proper $T[z[n]]$ that will provide a good estimation of $x[n]$. The conditional expectation $(E[x[n]/z[n]]$ is touted as a good estimate of $T[z[n]]$ [1].

According to the second approach, a predefined cost function $F[n]$ that characterizes the ISI is defined, see [5], [6], [8] and [9]. Minimizing this $F[n]$ with respect to the equalizer parameters will reduce the convolutional error. Minimization is performed with the gradient descent algorithm that searches for an optimal filter tap setting by moving in the direction of the negative gradient $-\nabla_c F\,[n]$ over the surface of the cost function in the equalizer filter tap space [7]. Thus the updated equation is given by [7]:

$$\underline{c}_{eq}[n+1] = \underline{c}_{eq}[n] + \mu \cdot \left(-\nabla_{c_{eq}} F\,[n]\right) = \underline{c}_{eq}[n] - \mu \frac{\partial F\,[n]}{\partial z\,[n]} \underline{y}^*\,[n] \quad (2.10)$$

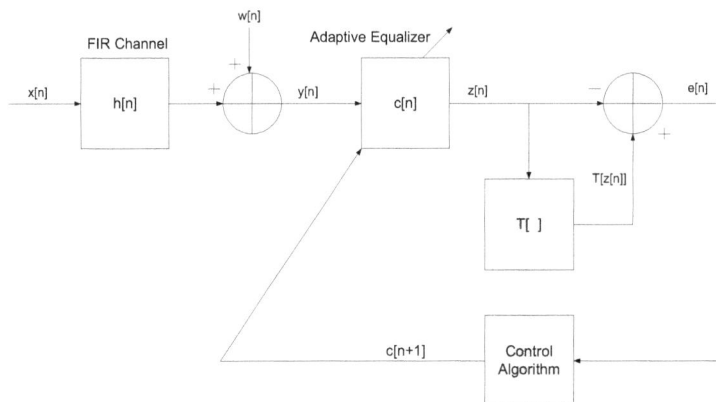

Figure 2.1: Block diagram of a baseband SISO communication system.

It should be pointed out, that the gradient of the function in (2.10) coincides with the partial derivative only if the parameter space (which is a vector space, in this case) is endowed with the standard Euclidean metric. If the space is a curved Riemannian manifold endowed with a different metric, the structure of the gradient may be very different.

The SIMO CASE

In this section we describe the SIMO system (Fig. **2.4**) where several serially blind adaptive equalizers are used. We use the following assumptions:

1.　The input sequence $x[n]$ belongs to a real or two independent quadrature carrier case constellation input with variance σ_x^2 where $x_1[n]$ and $x_2[n]$ are the real and imaginary parts of $x[n]$ respectively.

2.　The unknown subchannel $h^{(i)}[n]$ ($i = 1, 2, 3, ..., M$ where M is the number of subchannels) is a possibly nonminimum phase linear time-invariant filter. There is no common zero among all the subchannels.

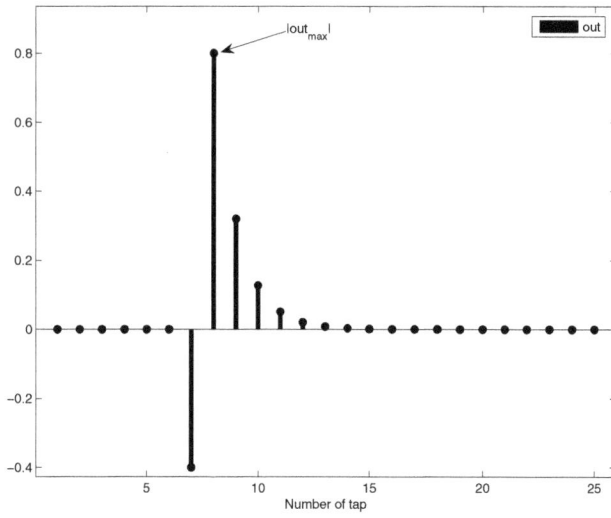

Figure 2.2: The output of $h[n] * \tilde{c}[n]$.

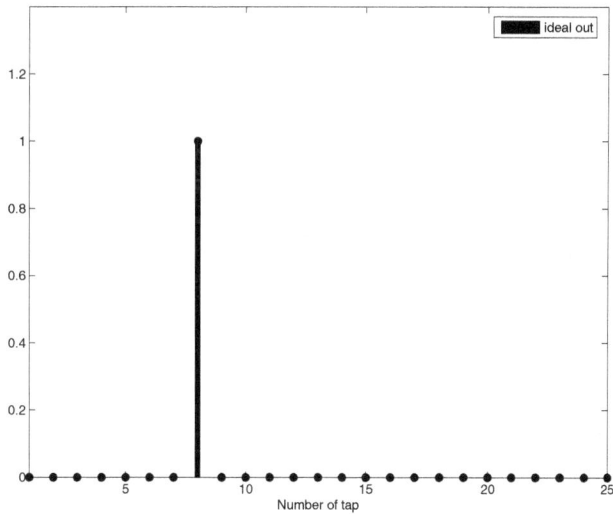

Figure 2.3: The ideal output of $h[n] * \tilde{c}[n]$ when $\tilde{c}[n] = c[n]$.

3. Each equalizer $c^{(i)}[n]$ $(i = 1, 2, 3, ..., M)$ is a tap-delay line.

4. The noise $w^{(i)}[n]$ $(i = 1, 2, 3, ..., M)$ is an additive Gaussian white noise with zero mean and variance $\sigma_w^2 = E[w[n]w^*[n]]$.

The sequence $x[n]$ is sent via M subchannels $h^{(i)}[n]$ $(i = 1, 2, 3, ..., M)$ and is corrupted at the output of each subchannel with additive noise $w^{(i)}[n]$. We may thus say, that the ith received signal $y^{(i)}[n]$ is the result of a linear convolution between the source signal $x[n]$ and the corresponding channel response $h^{(i)}[n]$, corrupted by an additive noise $w^{(i)}[n]$:

$$y^{(i)}[n] = x[n] * h^{(i)}[n] + w^{(i)}[n], \qquad i = 1, 2, 3, ..., M \qquad (2.11)$$

Now, by convolving $c^{(i)}[n]$ with the received sequence $y^{(i)}[n]$ and using (2.11), we obtain:

$$z^{(i)}[n] = y^{(i)}[n] * c^{(i)}[n] =$$

$$x[n] * h^{(i)}[n] * c^{(i)}[n] + w^{(i)}[n] * c^{(i)}[n], \qquad i = 1, 2, 3, ..., M$$
$$(2.12)$$

The equalizer's output is obtained from the sum of each subsystem's output by:

$$z[n] = \sum_{i=1}^{i=M} z^{(i)}[n] \qquad (2.13)$$

The overall impulse response of each subchannel is given by:

$$s^{(i)}[n] = c^{(i)}[n] * h^{(i)}[n], \qquad i = 1, 2, 3, ..., M \qquad (2.14)$$

where we may obtain according to (2.13) and [7] that the overall impulse response of the SIMO model is described by:

$$s[n] = \sum_{i=1}^{i=M} s^{(i)}[n] \qquad (2.15)$$

Perfect equalization according to [7] is described by the equation:

$$\sum_{i=1}^{i=M} s^{(i)}[n] = \delta(n-\tau)e^{j\phi} \qquad (2.16)$$

where δ is the Kronecker delta function. Please note that according to (2.16) not every individual subsystem must necessarily hold $s^{(i)}[n] = \delta(n-\tau)e^{j\phi}/M$ but the total sum of each subsystem's output must comply with (2.16). This outcome implies that each subsystem may have non perfect equalization performance even when the overall system reaches it, namely, complies with (2.16). This means that the requirement from each subsystem is relaxed. We assume here as was done in the previous section that $\tau = 0$ and $\phi = 0$. Thus, for the general case we can write (2.15) as:

$$s[n] = \delta[n] + \xi[n] \qquad (2.17)$$

where $\xi[n]$ stands for the error not having perfect equalization . Now, by using (2.13) and (2.17) we may write:

$$z[n] = x[n] + p[n] + \tilde{w}[n] \qquad (2.18)$$

where $p[n]$ is the convolutional noise $(p[n] = x[n] * \xi[n])$ and

$$\tilde{w}[n] = \sum_{i=1}^{i=M} w^{(i)}[n] * c^{(i)}[n] \qquad (2.19)$$

For the SIMO case, the update equation for each adaptive filter is quite similar to (2.10), (2.9) and is given by:

$$\underline{c}^{(i)}[n+1] = \underline{c}^{(i)}[n] + \mu^{(i)} \cdot (-\nabla_{c^{(i)}} F[n]) =$$

$$\underline{c}^{(i)}[n] - \mu^{(i)} \frac{\partial F[n]}{\partial z[n]} \underline{y}^{(i)*}[n] = \underline{c}^{(i)}[n] + \mu^{(i)} e[n] \underline{y}^{(i)*}[n] \qquad (2.20)$$

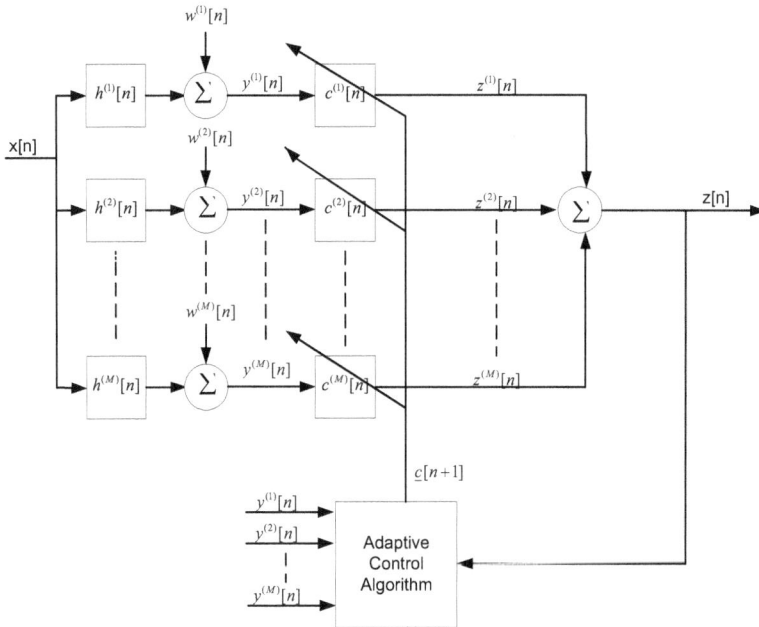

Figure 2.4: Block diagram of a baseband SIMO communication system.

where $\mu^{(i)}$ is the step-size parameter in the subchannel, $\underline{c}^{(i)}[n]$ is the equalizer vector where the input vector is $\underline{y}^{(i)}[n] = [y^{(i)}[n] \dots y^{(i)}[n - N + 1]]^T$ and N is the equalizer's tap length.

In the following chapters we consider only the SISO case.

ABBREVIATIONS

ISI=   Intersymbol Interference

SNR=   Signal to Noise Ratio

SISO=   Single Input Single Output

SIMO= Single Input Multiple Output

DISCLOSURE

Part of the information included in this chapter has been previously published in Signal, Image and Video Processing, DOI: 10.1007/s11760-011-0221-0 and is reused here with permission from Elsevier.

REFERENCES

[1] C. L. Nikias and A. P. Petropulu, Eds., *Higher-Order Spectra Analysis A Non-linear Signal Processing Framework*.   Prentice-Hall, 1993.

[2] M. Pinchas, "An analytical expression for the convergence time of adaptive blind equalizers," in *ICINCO-8th International Conference on Informatics in Control, Automation and Robotics*, 2011.

[3] ——, "A closed approximated formed expression for the achievable residual intersymbol interference obtained by blind equalizers," *Signal Processing Journal (Eurasip)*, vol. 90, pp. 1940–1962, 2010.

[4] M. Pinchas and B. Z. Bobrovsky, "A maximum entropy approach for blind deconvolution," *Signal Processing Journal (Eurasip)*, vol. 86, pp. 2913–2931, 2006.

[5] D. N. Godard, "Self recovering equalization and carrier tracking in two-dimensional data communication systems," *IEEE Transaction on Communications*, vol. COM-28, pp. 1867–1875, 1980.

[6] O. Shalvi and E. Weinstein, "New criteria for blind deconvolution of non-minimum phase systems (channels)," *IEEE Transaction on Information Theory*, vol. IT-36, pp. 312–321, 1990.

[7] A. K. Nandi, Ed., *Blind estimation using higher-order statistics*.   Boston: Kluwer Academic, 1999.

[8] M. Lazaro *et al.*, "Stochastic blind equalization based on pdf fitting using parzen estimator," *IEEE Trans. on Signal Processing*, vol. 53, pp. 696–704, 2005.

[9] G.-H. Im, C. J. Park, and H. C. Won, "A blind equalization with the sign algorithm for broadband access," *IEEE Comm. Letters*, vol. 5, pp. 70–72, 2001.

Chapter 3

THE COST FUNCTION APPROACH

MONIKA PINCHAS

Department of Electrical and Electronic Engineering, Ariel University Center of Samaria, Ariel 40700, ISRAEL

ABSTRACT

In this chapter the cost function approach for blind deconvolution is explained and the Bayesian estimation techniques are covered. We also show the relationship between the cost function and Bayesian approach and show a derivation from the cost function approach that was recently used successfully in literature.

In this chapter, we prove that for the real valued and two independent quadrature carrier cases, no blind adaptive equalizer can be found with perfect equalization performance from the residual Intersymbol Interference (ISI) point of view independent of the source signal statistics, where the error that is fed into the adaptive mechanism which updates the equalizer's taps can be expressed as a polynomial function of the equalized output of order three and where the gain between the equalized output and source signal is equal to one. In addition, we give in this chapter the entire mathematical basis that was missing in a recently presented paper where a new blind adaptive equalizer was introduced and shown to have an improved equalization performance compared with Godard's algorithm. We also show that the above mentioned blind adaptive equalizer was obtained by using a derivation from the cost function approach.

KEYWORDS

Blind deconvolution, Cost function approach, Bayesian estimation, Minimum Mean Square Error (MMSE) estimator, Bayesian approach, MAP (Maximum A-Posteriori) estimator, median estimator, conditional expectation, intersymbol interference (ISI), convolutional error

BAYESIAN ESTIMATION

In this section, we cover three different kinds of cost functions used in Bayesian estimation that give the MMSE (Minimum Mean Square Error), MAP (Maximum A-Posteriori) and median estimator . In addition, we also cover the main properties of the MMSE estimator. It should be pointed out that the MMSE estimator is used in the Bayesian approach (chapter 4). The reason for introducing the MMSE estimator in this chapter and not waiting for chapter 4 rests on the fact that the MMSE estimator and some of its properties are already described in the following section of this chapter.

We start our explanations by recalling the expression for the equalized output signal from chapter 2:

$$z[n] = x[n] + p[n] + \tilde{w}[n] \tag{3.1}$$

where $p[n]$ is the convolutional noise, $x[n]$ is the source signal and $\tilde{w}[n]$ is the output of the convolution between the channel noise and the equalizer. In the following we denote the sum of $(p[n] + \tilde{w}[n])$ as $\tilde{\tilde{w}}[n]$. Thus, we may rewrite (3.1) as:

$$z[n] = x[n] + \tilde{\tilde{w}}[n] \tag{3.2}$$

In this section we write $x[n]$, $z[n]$ and $\tilde{\tilde{w}}[n]$ as x, z and $\tilde{\tilde{w}}$ respectively for notation simplicity. In addition, we assume in this section that x

and z are real valued signals. We may denote $C\left(\widehat{x}\left(z\right),x\right)$ as a cost function that is a function of x and a function of the estimator $\widehat{x}\left(z\right)$ of x. Usually, the cost function is a function of the error arising between the estimator and the original signal. Thus, we may have the following cost function: $C\left(\epsilon\right)$ where $\epsilon = \widehat{x}\left(z\right) - x$. Now, the main idea here is to find the estimator $\widehat{x}\left(z\right)$ of x that will bring the following expression:

$$E\left[C\left(\widehat{x}\left(z\right),x\right)\right]$$
$$\text{or} \tag{3.3}$$
$$E\left[C\left(\epsilon\right)\right]$$

to minimum for every possible x and z where $E(\cdot)$ is the expectation operator.

As already mentioned earlier, we cover in this section three different kinds of cost functions. The first cost function is defined as:

$$C(\epsilon) = \epsilon^2 \tag{3.4}$$

which is described in Fig. (**3.1**). The second cost function is defined as:

$$C(\epsilon) = |\epsilon| \tag{3.5}$$

which is described in Fig. (**3.2**). The function $|(\cdot)|$ takes the absolute value of $(\cdot)$. The third cost function is defined as:

$$C(\epsilon) = 0 \quad \text{for} \quad |\epsilon| < \Delta \quad \text{and} \quad 1 \quad \text{for} \quad |\epsilon| > \Delta \tag{3.6}$$

which is described in Fig. (**3.3**). It should be pointed out that Δ is a positive small parameter. Next, we find the estimator that leads $E\left[C(\epsilon)\right]$ to minimum. We start with the cost function defined in (3.4).

$$E\left[\left(\widehat{x}\left(z\right) - x\right)^2\right] = E_z\left[E_x\left[\left(\widehat{x}\left(z\right) - x\right)^2/z\right]\right] \tag{3.7}$$

where E_z and E_x are the expectation with respect to z and x respectively. It should be pointed out that the estimator $\widehat{x}\left(z\right)$ that will lead

(3.7) to minimum is named as the MMSE. If we can find an estimator that leads the inner expectation $(E_x\left[\left(\widehat{x}\left(z\right)-x\right)^2/z\right])$ in (3.7) to minimum, then the same estimator will also lead the whole expression in (3.7) to minimum. Therefore, we concentrate on the inner expectation in (3.7).

$$E_x\left[\left(\widehat{x}\left(z\right)-x\right)^2/z\right] = E\left[\widehat{x}^2\left(z\right)/z\right] - 2E\left[\widehat{x}\left(z\right)x/z\right] + E\left[x^2/z\right] \quad (3.8)$$

Next we take the derivation of (3.8) with respect to $\widehat{x}\left(z\right)$ and set the output to zero:

$$\frac{\partial E_x\left[\left(\widehat{x}\left(z\right)-x\right)^2/z\right]}{\partial\widehat{x}\left(z\right)} = 2\widehat{x}\left(z\right) - 2E\left[x/z\right] = 0 \quad (3.9)$$

From (3.9), we obtain that the MMSE estimator is actually the conditional expectation:

$$\widehat{x}_{\mathrm{MMSE}} = \widehat{x}\left(z\right) = E\left[x/z\right] \quad (3.10)$$

The MMSE estimator has the following properties:
1.
$$E\left[\widehat{x}_{\mathrm{MMSE}}\right] = E\left[E\left[x/z\right]\right] = E\left[x\right]$$

$$\Downarrow \quad\quad\quad\quad (3.11)$$

$$E\left[\widehat{x}_{\mathrm{MMSE}} - x\right] = 0$$

which means that the MMSE estimator is unbiased.
2. The orthogonality principle:
The estimation error of the optimal estimator in the MMSE sense is orthogonal to any function of the uncorrelated measurements:

$$E\left[g\left(z\right)\left(\widehat{x}_{\mathrm{MMSE}} - x\right)\right] = 0 \quad (3.12)$$

Next, the estimator that leads $E[C(\epsilon)]$ to minimum where $C(\epsilon)$ is given in (3.5) is the median estimator defined by:

$$\int_{-\infty}^{\widehat{x}(z)} f(x/z)\, dx = \int_{\widehat{x}(z)}^{\infty} f(x/z)\, dx \qquad (3.13)$$

where $f(x/z)$ is the conditional probability density function (pdf) of the source signal given the measurement. In addition, the estimator that leads $E[C(\epsilon)]$ to minimum where $C(\epsilon)$ is given in (3.6) is the MAP estimator. The MAP estimator searches for the maximum point of $f(x/z)$.

COST FUNCTION APPROACH

We have already seen in the previous chapter, chapter 2, that in the field of blind equalization, there are two approaches for updating the equalizer's coefficients. According to one approach, a predefined cost function $F[n]$ (thus the name of cost function approach) that characterizes the intersymbol interference (ISI) is defined, see [1], [2], [3] and [4]. Minimizing this $F[n]$ with respect to the equalizer parameters will reduce the convolutional error. Minimization is performed with the gradient descent algorithm that searches for an optimal filter tap setting by moving in the direction of the negative gradient $-\nabla_c F[n]$ over the surface of the cost function in the equalizer filter tap space [5]. Thus, the updated equation is given by [5]:

$$\underline{c}_{eq}[n+1] = \underline{c}_{eq}[n] + \mu \cdot \left(-\nabla_{c_{eq}} F[n]\right) = \underline{c}_{eq}[n] - \mu \frac{\partial F[n]}{\partial z[n]} \underline{y}^*[n] \quad (3.14)$$

where $()^*$ is the conjugate operation, μ is the step-size parameter and $\underline{c}_{eq}[n]$ is the equalizer vector where the input vector is $\underline{y}[n] = [y[n]\ldots y[n-N+1]]^T$. The operator $()^T$ denotes the transpose of the function $()$ and N is the equalizer's tap length. According to the second approach, we may define some estimator for the input sequence $x[n]$, which is produced by the function $T[z[n]]$. Thus, the error signal is:

$$e\,[n] = T\,[z\,[n]] - z\,[n] \tag{3.15}$$

This error is fed into the adaptive mechanism which updates the equalizer's taps. In a compact filter vector notation, the updated equation for the taps of the equalizer looks quite similar to the LMS algorithm for the non-blind case [5]:

$$\underline{c}_{eq}[n+1] = \underline{c}_{eq}[n] + \mu \cdot e\,[n]\,\underline{y}^{*}\,[n] \tag{3.16}$$

According to [6], the conditional expectation $(E[x[n]/z[n]])$ is touted as a good estimate of $T[z[n]]$ which we have shown in the previous section to be the MMSE estimator.

Note that by (3.15), (3.16) and (3.14):

$$T\,(z\,[n]) = z\,[n] - \frac{\partial F\,[n]}{\partial z\,[n]} \tag{3.17}$$

thus, choosing the cost function $F'[n]$ results in a corresponding choice of $T(z[n])$ [5].

In the previous section, we have shown for the real valued case, three different kinds of cost functions used in the Bayesian estimation. In all those cases, the conditional probability density function (pdf) $f(x/z)$ has to be known. But since we do not have in hand the source pdf $f(x)$, we also do not have or know the conditional pdf $f(x/z)$. Thus, we cannot derive a closed-form analytical expression for any of the estimators presented in the previous section especially not for the MMSE estimator. It should be pointed out that for the general complex case we have the same problem of not knowing the conditional pdf. Therefore, we have in literature many blind equalizers based on the cost function approach while only few techniques based on the Bayesian approach (please refer to chapter 4) where in those cases, the conditional expectation is calculated by assuming that we have a uniformly distributed source signal $x[n]$ within $[-\sqrt{3}, +\sqrt{(3)}]$ [7], [8], [9] or the source pdf is approximated with the Edgeworth expansion or Maximum Entropy density technique [10], [11].

A DERIVATION OF THE COST FUNCTION APPROACH

In the cost function approach (as we have seen in the previous section), we are looking for a cost function $F[n]$ that characterizes the ISI . We try to minimize this $F[n]$ with respect to the equalizer parameters in order to reduce the convolutional error. Thus, each new cost function may lead to a new blind equalizer. But, alternatively according to (3.14), we could also search for such a function $\partial F[n]/\partial z[n]$ that leads to satisfying equalization performance from the residual ISI point of view. In order to find such a function $\partial F[n]/\partial z[n]$ that leads to satisfying equalization performance from the residual ISI point of view, we first must learn more of the nature of the function. For that reason we rewrite (3.17) and take the expectation operator to obtain:

$$E\left[T\left(z\left[n\right]\right) - z\left[n\right]\right] = -E\left[\frac{\partial F\left[n\right]}{\partial z\left[n\right]}\right] \qquad (3.18)$$

But when we substitute the conditional expectation $(E[x[n]/z[n]])$ for $T(z[n])$ into (3.18) and take into account (3.11) we obtain:

$$E\left[\frac{\partial F\left[n\right]}{\partial z\left[n\right]}\right] = 0 \qquad (3.19)$$

Now, suppose the function $\partial F[n]/\partial z[n]$ is a polynomial function of order "K". Then by (3.19) we know that if the function has only odd orders of the equalized output $z[n]$, (3.19) is complied assuming $E[z[n]] = 0$. Please note that in Godard's algorithm [1], the function

$\partial F[n]/\partial z[n]$ is a polynomial function of order three (please refer to Table (3.1)) where we have only the first and third order of the equalized output $z[n]$. But it should be pointed out that the algorithm was not obtained via the idea described above. In the following we will show some blind adaptive equalizers which were derived based on the idea described above.

We start with the blind adaptive equalizers presented in [12] valid for the 16QAM (Quadrature Amplitude Modulation) and 64QAM input case. Those equalizers were derived for the noiseless case but simulation results have shown that they work well even down to SNR=10 [dB] and SNR=20 [dB] for the 16QAM and 64QAM case respectively. In this paper [12], the function $\partial F[n]/\partial z[n]$ was chosen as a polynomial function where the coefficients were optimized with the mean square error (MSE) criteria [12]:

$$\frac{\partial F_1[n]}{\partial z_1[n]} = a_1 z_1^3[n] + b_1 z_1^5[n] - z_1[n] \qquad \text{for 16QAM}$$

$$\frac{\partial F_2[n]}{\partial z_1[n]} = a_2 z_1^3[n] + b_2 z_1^5[n] + d_2 z_1^7[n] - z_1[n] \qquad \text{for 64QAM}$$

$$(3.20)$$

where $z_1[n]$ is the real part of $z[n] = x[n] + p[n]$, a_1, b_1 and a_2, b_2, d_2 are the coefficients of the polynomial function $\partial F_1[n]/\partial z_1[n]$ and

$\partial F_2\left[n\right]/\partial z_1\left[n\right]$ respectively. The MSE was defined as [12]:

For 16QAM

$$MSE_1 = E\left[\left(T_1\left(z_1[n]\right) - x_1[n]\right)^2\right] =$$

$$E\left[\left(z_1[n] - \frac{\partial F_1\left[n\right]}{\partial z_1\left[n\right]} - x_1[n]\right)^2\right] \cong MSE_1^A + f\left(\sigma_p^2\right)$$

$$(3.21)$$

For 64QAM

$$MSE_2 = E\left[\left(T_2\left(z_1[n]\right) - x_1[n]\right)^2\right] =$$

$$E\left[\left(z_1[n] - \frac{\partial F_2\left[n\right]}{\partial z_1\left[n\right]} - x_1[n]\right)^2\right] \cong MSE_2^A + g\left(\sigma_p^2\right)$$

where $x_1[n] = \pm1, \pm3$ is for the 16QAM case while $x_1[n] = \pm1, \pm3, \pm5, \pm7$ is for the other case (64QAM). In chapter 9 we show that the MSE is a linear function of the ISI for the case where the gain between the equalized and input signal is equal to one as it is in our case (please see (3.21). Therefore, searching for a lower MSE is equivalent to searching a lower residual ISI. MSE_1^A and MSE_2^A are independent of the convolutional noise power but are a function of the source statistics and of the coefficients given in (3.20). $f\left(\sigma_p^2\right)$ and $g\left(\sigma_p^2\right)$ are two functions that depend on the convolutional noise power σ_p^2. The coefficients a_1, b_1, a_2, b_2 and d_2 were obtained by solving the following equations [12]:

$$\frac{\partial MSE_1^A}{\partial a_1} = 0; \qquad \frac{\partial MSE_1^A}{\partial b_1} = 0$$

$$(3.22)$$

$$\frac{\partial MSE_2^A}{\partial a_2} = 0; \qquad \frac{\partial MSE_2^A}{\partial b_2} = 0; \qquad \frac{\partial MSE_2^A}{\partial d_2} = 0$$

Next, we turn to another blind adaptive equalizer recently presented in [13] which was shown there having an improved equalization performance (faster convergence speed and lower residual ISI) compared with Godard's [1] algorithm for the 16QAM and 64QAM input case. It should be pointed out that this new blind equalizer was presented in [13] without any indication how it was obtained since it was not the purpose of that paper [13]. In the following we will first present again the above mentioned equalizer from [13] and then connect this equalizer to the approach of finding a function of $\partial F [n]/\partial z [n]$ that will lead to a low residual ISI rather than finding a cost function $F[n]$ that should be later minimized with the respect to the equalizer's co-efficients.

The blind adaptive equalizer is given by [13]:

$$c_m [n + 1] = c_m [n] - \mu_N \cdot$$
$$\left(\frac{E\left[(x_r [n])^4\right]}{E\left[(x_r [n])^6\right]} z_r^3 [n] + j \frac{E\left[(x_r [n])^4\right]}{E\left[(x_r [n])^6\right]} z_i^3 [n] - z [n] \right) y^* [n - m]$$

$$(3.23)$$

where μ_N is the step-size parameter , $x_r[n]$ and $z_r[n]$ are the real parts of $x[n]$ and $z[n]$ respectively. Next, we give the mathematical basis that leads to (3.23).

Recently [14], a closed-form approximated expression was obtained for the achievable residual ISI for the noiseless, real valued and two independent quadrature carrier cases applicable for the type of blind equalizers where the error that is fed into the adaptive mechanism which updates the equalizer's taps can be expressed as a polynomial function of the equalized output up to order three and where the gain between the equalized output and source signal is equal to one:

$$ISI = 10 \log_{10} (m_p) - 10 \log_{10} \left(\sigma_{x_r}^2\right) \qquad (3.24)$$

where $\sigma_{x_r}^2$ is the variance of the real part of the input sequence $x[n]$ and m_p is defined by:

$$m_p = \min\left[Sol_1^{mp_1}, Sol_2^{mp_1}\right] \quad \text{for} \quad Sol_1^{mp_1} > 0 \quad \text{and} \quad Sol_2^{mp_1} > 0$$

or

$$m_p = \max\left[Sol_1^{mp_1}, Sol_2^{mp_1}\right] \quad \text{for} \quad Sol_1^{mp_1} \cdot Sol_2^{mp_1} < 0$$

where

$$Sol_1^{mp_1} = \frac{-B_1 + \sqrt{B_1^2 - 4A_1 C_1 B}}{2A_1}; \quad Sol_2^{mp_1} = \frac{-B_1 - \sqrt{B_1^2 - 4A_1 C_1 B}}{2A_1}$$

$$(3.25)$$

$$A_1 = B\left(45\sigma_{x_r}^2 a_3^2 + 18\sigma_{x_r}^2 a_3 a_{12} + 6a_1 a_3 + 9\sigma_{x_r}^2 a_{12}^2 + 2a_1 a_{12}\right) - 2\left(3a_3 + a_{12}\right)$$

$$B_1 = B\left(12\left(\sigma_{x_r}^2\right)^2 a_3 a_{12} + 6\left(\sigma_{x_r}^2\right)^2 a_{12}^2 + 12\sigma_{x_r}^2 a_1 a_3 + 4\sigma_{x_r}^2 a_1 a_{12} + a_1^2 + 15E\left[x_r^4\right] a_3^2 + 2E\left[x_r^4\right] a_3 a_{12} + E\left[x_r^4\right] a_{12}^2\right) - 2\left(a_1 + 3\sigma_{x_r}^2 a_3 + \sigma_{x_r}^2 a_{12}\right)$$

$$C_1 = 2\left(\sigma_{x_r}^2\right)^2 a_1 a_{12} + \sigma_{x_r}^2 a_1^2 + 2E\left[x_r^4\right]\sigma_{x_r}^2 a_3 a_{12} + E\left[x_r^4\right]\sigma_{x_r}^2 a_{12}^2 + 2E\left[x_r^4\right] a_1 a_3 + E\left[x_r^6\right] a_3^2$$

$$B = \mu N \sigma_x^2 \sum_{k=0}^{k=R-1} |h(k)|^2$$

$$(3.26)$$

$x_r = x_r[n]$, R is the channel length, N is the equalizer's tap length and a_1, a_{12}, a_3 are properties of the chosen equalizer and found by:

$$Re\left(\frac{\partial F[n]}{\partial z[n]}\right) = \left(a_1(z_r) + a_3(z_r)^3 + a_{12}(z_r)(z_i)^2\right) \qquad (3.27)$$

where $Re(\cdot)$ is the real part of $(\cdot)$ and z_r, z_i are the real and imaginary parts of the equalized output $z[n]$ respectively. According to [14], the convolutional noise power $\sigma_p^2 = m_p$ tends to zero when C_1 given in (3.26) is zero. Thus, we are looking in the following for those coefficients a_1, a_{12}, a_3 that will lead C_1 given in (3.26) to minimum. In other words, we have to solve the following equations:

$$\frac{\partial C_1}{\partial a_1} = 2a_{12} \left(\sigma_{x_r}^2\right)^2 + 2a_1 \sigma_{x_r}^2 + 2E\left[x_r^4\right] a_3 = 0$$

$$\frac{\partial C_1}{\partial a_{12}} = 2\sigma_{x_r}^2 \left(E\left[x_r^4\right] a_3 + \sigma_{x_r}^2 a_1 + E\left[x_r^4\right] a_{12}\right) = 0 \qquad (3.28)$$

$$\frac{\partial C_1}{\partial a_3} = 2E\left[x_r^6\right] a_3 + 2E\left[x_r^4\right] a_1 + 2E\left[x_r^4\right] \sigma_{x_r}^2 a_{12} = 0$$

which can be written by assuming $\sigma_{x_r}^2 \neq 0$ as:

$$a_{12} \left(\sigma_{x_r}^2\right)^2 + a_1 \sigma_{x_r}^2 + E\left[x_r^4\right] a_3 = 0$$

$$E\left[x_r^4\right] a_3 + \sigma_{x_r}^2 a_1 + E\left[x_r^4\right] a_{12} = 0 \qquad (3.29)$$

$$E\left[x_r^6\right] a_3 + E\left[x_r^4\right] a_1 + E\left[x_r^4\right] \sigma_{x_r}^2 a_{12} = 0$$

Now, by subtracting the first line with the second line from (3.29) we obtain:

$$a_{12} \left(\left(\sigma_{x_r}^2\right)^2 - E\left[x_r^4\right]\right) = 0 \qquad (3.30)$$

Obviously, $\left(\sigma_{x_r}^2\right)^2 - E\left[x_r^4\right] = 0$ for some cases such as for the quadrature phase shift keying (QPSK) input constellation case while for other cases such as for the 16QAM input case, $\left(\sigma_{x_r}^2\right)^2 - E\left[x_r^4\right] \neq 0$. Therefore, the solution for (3.30) is $a_{12} = 0$. Now, substituting $a_{12} = 0$ back into (3.29) leads to:

$$E\left[x_r^4\right] a_3 + \sigma_{x_r}^2 a_1 = 0$$

$$\qquad (3.31)$$

$$E\left[x_r^6\right] a_3 + E\left[x_r^4\right] a_1 = 0$$

From (3.31) we obtain:

$$a_3 = -\frac{E\left[x_r^4\right]}{E\left[x_r^6\right]} a_1 \qquad (3.32)$$

By using (3.32) and (3.31) we have:

$$a_1 \left(\sigma_{x_r}^2 - \frac{E\left[x_r^4\right]}{E\left[x_r^6\right]} E\left[x_r^4\right] \right) = 0 \tag{3.33}$$

Since, the expression of $(\sigma_{x_r}^2 - (E\left[x_r^4\right]/E\left[x_r^6\right])E\left[x_r^4\right])$ is not zero for every real valued and two independent quadrature carrier cases source signal, the solution for (3.33) is $a_1 = 0$ which leads according to (3.32) to $a_3 = 0$. Thus, the conclusion here is that there does not exist an equalizer of order three for which we get $C_1 = 0$ for every real valued and two independent quadrature carrier cases input. If this is the case we may go back to (3.32) and substitute for a_1 any value especially the value of -1 thus obtaining:

$$a_3 = -\frac{E\left[x_r^4\right]}{E\left[x_r^6\right]}; \qquad a_1 = -1; \qquad a_{12} = 0 \tag{3.34}$$

By using (3.34) and (3.27) we obtain the function $\partial F\left[n\right]/\partial z\left[n\right]$ given in (3.23) for the real valued and two independent quadrature carrier cases.

According to (3.31) we could also choose another value for a_3:

$$a_3 = -\frac{\sigma_{x_r}^2}{E\left[x_r^4\right]} a_1 \tag{3.35}$$

The obvious question that may arise here is which value for a_3 will lead to a lower value for C_1, the option given in (3.35) or given in (3.32). In order to answer that question we substitute both options for a_3 given in (3.35) and (3.32) in the expression for C_1 (3.26) and take into account that $a_{12} = 0$. Thus, obtaining:

$$C_{1\_1} = \sigma_{x_r}^2 a_1^2 - 2\frac{\left(E\left[x_r^4\right]\right)^2}{E\left[x_r^6\right]} a_1^2 + \frac{\left(E\left[x_r^4\right]\right)^2}{E\left[x_r^6\right]} a_1^2$$
$$\tag{3.36}$$
$$C_{1\_2} = \sigma_{x_r}^2 a_1^2 - 2\sigma_{x_r}^2 a_1^2 + \frac{E\left[x_r^6\right]\sigma_{x_r}^4}{\left(E\left[x_r^4\right]\right)^2} a_1^2$$

where C_{1_1} and C_{1_2} describe the function C_1 with $a_{12} = 0$ and a_3 given in (3.32) and (3.35) respectively. We start our derivations by assuming that $C_{1\_1} < C_{1\_2}$. Thus, by using (3.36) we may write:

$$\sigma_{x_r}^2 a_1^2 - 2 \frac{(E[x_r^4])^2}{E[x_r^6]} a_1^2 + \frac{(E[x_r^4])^2}{E[x_r^6]} a_1^2 < \sigma_{x_r}^2 a_1^2 - 2\sigma_{x_r}^2 a_1^2 + \frac{E[x_r^6] \sigma_{x_r}^4}{(E[x_r^4])^2} a_1^2$$

(3.37)

Next, we divide both sides of (3.37) by a_1^2 and then multiply both sides of the obtained inequality by $E[x_r^6]$ and obtain:

$$-\frac{(E[x_r^6])^2 \sigma_{x_r}^4}{(E[x_r^4])^2} + 2\sigma_{x_r}^2 E[x_r^6] - (E[x_r^4])^2 < 0 \qquad (3.38)$$

Please note that the following equation:

$$-\frac{(E[x_r^6])^2 \sigma_{x_r}^4}{(E[x_r^4])^2} + 2\sigma_{x_r}^2 E[x_r^6] - (E[x_r^4])^2 = 0 \qquad (3.39)$$

is a second order equation with respect to $E[x_r^6]$ with a single solution of:

$$E[x_r^6] - \frac{(E[x_r^4])^2}{\sigma_{x_r}^2} \qquad (3.40)$$

Since $-\dfrac{\sigma_{x_r}^4}{(E[x_r^4])^2} < 0$ and there exists only a single solution (3.40) for (3.39) we conclude that the inequality given in 3.38 is always true for the real valued and two independent quadrature carrier cases for input sources having non constant amplitude. Thus, we may conclude that the option of using a_3 defined in (3.32) will lead to a lower value for C_1 for sources having non constant amplitude , namely, will lead to a lower value for the residual ISI compared with the option when a_3 is taken according to (3.35). For constant amplitude sources such as for the QPSK or binary phase shift keying (BPSK) case, both options for a_3 ((3.35) or (3.32)) will lead to the same value for C_1.

In this chapter, we discussed three different kinds of blind adaptive equalizers based on the cost function approach. The reader may find in Table **(3.1)** the expression for the function $\partial F\left[n\right]/\partial z\left[n\right]$ corresponding to each equalizer.

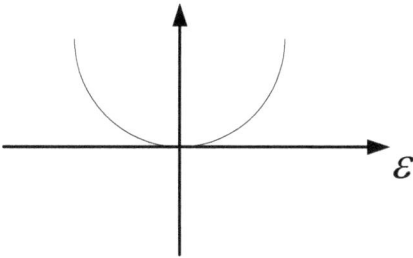

Figure 3.1: The value of the cost function as a function of ϵ

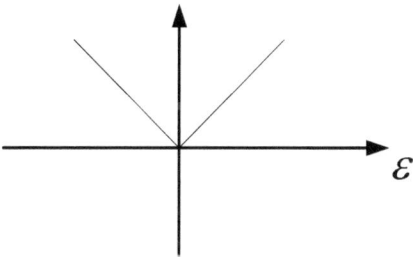

Figure 3.2: The value of the cost function as a function of ϵ

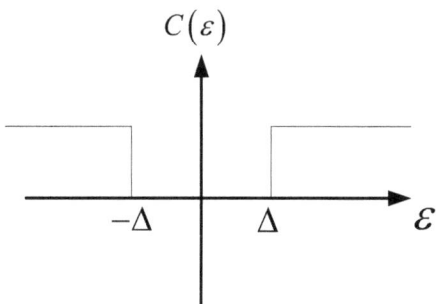

Figure 3.3: The value of the cost function as a function of ϵ

Chosen Equalizer	$\partial F[n]/\partial z[n]$
Godard [1]	$\left(|z[n]|^2 - \dfrac{E\left[|x[n]|^4\right]}{E\left[|x[n]|^2\right]} \right) z[n]$
Equalizer given in [12]	For the 16QAM case: $$-\left(\tfrac{10}{9}\right) z_r^3(n) - \left(\tfrac{-1}{9}\right) z_r^5(n)+$$ $$j\left(-\left(\tfrac{10}{9}\right) z_i^3(n) - \left(\tfrac{-1}{9}\right) z_i^5(n)\right) + z(n)$$ For the 64QAM case: $$a_2 z_r^3(n) + b_2 z_r^5(n) + d_2 z_r^7(n)+$$ $$j\left(a_2 z_i^3(n) + b_2 z_i^5(n) + d_2 z_i^7(n)\right) - z(n)$$ $a_2 = 0.16726,\ b_2 = -7.2482e{-}3,\ d_2 = 8.6752e{-}5$
Equalizer given in [13]	$\left(\dfrac{E\left[(x_r[n])^4\right]}{E\left[(x_r[n])^6\right]} z_r^3[n] + j \dfrac{E\left[(x_r[n])^4\right]}{E\left[(x_r[n])^6\right]} z_i^3[n] - z[n] \right)$

Table 3.1: $\partial F[n]/\partial z[n]$ as a function of the chosen equalizer.

ABBREVIATIONS

ISI=   Intersymbol Interference

SNR=   Signal to Noise Ratio

MMSE=   Minimum Mean Square Error

MAP=   Maximum A-Posteriori

pdf=   Probability Density Function

QAM= Quadrature Amplitude Modulation

MSE=   Mean Square Error

QPSK=   Quadrature Phase Shift Keying

BPSK=   Binary Phase Shift Keying

REFERENCES

[1] D. N. Godard, "Self recovering equalization and carrier tracking in two-dimensional data communication systems," *IEEE Transaction on Communications*, vol. COM-28, pp. 1867–1875, 1980.

[2] O. Shalvi and E. Weinstein, "New criteria for blind deconvolution of non-minimum phase systems (channels)," *IEEE Transaction on Information Theory*, vol. IT-36, pp. 312–321, 1990.

[3] M. Lazaro *et al.*, "Stochastic blind equalization based on pdf fitting using parzen estimator," *IEEE Trans. on Signal Processing*, vol. 53, pp. 696–704, 2005.

[4] G.-H. Im, C. J. Park, and H. C. Won, "A blind equalization with the sign algorithm for broadband access," *IEEE Comm. Letters*, vol. 5, pp. 70–72, 2001.

[5] A. K. Nandi, Ed., *Blind estimation using higher-order statistics.*   Boston: Kluwer Academic, 1999.

[6] C. L. Nikias and A. P. Petropulu, Eds., *Higher-Order Spectra Analysis A Nonlinear Signal Processing Framework.*   Prentice-Hall, 1993.

[7] S. Bellini, "Blind equalization," *Alta Freq.*, vol. 57, pp. 445–450, 1988.

[8] S. Fiori, "A contribution to (neuromorphic) blind deconvolution by flexible approximated bayesian estimation," *Signal Processing*, vol. 81, pp. 2131–2153, 2001.

[9] S. Haykin, Ed., *Bussgang Techniques for blind deconvolution and equalization, in: S. Haykin (Ed.), Blind Deconvolution.*   Prentice Hall, 1994.

[10] M. Pinchas and B. Z. Bobrovsky, "A novel hos approach for blind channel equalization," *IEEE Wireless Communication Journal*, vol. 6, 2007.

[11] M. Pinchas and B. Z. Bobrovsky, "A maximum entropy approach for blind deconvolution," *Signal Processing Journal (Eurasip)*, vol. 86, pp. 2913–2931, 2006.

[12] M. Pinchas, "A mse optimized polynomial equalizer for 16qam and 64qam constellation," *Signal, Image and Video Processing*, vol. 5 , No. 1, pp. 29–37, 2011.

[13] M. Pinchas, "What are the analytical conditions for which a blind equalizer will loose the converge state," *Signal, Image and Video Processing*, DOI: 10.1007/s11760-011-0221-0.

[14] M. Pinchas, "A closed approximated formed expression for the achievable residual intersymbol interference obtained by blind equalizers," *Signal Processing Journal (Eurasip)*, vol. 90, pp. 1940–1962, 2010.

Chapter 4

THE BAYESIAN APPROACH

MONIKA PINCHAS

Department of Electrical and Electronic Engineering, Ariel University Center of Samaria, Ariel 40700, ISRAEL

ABSTRACT

In this chapter the Bayesian approach for blind deconvolution is explained. We cover the different proposed expressions for the conditional expectation known in the literature and explain in detail the advantages and disadvantages of each obtained expression.

In the literature, the convolutional noise probability density function (pdf) is modeled as a Gaussian pdf. In this chapter, a new closed-form approximated expression is derived for the conditional expectation based on a new model for the convolutional noise pdf. The new derived expression for the conditional expectation, is based on the Maximum Entropy approach, Edgeworth expansion, Laplace integral method and is valid for the noiseless, real valued and two independent quadrature carrier case. According to simulation results carried out for signal to noise ratio (SNR) $SNR = 30$ [dB], the new derived expression leads to improved equalization performance for the 16QAM input constellation case.

KEYWORDS

Bayesian approach, conditional expectation, Maximum Entropy, Edgeworth expansion, Laplace integral, blind deconvolution, nonlinearity function, Shannon's entropy, Lagrange multipliers, orthogonal expansion, Hermite polynomials

THE CONDITIONAL EXPECTATION

In the previous chapters 3 and 2, we have seen that the non-linearity function $T[z[n]]$ should provide an estimation of the input signal $x[n]$. According to [1], the conditional expectation ($E[x[n]/z[n]]$, where $E[\cdot]$ stands for the expectation operation, $x[n]$ and $z[n]$ are the source and equalized output signal respectively) is touted as a good estimate of $T[z[n]]$. The conditional expectation ($E[x[n]/z[n]]$) is based on Bayes rule as will be shown in the following. Thus, when saying blind deconvolution according to the Bayesian approach, we actually mean that the blind deconvolution is based on the conditional expectation expression.

In this chapter we consider the noiseless, real valued and two independent quadrature carrier case. We start in this section with the real valued case while in the next section we turn to the two independent quadrature carriers one.

We recall the expression for the equalized output signal from chapter 2 for the noiseless case:

$$z[n] = x[n] + p[n] \tag{4.1}$$

where $p[n]$ is the convolutional noise and $x[n]$ is the source signal. The following assumptions are widely used in the literature [1], [2], [3], [4], [5], [6]:

1. The convolutional noise $p[n]$ is a zero mean, white Gaussian process with variance $\sigma_p^2 = E[p^2]$ where $E(\cdot)$ is the expectation operator.

2. The convolutional noise $p[n]$ and the source signal $x[n]$ are independent. Thus, $E[z[n]^2] = \sigma_z^2 = E[[x[n] + p[n]]^2] = E[x[n]^2] + E[p[n]^2] = \sigma_x^2 + \sigma_p^2$ where σ_x^2 and σ_p^2 are the variances of the source signal $x[n]$ and convolutional noise $p[n]$ respectively.

Some comments concerned the mentioned assumptions. According to [6], the described model for the convolutional noise $p[n]$ is applicable during the latter stages of the process where the process is close to optimality. In the early stages of the iterative deconvolution process, the Intersymbol interefence (ISI) is typically large with the result that the data sequence and the convolutional noise are strongly correlated and the convolutional noise sequence is more uniform than Gaussian [6], [7]. It should be pointed out, that good equalization performances were obtained in [2], [4] and [5] in spite of the fact that the described model for the convolutional noise $p[n]$ was used there. Based on these results ([2], [4],[5]), the described model for the convolutional noise $p[n]$ can be used (maybe not in the optimum way) also in the early stages where the "eye diagram" is still closed [2].

The conditional expectation $(E[x[n]/z[n]])$ is defined as:

$$E\left[x\left[n\right]/z\left[n\right]\right] = \int\limits_{-\infty}^{\infty} x f_{x/z}\left(x/z\right)dx \qquad (4.2)$$

where by Bayes rule we have:

$$f_{x/z}\left(x/z\right) = \frac{f_{z/x}\left(z/x\right)f_x\left(x\right)}{f_z\left(z\right)} = \frac{f_{z/x}(z/x)f_x(x)}{\int\limits_{-\infty}^{\infty} f_{z/x}(z/x)f_x(x)dx} \qquad (4.3)$$

By substituting (4.3) into (4.2) we obtain:

$$E\left[x[n]/z[n]\right] = \frac{\int\limits_{-\infty}^{\infty} x f_{z/x}(z/x) f_x(x) dx}{\int\limits_{-\infty}^{\infty} f_{z/x}(z/x) f_x(x) dx} \tag{4.4}$$

Based on our above made assumptions concerning the convolutional noise $p[n]$, we may write $f_{z/x}(z/x)$ as:

$$f_{z/x}(z/x) = \frac{1}{\sqrt{2\pi}\sigma_p} \exp\left(-\frac{(z-x)^2}{2\sigma_p^2}\right) \tag{4.5}$$

In order to evaluate (4.4) the source pdf has to be known. For a uniformly distributed source signal $x[n]$ within $[-\sqrt{3}, +\sqrt{(3)}]$, the conditional expectation (4.4) is given in [4], [8] and [9]:

$$E\left[x[n]/z[n]\right] = z[n] + \sigma_p \frac{\theta\left(\left(z[n]+\sqrt{3}\right)/\sigma_p\right) - \theta\left(\left(z[n]-\sqrt{3}\right)/\sigma_p\right)}{\tau\left(\left(z[n]-\sqrt{3}\right)/\sigma_p\right) - \tau\left(\left(z[n]+\sqrt{3}\right)/\sigma_p\right)} \tag{4.6}$$

where, by definition:

$$\theta(\gamma) = \frac{1}{\sqrt{2\pi}} \exp\left(-\frac{\gamma^2}{2}\right); \qquad \tau(\gamma) = \int\limits_{\gamma}^{\infty} \theta(\gamma)\, d\gamma = \frac{1}{2} erfc\left(\frac{\gamma}{\sqrt{2}}\right) \tag{4.7}$$

Three things should be pointed out when considering (4.6):

1. The expression for the conditional expectation (4.6) involves the variance of the convolutional noise, σ_p^2, which is an unknown parameter that changes with time.

2. The expression (4.6) is not easy to carry out when considering hardware implementation.

3. The expression (4.6) is only valid for uniformly distributed source

data $(x[n] \sim U[-\sqrt{(3)}, +\sqrt{(3)}])$.

According to [4], for a wide convolutional noise power range a suitable approximation for the conditional expectation given in (4.6) is the bilateral 'sigmoidal' function [6]

$$E[x[n]/z[n]] = a \tanh(\lambda z[n]) \tag{4.8}$$

with a and λ being properly chosen parameters. According to [4], in [6] a pair of values for a and λ is obtained by fitting the expression (4.8) with the actual estimator for a given convolutional distortion level. But as was pointed out in [4], the approximated expression (4.8) for the conditional expectation suffers from two problems:

1. It is clear that as an optimal constant value for σ_p^2 can not be found, a suitable pair of constant parameters a and λ can not determined too.

2. The approximation holds only for uniformly distributed source data.

In order to overcome the problem of having a pair of constant parameters a and λ, the author in [4] focused on deriving the best choice of the parameters a and λ. He proposed [4] to adapt the parameters a and λ through time by means of a stochastic gradient steepest descent (SGSD) algorithm applied to U where $\left(U = 0.5\left(\hat{b}(z[n]) - z[n]\right)^2\right)$:

$$\Delta a = -\eta_a \frac{\partial U}{\partial a} = -\eta_a \left[\hat{b}(z[n]) - z[n]\right] \frac{\hat{b}(z[n])}{a}$$

$$\Delta \lambda = -\eta_\lambda \frac{\partial U}{\partial \lambda} = -\eta_\lambda \left[\hat{b}(z[n]) - z[n]\right] \left[a^2 - \left(\hat{b}(z[n])\right)^2\right] \frac{z[n]}{a} \tag{4.9}$$

$$\hat{b}(z[n]) = a \tanh(\lambda z[n])$$

where η_λ and η_a are constant positive learning step-sizes. It should be pointed out that (4.9) already took into account (4.1), where the gain between the equalized output and input signal is equal to one for

$p[n] = 0$. Although (4.9) solved the problem of finding more suitable values for a and λ compared with the constant values proposed by [6], the expression for the conditional expectation proposed by [4] ($\hat{b}\,(z[n])$ in (4.9)) is suitable only for uniformly distributed source signals, thus it can not cope with a source having a general pdf shape. In order to derive a closed-form approximated expression for the conditional expectation ($E\,[x[n]/z[n]]$) valid for different source probability density functions, the input pdf has to be approximated with some literature known techniques. In the following we show two different approximations for the input pdf recently used for the blind deconvolution problem [2], [5]. In addition, we show the closed-form approximated expressions for the conditional expectation obtained from those approximations [2], [5]. We start with the Maximum Entropy approach used in [2] and then continue with the Edgeworth expansion technique used in [5].

According to [2], the maximum entropy density is typically obtained by maximizing Shannon's entropy:

$$H(x) = -\int f_x(x) \ln f_x(x) dx \qquad (4.10)$$

subject to some known moments constraints or equations of moments. Following Jumarie [10], we consider the arithmetic moments of the form:

$$\int x^i f_x(x) dx = m_i \qquad i = 0, 1, ..., K \qquad (4.11)$$

where m_i is the i-th moment of the random variable x with pdf $f_x(x)$. Using Lagrange's method to solve for the maximum entropy density leads according to Jumarie [10] and others [11]:

$$\widehat{f_x}(x) = \exp\left(\sum_{k=0}^{K} \lambda_k x^k\right) \qquad (4.12)$$

where λ_k, $k = 0, 1, 2, ..., K$ are the Lagrange multipliers. Now, by substituting (4.12) and (4.5) into (4.4) and using the Laplace integral method [2], [5], [12] we obtain [2]:

$$E\left[x[n]/z[n]\right] \cong \frac{z + \frac{g_1''(z)}{2g(z)}\left(\sigma_z^2 - \sigma_x^2\right) + \frac{g_1''''(z)}{8g(z)}\left(\sigma_z^2 - \sigma_x^2\right)^2}{1 + \frac{g''(z)}{2g(z)}\left(\sigma_z^2 - \sigma_x^2\right) + \frac{g''''(z)}{8g(z)}\left(\sigma_z^2 - \sigma_x^2\right)^2} \tag{4.13}$$

where

$$g(z) = \exp\left(\sum_{k=2}^{K} \lambda_k z^k\right)$$

$$g''(z) = \left\{\frac{d^2}{dx^2}\left[\exp\left(\sum_{k=2}^{K} \lambda_k x^k\right)\right]\right\}_{x=z}$$

$$g''''(z) = \left\{\frac{d^4}{dx^4}\left[\exp\left(\sum_{k=2}^{K} \lambda_k x^k\right)\right]\right\}_{x=z} \tag{4.14}$$

$$g_1''(z) = \left\{\frac{d^2}{dx^2}\left[x \exp\left(\sum_{k=2}^{K} \lambda_k x^k\right)\right]\right\}_{x=z}$$

$$g_1''''(z) = \left\{\frac{d^4}{dx^4}\left[x \exp\left(\sum_{k=2}^{K} \lambda_k x^k\right)\right]\right\}_{x=z}$$

and $d^4/dx^4(\cdot)$ and $d^2/dx^2(\cdot)$ are the fourth and second derivation of $(\cdot)$ respectively. The variance of the equalized output signal may be estimated by [2]:

$$\langle z^2 \rangle_n = (1 - \beta)\langle z^2 \rangle_{n-1} + \beta \cdot (z)_n^2 \tag{4.15}$$

where $\langle\rangle$ stands for the estimated expectation, $\langle z^2 \rangle_0 > 0$, and β is a positive stepsize parameter. The Lagrange multipliers λ_k are given by [2]:

$$m_{k-2}\left(k - 1\right)k + 2\lambda_k\; m_{2k-2}\; k^2 +$$

$$\sum_{L=2\;\; L\neq k}^{K} 2\lambda_L\; m_{k+L-2}\; kL = 0 \qquad \text{for}\;\; k = 2, 4, 6, ..., K \tag{4.16}$$

Please note that the Lagrange multipliers (4.16) derived in [2] are valid for the real valued and noiseless case where the input signal has a zero mean. In addition, they are not obtained via:

$$\int x^i \widehat{f}_x(x) dx = \int x^i \exp\left(\sum_{k=0}^{K} \lambda_k x^k\right) dx = m_i \qquad i = 0, 1, ..., N$$

(4.17)

which means that the derived expression for the conditional expectation (4.13) together with the Lagrange multiplies from (4.16) can not be considered to be a truly Maximum Entropy based derivation. It was already explained in [2], that it is not easy to carry out the integral in (4.17) to obtain the values for the desired Lagrange multipliers. As a matter of fact there are cases where an analytical solution of (4.17) for the Lagrange multipliers does not exist. Thus, trying to find the Lagrange multipliers via the integral (4.17) may only lead to approximated values for the Lagrange multipliers, which may cause degradation to the equalization performance. In order to obtain maximum equalization performance, the Lagrange multipliers (4.16) obtained in [2] were optimized in a mean square error (MSE) sense. Next we turn to the Edgeworth expansion technique.

We consider the one dimensional Edgeworth expansion with order up to four as following [5], [13], [14]:

$$f_x\left(x\right) \cong \frac{1}{\sqrt{2\pi}\sigma_x} \exp\left(-\frac{x^2}{2\sigma_x^2}\right)\left[1 + \frac{E\left[x^4\right] - 3\left(\sigma_x^2\right)^2}{4!\left(\sigma_x^2\right)^2}\left(3 - \frac{6x^2}{\sigma_x^2} + \frac{x^4}{\left(\sigma_x^2\right)^2}\right)\right]$$

(4.18)

It should be pointed out that the Edgeworth expansion is an orthogonal expansion using Hermite polynomials. This expansion is directly connected to the moments and cumulants of a pdf (the property which is lost in the Gauss-Hermite series) and, it is a true asymptotic expansion, so that the error of the approximation is controlled [5], [15]. Now, substituting (4.18) and (4.5) into (4.4) and using the Laplace integral method [2], [5], [12] we obtain [5]:

$$E\left[x[n]/z[n]\right] \cong \frac{A_1 \cdot z + B_1 \cdot z^3 + C_1 \cdot z^5}{1 + 3d_1(z) - 6q_1(z)z^2 + m_1(z)z^4} \tag{4.19}$$

$$A_1 = \frac{\sigma_x^2}{\sigma_z^2}\left[1 + 3\left(\frac{E[x^4]-3(\sigma_x^2)^2}{4!(\sigma_x^2)^2}\right)\left(1 - \frac{6(\sigma_z^2-\sigma_x^2)}{\sigma_z^2} + 5\left(1 - \frac{\sigma_x^2}{\sigma_z^2}\right)^2\right)\right]$$

$$B_1 = \left(\frac{1}{\sigma_z^2}\right)^3 \left(\frac{E[x^4]-3(\sigma_x^2)^2}{4!}\right)\left[-6 + \frac{10(\sigma_z^2-\sigma_x^2)}{\sigma_z^2}\right]$$

$$C_1 = \frac{\sigma_x^2}{(\sigma_z^2)^5}\left(\frac{E[x^4]-3(\sigma_x^2)^2}{4!}\right) \tag{4.20}$$

$$d_1(z) = \frac{E[z^4]-3(\sigma_z^2)^2}{4!(\sigma_z^2)^2} ; \qquad q_1(z) = \frac{E[z^4]-3(\sigma_z^2)^2}{4!(\sigma_z^2)^3} ; \tag{4.21}$$

$$m_1(z) = \frac{E[z^4]-3(\sigma_z^2)^2}{4!(\sigma_z^2)^4}$$

where $z = z[n]$ and $x = x[n]$. The moments of z in (4.21), (4.20) may be estimated by:

$$\langle z^2 \rangle_n = (1 - \beta_e)\langle z^2 \rangle_{n-1} + \beta_e \cdot (z)_n^2$$

$$\tag{4.22}$$

$$\langle z^4 \rangle_n = (1 - \beta_e)\langle z^4 \rangle_{n-1} + \beta_e \cdot (z)_n^4$$

where β_e is a small positive predefined parameter, $\langle z^2 \rangle_0 > 0$ and $\langle z^4 \rangle_0 > 0$.

The reader may find in Table (**4.1**), Table (**4.2**) a summary of the various expressions for the conditional expectation $E\left[x[n]/z[n]\right]$ discussed in this section.

ADAPTATION MECHANISM FOR BAYESIAN APPROACH

In this section, we extend the results obtained for the conditional expectation shown in the previous section to the two independent quadrature carrier case and describe the equalizer's adaptation scheme for Fiori's [4], Maximum Entropy [2] and Edgeworth Expansion [5] method for the noiseless, real valued and two independent quadrature carrier case.

According to [2], [3], [5], the conditional mean estimate of the complex datum x given the complex observation z can be written as:

$$E\left[x[n]/z[n]\right] = \hat{x}_1[n] + j\hat{x}_2[n] =$$

$$(4.23)$$

$$E\left[x_1[n]/z_1[n]\right] + jE\left[x_2[n]/z_2[n]\right] = T_1\left(z_1[n]\right) + jT_2\left(z_2[n]\right)$$

where

$$x[n] = x_1[n] + jx_2[n]; \qquad z[n] = z_1[n] + jz_2[n]$$

$$\hat{x}_1[n] = E\left[x_1[n]/z_1[n]\right]; \qquad \hat{x}_2[n] = E\left[x_2[n]/z_2[n]\right]$$

$$(4.24)$$

$$x_1[n] = Re(x[n]); \qquad x_2[n] = Imag(x[n])$$

$$z_1[n] = Re(z[n]); \qquad z_2[n] = Imag(z[n])$$

and $Re(\cdot)$, $Imag(\cdot)$ are the real and imaginary parts of $(\cdot)$ respectively. According to (4.23), real and imaginary parts of the data are to be estimated separately on the basis of the real and imaginary parts of the equalizer output. Now, by using (4.23) together with the different expressions for the conditional expectation from the previous section, we are ready to show the adaptation mechanism for the Maximum Entropy [2], Fiori's [4] and Edgeworth expansion [5] method.

The equalizer's taps for the Maximum Entropy approach [2], are updated only when $N_{\breve{s}} > \varepsilon$, where ε is a small positive parameter according to:

$$c_l(n+1) = c_l(n) - \mu W y^*(n-l) \qquad \text{with}$$

$$W = \left[\left(E\left[x_1/z_1\right] \left[\frac{(z_1[n] E\left[x_1/z_1\right])}{\langle (z_1)^2 \rangle_n} \right] + j E\left[x_2/z_2\right] \left[\frac{(z_2[n] E\left[x_2/z_2\right])}{\langle (z_2)^2 \rangle_n} \right] \right) - z[n] \right]$$
$$(4.25)$$

$$\breve{s} = 1, 2$$

$$(4.26)$$

$$N_{\breve{s}} = 1 + \frac{g''(z_{\breve{s}})}{2g(z_{\breve{s}})}\left(\sigma^2_{z_{\breve{s}}} - \sigma^2_{x_{\breve{s}}}\right) + \frac{g''''(z_{\breve{s}})}{8g(z_{\breve{s}})}\left(\sigma^2_{z_{\breve{s}}} - \sigma^2_{x_{\breve{s}}}\right)^2$$

where $E\left[x_{\breve{s}}/z_{\breve{s}}\right]$ is given in (4.13), $g''(z_{\breve{s}})$, $g(z_{\breve{s}})$, $g''''(z_{\breve{s}})$ are given in (4.14), $\sigma^2_{x_{\breve{s}}} = E\left[x_{\breve{s}}^2[n]\right]$, $\sigma^2_{z_{\breve{s}}} = E\left[z_{\breve{s}}^2[n]\right]$, μ is a positive step-size parameter and l stands for the l-th tap of the equalizer. The factor $[(z_{\breve{s}}(n) E\left[x_{\breve{s}}/z_{\breve{s}}\right])/\langle(z_{\breve{s}})^2\rangle_n]$ in (4.25) is a small gain correction which is needed since conditional mean estimators tend to have smaller gains [3].

The equalizer's taps for the Edgeworth expansion approach [5], are updated only when $\tilde{N}_{\breve{s}} > \varepsilon_e$, where ε_e is a small positive parameter according to:

$$c_l(n+1) = c_l(n) - \mu_e \tilde{W} y^*(n-l) \qquad \text{with}$$

$$\tilde{W} = \left[\left(E\left[x_1/z_1\right] \left[\frac{(z_1[n] E\left[x_1/z_1\right])}{\langle (z_1)^2 \rangle_n} \right] + j E\left[x_2/z_2\right] \left[\frac{(z_2[n] E\left[x_2/z_2\right])}{\langle (z_2)^2 \rangle_n} \right] \right) - z[n] \right]$$
$$(4.27)$$

$$\tilde{N}_{\breve{s}} = 1 + 3d_1(z_{\breve{s}}) - 6q_1(z_{\breve{s}})z_{\breve{s}}^2 + m_1(z_{\breve{s}})z_{\breve{s}}^4 \qquad (4.28)$$

where μ_e is a positive step-size parameter, $E\left[x_{\breve{s}}/z_{\breve{s}}\right]$ is given in (4.19), $\sigma^2_{x_{\breve{s}}} = E\left[x_{\breve{s}}^2[n]\right]$, $\sigma^2_{z_{\breve{s}}} = E\left[z_{\breve{s}}^2[n]\right]$, $d_1(z_{\breve{s}})$, $q_1(z_{\breve{s}})$ and $m_1(z_{\breve{s}})$ are given (4.21). Note, that a small gain correction $[(z_{\breve{s}}(n) E\left[x_{\breve{s}}/z_{\breve{s}}\right])/\langle(z_{\breve{s}})^2\rangle_n]$ is also needed here (4.27).

The equalizer's taps for Fiori's [4] approach, are updated according to:

$$c_l\,(n+1) = c_l\,(n) + \mu_F \cdot \left(\tilde{c}_1 \cdot \hat{b}(z_1[n]) + j\tilde{c}_2 \cdot \hat{b}(z_2[n]) - z[n] \right) y^*\,(n-l)$$

$$c \longleftarrow \frac{|\kappa|c}{\|c\|}$$

(4.29)

with

$$\hat{b}(z_\delta[n]) = \frac{a_\delta}{\tilde{c}_\delta} \tanh(\lambda_s z_\delta[n])$$

$$\Delta a_\delta = -\eta_a \left[\tilde{c}_\delta \cdot \hat{b}(z_\delta[n]) - z_\delta[n] \right] \frac{\tilde{c}_\delta \cdot \hat{b}(z_\delta[n])}{a_\delta}$$

$$\Delta \lambda_\delta = -\eta_\lambda \left[\tilde{c}_\delta \cdot \hat{b}(z_\delta[n]) - z_\delta[n] \right] \left[a_\delta^2 - \tilde{c}_\delta^2 \cdot \hat{b}^2(z_\delta[n]) \right] \frac{z_\delta[n]}{a_\delta}$$

(4.30)

where μ_F, η_a and η_λ are again step-size parameters and $|\kappa|$ ia a non-null constant that provides an amplification of the filter output signal with a factor $|\kappa|$. In the following some insights and comments are given concerning the Maximum Entropy [2], Edgeworth expansion [5], Fiori's [4] approach and (4.6).

The Edgeworth expansion, which was used for approximating the source pdf, turns to the Gaussian distribution when the source signal becomes Gaussian. Thus, it is likely to have better equalization performance with the Edgeworth expansion method [5] compared with Fiori's algorithm [4] for nearly Gaussian sources. Since the Edgeworth expansion was used only up to order four, while the Maximum Entropy [2] method may use λ_k with order higher than four, the equalization performance of the Maximum Entropy [2] approach may be better for non-Gaussian sources compared with the Edgeworth expansion algorithm [5]. Indeed, it was shown [16], that for the 64QAM (quadrature amplitude modulation) input case, the Maximum Entropy [2] method achieves improved equalization performance over 15 [dB] in the residual ISI compared with the Edgeworth expansion algorithm [5]. The maximum Entropy [2] and Edgeworth expansion method [5] are based on approximated expressions for the conditional expectation that are suitable for a wider range of source probability density function compared with Bellini's [8], Fiori's [4] or Haykin's [9] expression. According to simulation results shown in [2], the Maximum Entropy method [2] outperforms Bellini's [8], Fiori's [4] and Haykin's [9] based equalizers. Even when using a uniformly distributed source signal within $[-\sqrt{(3)}, +\sqrt{(3)}]$, the Maximum Entropy approach [2] outperforms the equalizer based on the exact expression for the conditional expectation (4.6) given in [8], [9] since in the Maximum Entropy approach [2], the Lagrange multipliers are optimized in the MSE sense. According to simulation results shown in [5], the Edgeworth expansion method outperforms Fiori's [4] approach for blind equalization.

A NEW MODEL FOR THE CONVOLUTIONAL NOISE PDF

In this section we show that improved equalization performance can be obtained when a nearly Gaussian distribution is chosen for the convolutional noise pdf instead of the Gaussian model proposed in the literature. The improved equalization performance is shown with a 16QAM constellation input sent via the channel defined in [17].

In the literature, the convolutional noise is often assumed to be a Gaussian process. Thus the convolutional noise pdf is assumed to have a Gaussian distribution. However, according to [6], the described model for the convolutional noise $p[n]$ is applicable during the latter stages of the process where the process is close to optimality. In the early stages of the iterative deconvolution process, the ISI is typically large with the result that the data sequence and the convolutional noise are strongly correlated and the convolutional noise sequence is more uniform than Gaussian [6], [7]. Thus, it is reasonable to search for an-other model for the convolutional noise pdf that is Gaussian only at at the latter stages of the process where the process is close to optimality and is non-Gaussian in the early stages of the iterative deconvolution process. Let us consider for a moment the real valued case where we use the Edgeworth expansion up to order four [5], [13], [14] for the convolutional noise pdf.

$$f_p\left(p\right) \cong \frac{1}{\sqrt{2\pi}\sigma_p} \exp\left(-\frac{p^2[n]}{2\sigma_p^2}\right) \left[1 + \frac{E\left[p^4[n]\right] - 3\left(\sigma_p^2\right)^2}{4!\left(\sigma_p^2\right)^2}\left(3 - \frac{6p^2[n]}{\sigma_p^2} + \frac{p^4[n]}{\left(\sigma_p^2\right)^2}\right)\right]$$

(4.31)

where

$$\frac{E\left[p^4[n]\right] - 3\left(\sigma_p^2\right)^2}{4!\left(\sigma_p^2\right)^2}$$

(4.32)

is zero for a Gaussian signal with zero mean. Thus, the Edgeworth ex-pansion (4.31) turns to the Gaussian distribution if $p[n]$ is a Gaussian signal. Since $p[n] = z[n] - x[n]$ and the input signal $x[n]$ is not avail-able (blind mode), the expression for $E[p^4[n]]$ can not be calculated. Since we wish to have a non Gaussian distribution for the convolu-tional noise pdf for the early stages of the deconvolution process we set $E[p^4[n]] = 6\sigma_p^4$ (found experimentally for the 16QAM source sent via a channel defined in [17]) which leads the expression of (4.32) to be equal to 1/8. In the latter stages of the iterative deconvolution process where the process is close to optimality, σ_p^2 is supposed to be very small. Therefore, any contribution of $O(\sigma_p^4)$ ($O(L)$ is defined as $\lim_{L\to 0}(O(L)/L) = r_{const}$ and "r_{const}" is a constant) to the final resid-ual ISI should be negligible. In the following we derive a closed-form approximated expression for the conditional expectation based on the Maximum Entropy approach with two Lagrange multipliers (for the 16QAM case) and based on the new model for the convolutional noise pdf (4.31).

A NEW EXPRESSION FOR THE CONDITIONAL EXPECTATION

In this subsection we start our derivations for the real valued case and then we will turn to the 16QAM constellation input (complex case). It should be pointed out that our derivations are valid for the noiseless case. Let us recall the following expression for the conditional expectation:

$$E\left[x[n]/z[n]\right] = \frac{\int\limits_{-\infty}^{\infty} x f_{z/x}(z/x) f_x(x) dx}{\int\limits_{-\infty}^{\infty} f_{z/x}(z/x) f_x(x) dx} \tag{4.33}$$

where we use:

$$f_{z/x}(z/x) = \frac{1}{\sqrt{2\pi}\sigma_p} \exp\left(-\frac{p^2[n]}{2\sigma_p^2}\right) \left[1 + \frac{E\left[p^4[n]\right] - 3\left(\sigma_p^2\right)^2}{4!\left(\sigma_p^2\right)^2}\left(3 - \frac{6p^2[n]}{\sigma_p^2} + \frac{p^4[n]}{\left(\sigma_p^2\right)}\right)\right] \tag{4.34}$$

and

$$f_x(x) \cong \exp\left(\lambda_0 + \lambda_2 x^2 + \lambda_4 x^4\right) \tag{4.35}$$

Substituting (4.34) and (4.35) into (4.33) yields to:

$$E\left[x[n]/z[n]\right] \cong \frac{\int\limits_{-\infty}^{\infty} \widetilde{g}_1(x) \exp\left(-\psi(x)/\rho\right) dx}{\int\limits_{-\infty}^{\infty} \widetilde{g}(x) \exp\left(-\psi(x)/\rho\right) dx} \tag{4.36}$$

where

$$\widetilde{g}(x) = \exp\left(\lambda_2 x^2 + \lambda_4 x^4\right)\left[1 + \frac{E[p^4[n] - 3\sigma_p^4}{4!\sigma_p^4}\left(3 - \frac{6p^2[n]}{\sigma_p^2} + \frac{p^4[n]}{\sigma_p^4}\right)\right]$$

$$\widetilde{g}_1(x) = x\widetilde{g}(x); \qquad \psi(x) = p^2[n] = (z-x)^2 \qquad \rho = 2\sigma_p^2 \tag{4.37}$$

Solving (4.36) with the Laplace integral method [12] as was done in [2], [5] we obtain:

$$E\left[x[n]/z[n]\right] \cong \frac{z + \dfrac{\tilde{g}_1''(z)}{2\tilde{g}(z)}\left(\sigma_z^2 - \sigma_x^2\right) + \dfrac{\tilde{g}_1''''(z)}{8\tilde{g}(z)}\left(\sigma_z^2 - \sigma_x^2\right)^2}{1 + \dfrac{\tilde{g}''(z)}{2\tilde{g}(z)}\left(\sigma_z^2 - \sigma_x^2\right) + \dfrac{\tilde{g}''''(z)}{8\tilde{g}(z)}\left(\sigma_z^2 - \sigma_x^2\right)^2} \qquad (4.38)$$

where

$$\tilde{g}(z) = \exp\left(\lambda_2 z^2 + \lambda_4 z^4\right)\left[1 + 3\frac{E[p^4[n] - 3\sigma_p^4]}{4!\sigma_p^4}\right] \qquad (4.39)$$

$$\tilde{g}''(z) = \exp\left(\lambda_2 z^2 + \lambda_4 z^4\right) \cdot$$

$$\left[\left(1 + \frac{1}{8}\frac{E[p^4[n] - 3\sigma_p^4]}{\sigma_p^4}\right)\left(2\lambda_2 + 12z^2\lambda_4 + 4z^2\lambda_2^2 + 16z^4\lambda_2\lambda_4 + 16z^6\lambda_4^2\right) - \right.$$

$$\left. \frac{1}{2}\frac{E[p^4[n] - 3\sigma_p^4]}{\sigma_p^6}\right]$$

$$(4.40)$$

$$\widetilde{g}''''(z) = \exp\left(\lambda_2 z^2 + \lambda_4 z^4\right) \cdot$$

$$\left[\left(1 + \frac{1}{8}\frac{E[p^4[n]] - 3\sigma_p^4}{\sigma_p^4}\right)\left(24\lambda_4 + 336z^2\lambda_2\lambda_4 + 816z^4\lambda_4^2 + \right.\right.$$

$$12\lambda_2^2 + 48z^2\lambda_2^3 + 480z^4\lambda_2^2\lambda_4 + 1344z^6\lambda_2\lambda_4^2 + 1152z^8\lambda_4^3 + 16z^4\lambda_2^4 +$$

$$\left.128z^6\lambda_2^3\lambda_4 + 384z^8\lambda_2^2\lambda_4^2 + 512z^{10}\lambda_2\lambda_4^3 + 256z^{12}\lambda_4^4\right)$$

$$-3\frac{E[p^4[n]] - 3\sigma_p^4}{\sigma_p^6}\left(2\lambda_2 + 12z^2\lambda_4 + 4z^2\lambda_2^2 + 16z^4\lambda_2\lambda_4 + 16z^6\lambda_4^2\right) +$$

$$\left.\frac{E[p^4[n]] - 3\sigma_p^4}{\sigma_p^8}\right]$$

$$(4.41)$$

$$\widetilde{g}_1''(z) = \exp\left(\lambda_2 z^2 + \lambda_4 z^4\right) \cdot$$

$$\left[\left(1 + \frac{1}{8}\frac{E[p^4[n]] - 3\sigma_p^4}{\sigma_p^4}\right)\left(6z\lambda_2 + 20z^3\lambda_4 + 4z^3\lambda_2^2 + 16z^5\lambda_2\lambda_4 + 16z^7\lambda_4^2\right) - \right.$$

$$\left.\frac{1}{2}z\left(\frac{E[p^4[n]] - 3\sigma_p^4}{\sigma_p^6}\right)\right]$$

$$(4.42)$$

$$\tilde{g}_1^{''''}(z) = \exp\left(\lambda_2 z^2 + \lambda_4 z^4\right) \cdot$$

$$\left[\left(1 + \frac{1}{8}\frac{E[p^4[n]] - 3\sigma_p^4}{\sigma_p^4}\right)(120 z \lambda_4 +\right.$$

$$60 z \lambda_2^2 + 720 z^3 \lambda_2 \lambda_4 + 1392 z^5 \lambda_4^2 + 80 z^3 \lambda_2^3 +$$

$$672 z^5 \lambda_2^2 \lambda_4 + 1728 z^7 \lambda_2 \lambda_4^2 + 1408 z^9 \lambda_4^3 +$$

$$16 z^5 \lambda_2^4 + 128 z^7 \lambda_2^3 \lambda_4 + 384 z^9 \lambda_2^2 \lambda_4^2 + 512 z^{11} \lambda_2 \lambda_4^3 + 256 z^{13} \lambda_4^4) -$$

$$3\left(\frac{E[p^4[n]] - 3\sigma_p^4}{\sigma_p^6}\right)(6 z \lambda_2 + 20 z^3 \lambda_4 + 4 z^3 \lambda_2^2 + 16 z^5 \lambda_2 \lambda_4 + 16 z^7 \lambda_4^2) +$$

$$\left. z\left(\frac{E[p^4[n]] - 3\sigma_p^4}{\sigma_p^8}\right)\right]$$

$$(4.43)$$

and the variance of the equalized output signal σ_z^2 may be estimated by [2]:

$$\langle z^2 \rangle_n = (1 - \beta_{edg})\langle z^2 \rangle_{n-1} + \beta_{edg} \cdot (z)_n^2 \qquad (4.44)$$

where $\langle\rangle$ stands for the estimated expectation, $\langle z^2 \rangle_0 > 0$, and β_{edg} is a positive stepsize parameter.

Next the new approximated expression for the conditional expectation (4.38) is tested via simulation. We use the 16QAM input case and the channel defined in [17]: **Channel1** (initial ISI $= 0.44$):
$h_n = 0$ for $n < 0$; -0.4 for $n = 0$ $0.84 \cdot 0.4^{n-1}$ for $n > 0$.

An equalizer with 13 taps is used in the simulation where the equalizer is initialized by setting the center tap equal to one and all others to zero. The equalizer's taps are updated only when $\check{N}_{\check{s}} > \varepsilon$, where ε is a small positive parameter according to:

$$c_l(n+1) = c_l(n) - \mu_{edg}\breve{W}y^*(n-l) \qquad \text{with}$$

$$\breve{W} = \left[\left(E[x_1/z_1]\left[\frac{(z_1[n]E[x_1/z_1])}{\langle(z_1)^2\rangle_n}\right] + jE[x_2/z_2]\left[\frac{(z_2[n]E[x_2/z_2])}{\langle(z_2)^2\rangle_n}\right]\right) - z[n]\right]$$

(4.45)

$$\breve{s} = 1, 2$$

(4.46)

$$\breve{N}_{\breve{s}} = 1 + \frac{g''(z_{\breve{s}})}{2g(z_{\breve{s}})}\left(\sigma_{z_{\breve{s}}}^2 - \sigma_{x_{\breve{s}}}^2\right) + \frac{g''''(z_{\breve{s}})}{8g(z_{\breve{s}})}\left(\sigma_{z_{\breve{s}}}^2 - \sigma_{x_{\breve{s}}}^2\right)^2$$

where $E[x_{\breve{s}}/z_{\breve{s}}]$ is given in (4.38), $g''(z_{\breve{s}})$, $g(z_{\breve{s}})$, $g''''(z_{\breve{s}})$ are given in (4.40), (4.39) and (4.41) respectively, $\sigma_{x_{\breve{s}}}^2 = E[x_{\breve{s}}^2[n]]$, $\sigma_{z_{\breve{s}}}^2 = E[z_{\breve{s}}^2[n]]$, μ_{edg} is a positive stepsize parameter and l stands for the l-th tap of the equalizer. In the simulation we denoted as μ_G, μ_{ent} and μ_{edg} as the step-size parameter for Godard's [18], Maximum Entropy [2] and Edgeworth (4.45) approach respectively. β_{ent} and β_{edg} are additional stepsize parameters for the Maximum Entropy [2] and Edgeworth (4.45) approach respectively. Fig. (**4.1**) shows the simulated performance of the Maximum Entropy [2], Edgeworth (4.45) and Godard's equalization method for the 16QAM input case, namely the ISI as a function of iteration number for SNR = 30 [dB]. According to Fig. (**4.1**), the new model for the convolutional noise pdf indeed helps to improve the equalization performance. It should be pointed out that we used $E[p^4[n]] = 6\sigma_p^4$ which was found experimentally for the 16QAM source and for the channel defined in [17]. This implies that for other types of channels and input constellations, a different substitution for $E[p^4[n]]$ may be needed. Anyway, the main purpose in this subsection was to show that other models than the Gaussian distribution for the convolutional noise pdf may lead to improved equalization performance. The simulation results gave a good indication of it.

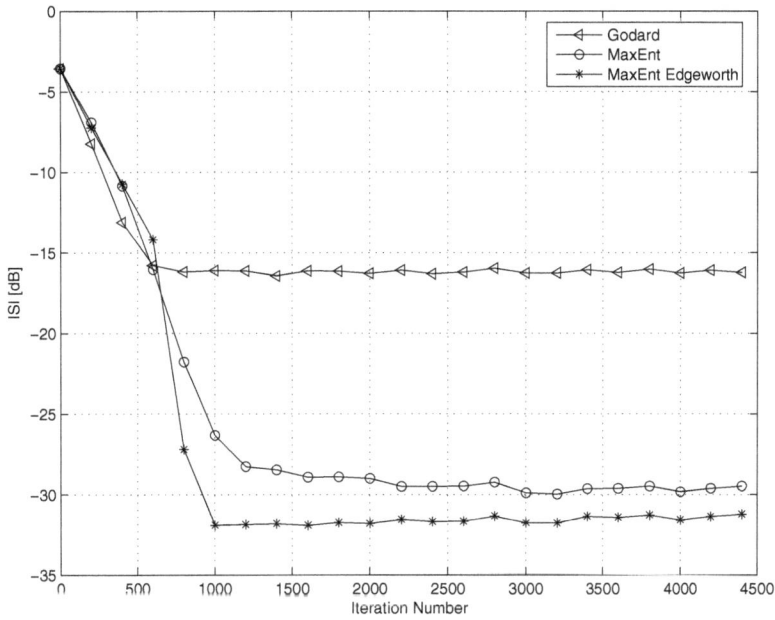

Figure 4.1: Equalization performance comparison between the equalization methods for the 16QAM source input going through channel1. The averaged results were obtained in 100 Monte Carlo trials for SNR = 30 [dB]. The equalizer's length was set to 13, $\mu_G = 0.0001$, $\mu_{ent} = 0.0004$, $\beta_{ent} = 0.0002$, $\mu_{edg} = 0.0005$, $\beta_{edg} = 0.0005$

Chosen Equalizer	$E[x[n]/z[n]]$
Given in [6]	The bilateral 'sigmoidal' function $$a \tanh\left(\lambda z[n]\right)$$ where a and λ are obtained by fitting the above expression with the actual estimator for a given convolutional distortion le⁻
Given in [4], [8], [9]	For a uniformly distributed source signal $x[n]$ within $[-\sqrt{3}, +\sqrt{(3)}]$: $$E\left[x[n]/z[n]\right] = z[n] + \sigma_p \frac{\theta\left(\left(z[n]+\sqrt{3}\right)/\sigma_p\right) - \theta\left(\left(z[n]-\sqrt{3}\right)/\sigma_p\right)}{\tau\left(\left(z[n]-\sqrt{3}\right)/\sigma_p\right) - \tau\left(\left(z[n]+\sqrt{3}\right)/\sigma_p\right)}$$ $$\theta\left(\gamma\right) = \frac{1}{\sqrt{2\pi}} \exp\left(-\frac{\gamma^2}{2}\right); \quad \tau\left(\gamma\right) = \int_{\gamma}^{\infty} \theta\left(\gamma\right) d\gamma = \frac{1}{2} erfc\left(\right.$$
Given in [4]	$$\hat{b}(z_{\breve{s}}[n]) = \frac{a_{\breve{s}}}{\tilde{c}_{\breve{s}}} \tanh(\lambda_{\breve{s}} z_{\breve{s}}[n])$$ $$\Delta a_{\breve{s}} = -\eta_a \left[\tilde{c}_{\breve{s}} \cdot \hat{b}(z_{\breve{s}}[n]) - z_{\breve{s}}[n]\right] \frac{\tilde{c}_{\breve{s}} \cdot \hat{b}(z_{\breve{s}}[n])}{a_{\breve{s}}}$$ $$\Delta \lambda_{\breve{s}} = -\eta_\lambda \left[\tilde{c}_{\breve{s}} \cdot \hat{b}(z_{\breve{s}}[n]) - z_{\breve{s}}[n]\right] \left[a_{\breve{s}}^2 - \tilde{c}_{\breve{s}}^2 \cdot \hat{b}^2(z_{\breve{s}}[n])\right] \frac{z_{\breve{s}}}{a}$$

Table 4.1: $E[x[n]/z[n]]$ as a function of the chosen equalizer for the real valued case.

Chosen Equalizer Continued	$E[x[n]/z[n]]$
Given in [2]	$E\left[x[n]/z[n]\right] \cong \dfrac{z + \dfrac{g_1''(z)}{2g(z)}\left(\sigma_z^2 - \sigma_x^2\right) + \dfrac{g_1''''(z)}{8g(z)}\left(\sigma_z^2 - \sigma_x^2\right)^2}{1 + \dfrac{g''(z)}{2g(z)}\left(\sigma_z^2 - \sigma_x^2\right) + \dfrac{g''''(z)}{8g(z)}\left(\sigma_z^2 - \sigma_x^2\right)^2}$ where $g(z),\ g''(z),\ g''''(z),\ g_1''(z),\ g_1''''(z)$ are given in (4.14) The variance of the equalized output signal may be estimated by (4.15)
Given in [5]	$E\left[x[n]/z[n]\right] \cong \dfrac{A_1 \cdot z + B_1 \cdot z^3 + C_1 \cdot z^5}{1 + 3d_1(z) - 6q_1(z)z^2 + m_1(z)z^4}$ where $A_1,\ B_1,\ C_1$ are given in (4.20) and $d_1(z),\ q_1(z),\ m_1(z)$ are given in (4.21)

Table 4.2: $E[x[n]/z[n]]$ as a function of the chosen equalizer for the real valued case.

ABBREVIATIONS

ISI=   Intersymbol Interference

SNR=   Signal to Noise Ratio

SGSD= Stochastic Gradient Steepest Descent

pdf=   Probability Density Function

QAM=   Quadrature Amplitude Modulation

MSE=   Mean Square Error

REFERENCES

[1] C. L. Nikias and A. P. Petropulu, Eds., *Higher-Order Spectra Analysis A Non-linear Signal Processing Framework.*   Prentice-Hall, 1993.

[2] M. Pinchas and B. Z. Bobrovsky, "A maximum entropy approach for blind deconvolution," *Signal Processing Journal (Eurasip)*, vol. 86, pp. 2913–2931, 2006.

[3] S. Bellini, "Bussgang techniques for blind equalization," in *IEEE Global Telecommunication Conference Records*, 1986, pp. 1634–1640.

[4] S. Fiori, "A contribution to (neuromorphic) blind deconvolution by flexible approximated bayesian estimation," *Signal Processing*, vol. 81, pp. 2131–2153, 2001.

[5] M. Pinchas, "A novel hos approach for blind channel equalization," *IEEE Wireless Communication Journal*, vol. 6, 2007.

[6] S. Haykin, Ed., *Adaptive Filter Theory.*   Prentice-Hall, Englewood cliffs,NJ, 1991.

[7] R. Godfrey and F. Rocca, "Zero memory non-linear deconvolution," *Geophys. Prospect.*, vol. 29, pp. 189–228, 1981.

[8] S. Bellini, "Blind equalization," *Alta Freq.*, vol. 57, pp. 445–450, 1988.

[9] S. Haykin, *Bussgang Techniques for blind deconvolution and equalization, in: S. Haykin (Ed.), Blind Deconvolution.*   Prentice Hall, 1994.

[10] G. Jumarie, "Nonlinear filtering. a weighted mean squares approach and a bayesian one via the maximum entropy principle," *Signal Processing*, vol. 21, pp. 323–338, 1990.

[11] A. Papoulis, Ed., *Probability, Random Variables, and Stochastic Processes*, second international ed.   Kogakusha: McGraw-Hill, 1984.

[12] S. A. Orszag and C. M. Bender, Eds., *Advanced Mathematical Methods for Scientist Engineers, International Series in Pure and Applied Mathematics*. McDraw-Hill, 1978.

[13] S. A. Assaf and L. D. Zirkle, "Approximate analysis of nonlinear stochastic systems," *International Journal of Control*, vol. 23, pp. 477–492, 1976.

[14] D. C. C. Bover, "Moment equation methods for nonlinear stochastic systems," *Journal of Mathematical Analysis and Applications*, vol. 65, pp. 306–320, 1978.

[15] H. Cramer, Ed., *Mathematical Methods of Statistics*.   Princeton University Press, 1951.

[16] M. Pinchas, Ed., *Blind Equalizers By Techniques Of Optimal Non-Linear Filtering Theory*.   ISBN 978-3-639-15530-3, VDM Verlagsservice gesellschaft mbH, 2009.

[17] O. Shalvi and E. Weinstein, "New criteria for blind deconvolution of non-minimum phase systems (channels)," *IEEE Transaction on Information Theory*, vol. IT-36, pp. 312–321, 1990.

[18] D. N. Godard, "Self recovering equalization and carrier tracking in two-dimensional data communication systems," *IEEE Transaction on Communications*, vol. COM-28, pp. 1867–1875, 1980.

Chapter 5

ADVANTAGES AND DISADVANTAGES OF EACH APPROACH

MONIKA PINCHAS

Department of Electrical and Electronic Engineering, Ariel University Center of Samaria, Ariel 40700, ISRAEL

ABSTRACT

In this chapter we explain the advantages and disadvantages of the Bayesian and Cost function approach. In doing so we also explain what is the meaning of higher order statistics (HOS) which is widely used in blind equalization, the reason why using HOS ($order \geq 3$) in blind equalization and why HOS can not be applied for Gaussian input signals.

KEYWORDS

Bayesian approach, Cost function approach, higher order statistics (HOS), moments, cumulants, blind deconvolution, inverse filter, Maximum Entropy density approximation technique, Edgeworth expansion, nonlinearity function

HOS: MOMENTS AND CUMULANTS

HOS measures are extensions of second-order measures (such as the autocorrelation function and power spectrum) to higher orders. The reason for using HOS (*order* ≥ 3) in blind equalization is due to the fact that second order statistics (SOS) (autocorrelations or power spectra [1]) based methods cannot be used if the channel is nonminimum-phase since these methods are blind to the phase of the channel [2], whereas phase information is preserved in statistics of order higher than two [3].

There are two classes of HOS-based methods for blind equalization:
1. Explicitly used high-order statistics include the inverse filter criteria (IFC) based algorithm [4], [5], [6], [7], [8]; the super-exponential algorithm [9]; the polyspectra-based algorithm [10]; and the eigenvector approach [11], [12].
2. Implicitly used high-order statistics. These methods, which implicitly use higher-order moments are also known as Bussgang-type algorithms and include the Sato [13] and Godard [14] algorithms as specials cases. Here, the training sequence in the non-blind architecture is replaced by a nonlinear estimate of the channel input. The nonlinearity is designed to minimize a cost function that is ***implicitly*** based on HOS according to one approach ([13], [14]), or the nonlinearity is calculated directly according to the Bayes rules [1], [15], [16], [17], [18], [19], [20].

Moments are statistical measures which characterize signal properties. According to [3], when signals are non-Gaussian the first two moments do not define their probability density function (pdf) and consequently HOS, namely of order greater than two, can reveal other information about them than SOS alone can. Ideally the entire pdf is needed to characterize a non-Gaussian signal. In practice this is not available but the pdf may be characterized by its moments [3]. For practical purposes, the knowledge of moments may be considered equivalent to the knowledge of the pdf [3]. In practice in HOS we usually use the cumulants rather than the moments. The n-th order cumulant is a function of the moments of orders up to (and including) n. Cumulants and moments are different though clearly related (as can be seen in the following). Cumulants are not directly estimable by summatory or integrative processes, and to find them it is necessary either to derive them from the characteristic function or to find the moments first [3]. For zero-mean distributions, the first three central moments and the corresponding cumulants are identical but they begin to differ from order four [3]:

$$C_1 = \mu_1 = 0; \quad C_2 = \mu_2; \quad C_3 = \mu_3; \quad C_4 = \mu_4 - 3\mu_2^4 \qquad (5.1)$$

where C_v $(v = 1, 2, 3, ...)$ is the cumulant and μ_v is the central moment. Let the cumulative distribution function (cdf) of x be denoted by $F(x)$ [3]. The central moment (about the mean) of order v of x is defined by [3]:

$$\mu_v = \int_{-\infty}^{\infty} (x - m)^v \, dF \qquad (5.2)$$

where m, the mean of x, is given by $\int_{-\infty}^{\infty} x dF$, $\mu_0 = 1$ and $\mu_1 = 0$. For zero-mean Gaussian distribution, $C_1 = 0$, $C_2 = \sigma^2$ (variance), and $C_v = 0$ for $v > 2$. Thus HOS are applicable when we are dealing with non-Gaussian signals and we may say that HOS "measure" non-Gaussianity. In the following we derive the cumulants via the characteristic function following [9]. Let $x_1, x_2, ..., x_n$ be a set of real/complex random variables possessing the joint characteristic function:

$$\phi(\omega) = E\left\{ \exp\left(j \sum_{i=1}^{n} \omega_i x_i \right) \right\} \qquad (5.3)$$

where $j = \sqrt{-1}$ and $E\{\cdot\}$ stands for the expectation operation. The joint cumulant of $x_{n_1}, x_{n_2}, ..., x_{n_m}$, $n_i \epsilon \{1, 2, ..., n\}$ is defined by [9]:

$$cum\left(x_{n_1}; x_{n_2}; x_{n_3} \cdots; x_{n_m}\right) = (-j)^m \frac{\partial^m \ln \phi\left(\omega\right)}{\partial \omega_{n_1} \cdots \partial \omega_{n_m}\big|_{\omega=0}} \qquad (5.4)$$

where $ln\left(\cdot\right)$ stands for the natural logarithm. Thus, if $x_1, x_2 \cdots$ are zero-mean random variables then [9]:

$$cum\left(x_1\right) = 0$$

$$cum\left(x_1; x_2\right) = E\left\{x_1 x_2\right\}$$

$$cum\left(x_1; x_2; x_3\right) = E\left\{x_1 x_2 x_3\right\}$$

$$cum\left(x_1; x_2; x_3; x_4\right) = E\left\{x_1 x_2 x_3 x_4\right\} - cum\left(x_1; x_2\right)cum\left(x_3; x_4\right) -$$

$$cum\left(x_1; x_3\right)cum\left(x_2; x_4\right) - cum\left(x_1; x_4\right)cum\left(x_2; x_3\right)$$

$$(5.5)$$

THE BAYESIAN AND COST FUNCTION APPROACH

In this section we describe the advantages and disadvantages of each approach. It should be pointed out that for both approaches (Bayesian and cost function), we adopt the LMS type algorithm for the adaptation of the equalizer's taps. It is well known that the LMS type algorithm is simple but may lead to slow convergence and poor performance for hostile scenarios, whereas a least-square type algorithm is more computationally complex but has a better performance.

In the cost function approach, a predefined cost function $F(n)$ that characterizes the intersymbol interference (ISI) is defined, see [8], [14], [21], [22] and [23]. Minimizing this $F(n)$ with respect to the equalizer parameters will reduce the convolutional error. A predefined cost function may be designed for special input cases such as for the 16QAM or 64QAM case only [21]. For those special input cases, the equalizer may reach perfect or close to perfect equalization performance in the residual ISI point of view. But when the input signal is different than for the designed one, the equalization performance (the residual ISI) may degrade very much. Namely, the system might be left with a relative high residual ISI that might not meet the system designers requirements. On the other hand, the Bayesian approach which is based on the conditional expectation of the input signal given the equalized output signal, is a general solution that should work for every (or almost every) input case. Thus, should lead to satisfying equalization performance for every (or almost every) input signal. Unfortunately, in the literature, we do not have yet a closed-form expression for the conditional expectation of the input signal given the equalized output signal valid for the general input complex case. But, we may find a closed-form approximated expression for the conditional expectation [19] of the input signal given the equalized output signal that is valid for the real valued and two independent quadrature carrier case [19]. The equalizer, based on the closed-form approximated expression for the conditional expectation [19], leads to perfect equalization performance in the residual ISI point of view [19] for all input signals belonging to the real valued and two independent quadrature carrier case.

As already discussed in the previous section, HOS can not be applied to Gaussian input signals since the HOS of Gaussian signals vanish. Therefore, equalizers based on cost functions that are based on HOS can not recover Gaussian input signals. Let us go back to the Bayesian approach. According to [20] and [19], the conditional expectation is based on some approximation for the input pdf. In [20], the input pdf is approximated with the Edgeworth expansion for the real valued case:

$$f_x(x) \cong \frac{1}{\sqrt{2\pi}\sigma_x} \exp\left(-\frac{x^2}{2\sigma_x^2}\right) \left[1 + \frac{E[x^4] - 3\left(\sigma_x^2\right)^2}{4!\left(\sigma_x^2\right)^2}\left(3 - \frac{6x^2}{\sigma_x^2} + \frac{x^4}{\left(\sigma_x^2\right)^2}\right)\right]$$
$$(5.6)$$

while in [19], the Maximum Entropy density approximation technique is used for defining the input pdf for the real valued case as:

$$f_x(x) \cong \exp\left(\sum_{k=0}^{K} \lambda_k x^k\right) \qquad (5.7)$$

where x is the input signal, σ_x^2 is the variance of the input signal, $E[(\cdot)]$ stand for the expectation operation, λ_k, $k = 0, 1, 2, ..., K$ are the Lagrange multipliers. Now, when the input signal is Gaussian, the expression of $(E[x^4] - 3\left(\sigma_x^2\right)^2)/4!\left(\sigma_x^2\right)^2$ will tend to zero and leave the approximated input pdf (5.6) with the expected Gaussian input pdf. In both cases (5.7) and (5.6), the approximated input pdf is close related to the Gaussian pdf. Therefore, the conditional expectation based on [20] or [19] is valid even when the input signal is Gaussian. This means that if the input signal is nearly Gaussian, the Bayesian approach will outperform (in the residual ISI point of view) any equalization method based on a cost function that is based on HOS.

Next we turn to the computational complexity issue. The computational complexity of blind adaptive equalizers based on the Bayesian approach [19], [20] is relative high compared to the cost function approach presented by [14], [21].

ABBREVIATIONS

ISI=   Intersymbol Interference

SNR=   Signal to Noise Ratio

SOS=   Second Order Statistics

pdf=   Probability Density Function

HOS=   Higher Order Statistics

cdf= Cumulative Distribution Function

REFERENCES

[1] S. Haykin, Ed., *Adaptive Filter Theory.*   Prentice-Hall, Englewood cliffs,NJ, 1991.

[2] C.-Y. Chi, C.-Y. Chen, C.-H. Chen, and C.-C. Feng, "Batch processing algorithms for blind equalization using higher-order statistics," *IEEE Signal Processing Magazine*, pp. 25–49, 2003.

[3] A. K. Nandi, Ed., *Blind estimation using higher-order statistics.*   Boston: Kluwer Academic, 1999.

[4] J. A. Cadzow, "Blind deconvolution via cumulant extrema," *IEEE Signal Processing Mag.*, vol. 13, pp. 24–42, 1996.

[5] C. Y. Chi and M. C. Wu, "Inverse filter criteria for blind deconvolution and equalization using two cumulants," *Signal Processing*, vol. 43, pp. 55–63, 1995.

[6] ——, "A unified class of inverse filter criteria using two cumulants for blind deconvolution and equalization," in *Proc. IEEE Int. Conf. Acoustics, Speech, Signal Processing*, Detroit, MI, 1995, pp. 1960–1963.

[7] D. L. Donoho, *On minimum entropy deconvolution*, ser. In Applied Time Series Analysis II, D. F. Findly, Ed.   New York: Academic, 1981.

[8] O. Shalvi and E. Weinstein, "New criteria for blind deconvolution of nonminimum phase systems (channels)," *IEEE Transaction on Information Theory*, vol. IT-36, pp. 312–321, 1990.

[9] ——, "Super-exponential methods for blind deconvolution," *IEEE Trans. on Information Theory*, vol. 39, pp. 504–519, 1993.

[10] D. Hatzinakos and C. L. Nikias, "Blind equalization using a tricespectrum based algorithm," *IEEE Transaction on Comm.*, vol. 39, pp. 669–682, 1991.

[11] B. Jellonek, D. Boss, and K. D. Kammeyer, "Generalized eigenvector algorithm for blind equalization," *EURASIP Signal Processing*, pp. 237–264, 1997.

[12] B. Jellonek and K. D. Kammeyer, "A closed-form solution to blind equalization," *EURASIP Signal Processing*, vol. 36, pp. 251–259, 1994.

[13] Y. Sato, "A method of self-recovering equalization for multilevel amplitude-modulation systems," *IEEE Transaction on Communications*, vol. COM-23, pp. 679–682, 1975.

[14] D. N. Godard, "Self recovering equalization and carrier tracking in two-dimensional data communication systems," *IEEE Transaction on Communications*, vol. COM-28, pp. 1867–1875, 1980.

[15] S. Bellini, "Bussgang techniques for blind equalization," in *IEEE Global Telecommunication Conference Records*, 1986, pp. 1634–1640.

[16] ——, "Blind equalization," *Alta Freq.*, vol. 57, pp. 445–450, 1988.

[17] S. Fiori, "A contribution to (neuromorphic) blind deconvolution by flexible approximated bayesian estimation," *Signal Processing*, vol. 81, pp. 2131–2153, 2001.

[18] ——, "A fast fixed-point neural blind deconvolution algorithm," *IEEE Transaction on Neural Networks*, vol. 15, pp. 455–459, 2004.

[19] M. Pinchas and B. Z. Bobrovsky, "A maximum entropy approach for blind deconvolution," *Signal Processing Journal (Eurasip)*, vol. 86, pp. 2913–2931, 2006.

[20] ——, "A novel hos approach for blind channel equalization," *IEEE Wireless Communication Journal*, vol. 6, 2007.

[21] M. Pinchas, "A mse optimized polynomial equalizer for 16qam and 64qam constellation," *Signal, Image and Video Processing*, vol. 5 , No. 1, pp. 29–37, 2011.

[22] M. Lazaro *et al.*, "Stochastic blind equalization based on pdf fitting using parzen estimator," *IEEE Trans. on Signal Processing*, vol. 53, pp. 696–704, 2005.

[23] G.-H. Im, C. J. Park, and H. C. Won, "A blind equalization with the sign algorithm for broadband access," *IEEE Comm. Letters*, vol. 5, pp. 70–72, 2001.

Chapter 6

EQUALIZATION PERFORMANCE ANALYSIS

MONIKA PINCHAS

Department of Electrical and Electronic Engineering, Ariel University Center of Samaria, Ariel 40700, ISRAEL

ABSTRACT

Recently, a closed-form approximated expression was proposed for the achievable residual intersymbol interference (ISI) valid for the real valued and two independent quadrature carrier case and for type of blind equalizers where the error that is fed into the adaptive mechanism which updates the equalizer's taps can be expressed as a polynomial function of order three of the equalized output. Thus the recently proposed expression for the achievable residual ISI can not be applied for input constellations such as the 32QAM or V29 case. In this chapter we propose for the noiseless case, a new closed-form approximated expression for the residual ISI that depends on the step-size parameter, equalizer's tap length, input signal statistics, channel power and that is valid for the general case of input constellations. The new closed-form approximated expression for the residual ISI is applicable for type of blind equalizers where the error that is fed into the adaptive mechanism which updates the equalizer's taps can be expressed as a polynomial function of order three of the equalized output like in Godard's algorithm. Since the channel power is measurable or can be calculated if the channel coefficients are given, there is no need anymore to carry out any simulation with various step-size parameters in order to reach the required residual ISI.

KEYWORDS

Blind deconvolution, intersymbol interference (ISI), convergence speed, perfect equalization, polynomial function, convolutional noise power, residual ISI, step-size parameter, equalizer's tap-length, channel power

INTRODUCTION

The main concern in blind adaptive equalizers is having a low residual ISI where the eye diagram is considered to be open with a fast convergence speed. Fast convergence speed may be obtained by increasing the step-size parameter. But increasing the step-size parameter may lead to a higher residual ISI which might not meet any more the system's requirements. Recently [1], [2], closed-form approximated expressions were proposed for the achievable residual ISI as a function of the step-size parameter, equalizer's tap-length, input signal statistics and channel power. These approximated expressions are valid for the real valued and two independent quadrature carrier case and for type of blind equalizers where the error that is fed into the adaptive mechanism which updates the equalizer's taps can be expressed as a polynomial function of order three of the equalized output. The main problem of the proposed expressions for the residual ISI [1], [2] is that they are valid only for the real valued and two independent quadrature carrier case such as the 8 PAM (pulse amplitude modulation), 16QAM, 64QAM or 256QAM constellation input. They are not applicable for the general case of complex input signals such as the 32QAM, 128QAM or V29 input constellation.

In this chapter, we propose for the noiseless case a closed-form approximated expression for the achievable residual ISI as a function of the step-size parameter, equalizer's tap length, input signal statistics and channel power that is valid for the real as well as for the general case of complex input signals such as the 32QAM or V29 input constellation. This new proposed expression is valid for type of blind equalizers where the error that is fed into the adaptive mechanism which updates the equalizer's taps can be expressed as a polynomial function of order three of the equalized output. Based on the new proposed expression for the achievable residual ISI, a new closed-form approximated expression is proposed for the above mentioned case, that indicates if the chosen blind adaptive equalizer leads to perfect equalization performance from the residual ISI point of view. This new closed-form approximated expression can be a useful tool for equalization performance comparison between blind adaptive equalizers.

In this chapter, we adopt the LMS type algorithm as was done in [1], [2] for the adaptation of the equalizer's taps. It should be pointed out that the LMS type algorithm is simple but may lead to slow convergence and poor performance for hostile scenarios, whereas a least-square type algorithm is more computationally complex but has a better performance.

THE ACHIEVABLE RESIDUAL ISI

In this section we derive a closed-form approximated expression for the expected residual ISI as a function of the equalizer's tap length, constellation input statistics, step-size parameter and channel power. In addition we derive an expression that indicates how low the residual ISI has to be in order that the simulated results will be close to those obtained from our new proposed expression for the achievable residual

ISI.

Let us recall from chapter 2 the expression for the equalized output signal for the noiseless case:

$$z[n] = x[n] + p[n] \tag{6.1}$$

where $p[n]$ is the convolutional noise and $x[n]$ is the source signal.

Theorem: For the following assumptions:

1. The convolutional noise $p[n]$, is a zero mean, white Gaussian process with variance $\sigma_p^2 = E[p[n]p^*[n]]$ where $E[\cdot]$ is the expectation operator and $()^*$ is the conjugate operation.

2. The source signal $x[n]$ is a signal with known variance ($\sigma_x^2 = E[x[n]x^*[n]]$) and higher moments.

3. The convolutional noise $p[n]$ and the source signal $x[n]$ are independent. Thus, $\sigma_z^2 = E[z[n]z^*[n]] = E[(x[n] + p[n])(x[n] + p[n])^*] = E[x[n]x^*[n]] + E[p[n]p^*[n]]$

4. No noise is added.

5. $\dfrac{\partial F(z[n])}{\partial z[n]}$ can be expressed as a polynomial function of the equalized output namely as $P[z[n]]$ of order three as defined in (6.5).

6. The gain between the source and equalized output signal is equal to one. Namely, $|\tilde{s}|_{max}^2 = 1$ where $\tilde{s}$ is given in (2.4).

The residual ISI expressed in dB units may be defined for $|\tilde{s}|_{max}^2 = 1$ and $P[z[n]]$ of order 3 as:

$$ISI = 10\log_{10}(m_p) - 10\log_{10}(\sigma_x^2) \tag{6.2}$$

where m_p is defined by:

$$m_p = \min\left[Sol_1^{mp1}, Sol_2^{mp1}\right] \quad \text{for} \quad Sol_1^{mp1} > 0 \quad \text{and} \quad Sol_2^{mp1} > 0$$

or

$$m_p = \max\left[Sol_1^{mp1}, Sol_2^{mp1}\right] \quad \text{for} \quad Sol_1^{mp1} \cdot Sol_2^{mp1} < 0$$

where

$$Sol_1^{mp1} = \frac{-B_1+\sqrt{B_1^2-4A_1C_1B}}{2A_1}; \quad Sol_2^{mp1} = \frac{-B_1-\sqrt{B_1^2-4A_1C_1B}}{2A_1}$$

$$\tag{6.3}$$

$$A_1 = B\left(27\sigma_x^2 a_3^2 + 6a_1 a_3\right) - 6a_3$$

$$B_1 = B\left(8\sigma_x^2 a_1 a_3 + a_1^2 + 9E\left[|x[n]|^4\right]a_3^2\right) - 2\left(2a_3\sigma_x^2 + a_1\right)$$

$$\tag{6.4}$$

$$C_1 = \sigma_x^2 a_1^2 + 2E\left[|x[n]|^4\right]a_1 a_3 + E\left[|x[n]|^6\right]a_3^2$$

$$B = \mu N\sigma_x^2 \sum_{k=0}^{k=R-1} |h(k)|^2$$

R is the channel length, N is the equalizer's tap length, μ is the step-size parameter, a_1 and a_3 are the property of the chosen blind adaptive equalizer via:

$$\frac{\partial F(z[n])}{\partial z[n]} = a_1 z[n] + a_3 |z[n]|^2 z[n] \tag{6.5}$$

Proof :

We begin our proof by first recalling the update equation of the equalizer and using the assumption that $\partial F(z[n])/\partial z[n]$ can be expressed as a polynomial function of the equalized output namely as $P[z[n]]$.

$$\underline{c}_{eq}[n+1] = \underline{c}_{eq}[n] - \mu\frac{\partial F(z[n])}{\partial z[n]}\underline{y}^*[n] = \underline{c}_{eq}[n] - \mu P[z[n]]\underline{y}^*[n] \tag{6.6}$$

where $(\cdot)^*$ denotes the conjugate operation of $(\cdot)$, $\underline{y}[n] = [y[n] \ldots y[n - N + 1]]^T$ and the operator $()^T$ denotes for transpose of the function $()$. Next we multiply both sides of (6.6) with the horizontal vector:

$(y[n - 0] \quad y[n - 1] \quad y[n - 2] \quad \cdot \quad \cdot \quad \cdot \quad y[n - N + 1])$ as was done in [1] and obtain:

$$\sum_{m=0}^{m=N-1} c_{n+1}[m]\, y\,[n - m] = \sum_{m=0}^{m=N-1} c_n[m]\, y\,[n - m] - \tag{6.7}$$

$$\mu P(z[n]) \sum_{m=0}^{m=N-1} y\,[n - m]\, y^*\,[n - m]$$

which can be written as:

$$z_{n+1} - z_n = -\mu P[z[n]] \sum_{m=0}^{m=N-1} y\,[n - m]\, y^*\,[n - m] \tag{6.8}$$

where $z_{n+1} = \sum_{m=0}^{m=N-1} c_{n+1}[m]\, y\,[n - m]$ and $z_n = \sum_{m=0}^{m=N-1} c_n[m]\, y\,[n - m] = z[n]$. According to [1], $x[n+1]$ may be equal to $x[n]$. For that case ($x[n+1] = x[n]$ and having $h[n+1] = h[n]$), we obtain by (6.1) that $z_{n+1} = z[n + 1] = x[n] + p[n + 1]$. By using (6.1) we may write (6.8) as:

$$z_{n+1} - z_n = p\,[n + 1] - p\,[n] = \Delta p = -\mu P[z[n]] \sum_{m=0}^{m=N-1} y\,[n - m]\, y^*\,[n - m] \tag{6.9}$$

According to [1], if we have a real function $g\,(p_r)$ (p_r is the real part of p[n]) then we may write by Tailor expansion [3] that

$$\Delta g = \frac{\partial g}{\partial p_r} \Delta p_r + \frac{1}{2} \frac{\partial^2 g}{\partial^2 p_r} (\Delta p_r)^2 + O((\Delta p_r)^3) \tag{6.10}$$

where $O(q)$ is defined as $\lim_{q \to 0}(O(q)/q) = r_{const}$ and "r_{const}" is a constant. Next we define $g\,(p_r) = p_r^2$ as was done in [1]. Thus we have by using (6.10):

$$\Delta p_r^2 \cong 2 p_r \Delta p_r + (\Delta p_r)^2 \tag{6.11}$$

Next, we wish to have an expression for $\Delta(pp^*)$. By using (6.10) we may define $\Delta(pp^*)$ as:

$$\Delta(pp^*) = \Delta(p_r^2 + p_i^2) = \Delta(p_r^2) + \Delta(p_i^2) \cong$$

$$2p_r\Delta p_r + (\Delta p_r)^2 + 2p_i\Delta p_i + (\Delta p_i)^2$$

(6.12)

where p_i is the imaginary part of $p[n]$. Next we use the following equalities:

$$p_r = \frac{p[n] + p^*[n]}{2}; \qquad p_i = \frac{p[n] - p^*[n]}{2}(-j)$$

$$\Delta p_r = \frac{\Delta p + \Delta p^*}{2}; \qquad \Delta p_i = \frac{\Delta p - \Delta p^*}{2}(-j)$$

(6.13)

By using (6.13) we may write (6.12) as:

$$\Delta(pp^*) \cong p[n]\Delta p^* + p^*[n]\Delta p + \Delta p\Delta p^*$$

(6.14)

Now by using (6.14) and (6.9) we may write:

$$\Delta(pp^*) \cong -\mu(p[n](P[z[n]])^* + p^*[n]P[z[n]]) \cdot$$

$$\sum_{m=o}^{m=N-1} y[n-m]y^*[n-m] + \mu^2(P[z[n]])^*(P[z[n]]) \cdot$$

(6.15)

$$\sum_{m=o}^{m=N-1} y[n-m]y^*[n-m] \sum_{m=o}^{m=N-1} y(n-m)y^*[n-m]$$

Our next step is applying the expectation operator on both sides of (6.15). Thus we obtain:

$$E[\Delta(pp^*)] \cong -\mu E[(p[n](P[z[n]])^* + p^*[n]P[z[n]])] \cdot$$

$$E\left[\sum_{m=o}^{m=N-1} y[n-m]y^*[n-m]\right] + \mu^2 E[(P[z[n]])^*(P[z[n]])] \cdot$$

$$E\left[\sum_{m=o}^{m=N-1} y[n-m]y^*[n-m] \sum_{m=o}^{m=N-1} y(n-m)y^*[n-m]\right]$$

(6.16)

By using (6.1) and (6.5) we obtain:

$$E\left[-\mu\left(p[n]\left(P\left[z[n]\right]\right)^* + p^*[n]P\left[z[n]\right]\right)\right] E\left[\sum_{m=o}^{m=N-1} y[n-m]y^*[n-m]\right] \cong$$

$$-2B\left(\sigma_p^2 a_1 + E\left[|p|^4\right]a_3 + 2\sigma_p^2\sigma_x^2 a_3\right)$$

$$E\left[(P\left[z[n]\right])^*\left(P\left[z[n]\right]\right)\right] \cong E\left[|p[n]|^6\right]a_3^2 + 9E\left[|p[n]|^4\right]\sigma_x^2 a_3^2 +$$
$$2E\left[|p[n]|^4\right]a_1 a_3 +$$
$$9E\left[|x[n]|^4\right]\sigma_p^2 a_3^2 + 8\sigma_p^2\sigma_x^2 a_1 a_3 + \sigma_p^2 a_1^2 + E\left[|x[n]|^6\right]a_3^2 +$$
$$2E\left[|x[n]|^4\right]a_1 a_3 + \sigma_x^2 a_1^2$$

$$E\left[\sum_{m=o}^{m=N-1} y[n-m]y^*[n-m]\sum_{m=o}^{m=N-1} y(n-m)y^*[n-m]\right] \cong B^2$$
$$(6.17)$$

where B is given in (6.4) and $\sigma_p^2 = E[p[n]p^*[n]]$. Thus by using (6.17) and (6.16) we may rewrite (6.16) as:

$$E\left[\Delta\left(pp^*\right)\right] \cong \left(15B^2 a_3^2\right)m_p^3 + \left(B^2\left(27\sigma_x^2 a_3^2 + 6a_1 a_3\right) - 6Ba_3\right)m_p^2 +$$

$$\left(B^2\left(8\sigma_x^2 a_1 a_3 + a_1^2 + 9E\left[|x[n]|^4\right]a_3^2\right) - 2B\left(2a_3\sigma_x^2 + a_1\right)\right)m_p +$$

$$B^2\left(\sigma_x^2 a_1^2 + 2E\left[|x[n]|^4\right]a_1 a_3 + E\left[|x[n]|^6\right]a_3^2\right)$$
$$(6.18)$$

where $m_p = E\left[p[n]p^*[n]\right]$. In the latter stages were the algorithm has converged we may write that $E\left[\Delta\left(pp^*\right)\right] \cong 0$. Therefore, substituting $E\left[\Delta\left(pp^*\right)\right] \cong 0$ into (6.18) and dividing by B (for $B \neq 0$) leads to:

$$D_1 m_p^3 + A_1 m_p^2 + B1m_p + BC_1 \cong 0 \qquad (6.19)$$

where A_1, B_1, C_1 are given in (6.4) and D_1 is given by:

$$D_1 = 15Ba_3^2 \qquad (6.20)$$

In the following we start to derive the relation between the convolutional noise power m_p and ISI. By using (6.1), (2.4) and (2.5) for the noiseless case, we may write:

$$z[n] = \tilde{c}[n] * h[n] * x[n] = \tilde{s}[n] * x[n]$$

$$\Downarrow \qquad\qquad (6.21)$$

$$E[z[n]z[n]^*] = \sigma_x^2 \sum_{\tilde{m}} |\tilde{s}(\tilde{m})|^2$$

Now we use the relation of $\sigma_p^2 = \sigma_z^2 - \sigma_x^2$ (assumption 3) together with (6.21) to obtain:

$$\sigma_p^2 = \sigma_z^2 - \sigma_x^2 = \sigma_x^2 \left[\sum_{\tilde{m}} |\tilde{s}(\tilde{m})|^2 - 1 \right] \qquad (6.22)$$

which by the help of (2.7) may be expressed for $|\tilde{s}|_{max}^2 = 1$ as:

$$\sigma_p^2 = \sigma_x^2 \cdot ISI \qquad \text{for} \quad |\tilde{s}|_{max}^2 = 1 \qquad (6.23)$$

For the case of $|D_1 m_p^3| \ll |A_1 m_p^2|$ (a case that fits an easy channel, namely a channel where the ISI is relatively low (but the eye diagram is still very closed)) we may neglect the product of $D_1 m_p^3$ in (6.19) and write that:

$$A_1 m_p^2 + B_1 m_p + BC_1 \cong 0 \qquad (6.24)$$

Next we solve (6.24) (which is a second order equation for m_p) under the assumption that $\Delta \geq 0$ where Δ is defined as:

$$\Delta = B_1^2 - 4A_1 C_1 B \qquad (6.25)$$

Assuming $\Delta \geq 0$ leads to one or two possible solutions for m_p named as Sol_1^{mp1} and Sol_2^{mp1} which are given in (6.3). This completes our *Proof*.

   Next we find an expression that shows how low the ISI has to be in order for (6.2) to be valid.

We neglected in our derivations the product of $D_1 m_p^3$ in (6.19) assuming that in the final stages of the deconvolutional process the ISI is relative low. As a matter of fact we assumed that

$$|D_1 m_p^3| \ll |A_1 m_p^2| \tag{6.26}$$

Since $B > 0$ and $a_3^2 > 0$, we have according to (6.20) that $D_1 > 0$. Since $m_p > 0$ we may write (6.26) as:

$$D_1 m_p \ll |A_1| \tag{6.27}$$

Now by using (6.4) , (6.20), (6.23) and (6.27) we obtain:

$$ISI \ll \frac{|A_1|}{\sigma_x^2 D_1}$$

$$\tag{6.28}$$

$$ISI \ll \frac{|B\left(27\sigma_x^2 a_3^2 + 6a_1 a_3\right) - 6a_3|}{\sigma_x^2 15 B a_3^2}$$

According to (6.28), our new proposed expression for the residual ISI (6.2) is valid if the condition in (6.28) is fulfilled.

A NEW TOOL FOR EQUALIZATION PERFORMANCE COMPARISON

In this section, we show a very useful tool for equalization performance comparison for blind adaptive equalizers where the error that is fed into the adaptive mechanism which updates the equalizer's taps is expressed as a polynomial function of order three of the equalized output.

According to (6.24), we see that the convolutional noise power m_p does not converge in the steady state approximately to zero unless $C_1 = 0$. C_1 depends on the constellation input statistics and on the algorithm itself via a_1 and a_3. This implies that there might be some kind of equalization methods for which we will be left with a residual ISI dependent on the constellation input statistics and on the other hand, there might be other algorithms that might reach $m_p \to 0$ in the steady state. Therefore, C_1 can be very useful in telling us which equalization method leads to perfect equalization performance ($m_p \to 0$) from the residual ISI point of view. Let us consider for a moment Shalvi and Weinstein's [4] algorithm where $a_1 = 0$ (please see (6.46)). According to C_1 we see that Shalvi and Weinstein's [4] algorithm will never reach perfect equalization performance from the residual ISI point of view while Godard's algorithm [5] with a_1 and a_3 given in (6.44) will reach $m_p \to 0$ for the QPSK (quadrature phase shift keying) input case.

In the following we show that for the polynomial function defined in (6.5), we can not find $a_1 \neq 0$ and $a_3 \neq 0$ that lead C_1 to zero independent to the source signal. Let us recall C_1:

$$C_1 = \sigma_x^2 a_1^2 + 2E\left[|x[n]|^4\right] a_1 a_3 + E\left[|x[n]|^6\right] a_3^2 \qquad (6.29)$$

Next we take the partial derivation of C_1 with respect to a_1 and a_3 respectively and set the outcome to zero. Thus we obtain:

$$\frac{\partial C_1}{\partial a_1} = 2a_1\sigma_x^2 + 2a_3 E\left[|x[n]|^4\right] = 0$$

$$\frac{\partial C_1}{\partial a_3} = 2a_1 E\left[|x[n]|^4\right] + 2a_3 E\left[|x[n]|^6\right] = 0 \qquad (6.30)$$

From the first line of (6.30) we obtain:

$$a_1 = -\frac{E\left[|x[n]|^4\right]}{\sigma_x^2} a_3 \qquad (6.31)$$

By substituting (6.31) into the second line of (6.30) we obtain:

$$a_3\left(E\left[|x[n]|^6\right] - \frac{\left(E\left[|x[n]|^4\right]\right)^2}{\sigma_x^2}\right) = 0 \qquad (6.32)$$

Thus according to (6.32), we may conclude that we can not find $a_1 \neq 0$ and $a_3 \neq 0$ that lead C_1 to zero independent to the source signal. In addition we may see according to (6.32) that perfect equalization performance is obtained from the residual ISI point of view for $a_1 \neq 0$ and $a_3 \neq 0$ when $E\left[|x[n]|^6\right] = \left(E\left[|x[n]|^4\right]\right)^2 / \sigma_x^2$.

The expression for a_1 (6.31) was obtained from the first line of (6.30). But, we could also use the second line of (6.30) in order to derive a_1 where in that case we would have:

$$a_1 = -\frac{E\left[|x[n]|^6\right]}{E\left[|x[n]|^4\right]} a_3 \tag{6.33}$$

It should be pointed out that the expression for a_1 (6.31) corresponds to Godard's [5] expression for a_1 for $a_3 = 1$ (please see (6.44). But even if $a_3 \neq 1$ it is still Godard's [5] algorithm because in that case we might say that we scaled the step-size parameter accordingly. In order to see this we recall the equalizer's update equation:

$$c_m[n+1] = c_m[n] - \mu\left(a_1 z[n] + a_3 |z[n]|^2 z[n]\right) y^*[n-m] \tag{6.34}$$

Now we substitute (6.31) into (6.34) and obtain:

$$c_m[n+1] = c_m[n] - \mu\left(\left(-\frac{E\left[|x[n]|^4\right]}{\sigma_x^2} a_3\right) z[n] + a_3 |z[n]|^2 z[n]\right) y^*[n-m] \tag{6.35}$$

Next we define $a_3 = \tilde{a}_3 q$ where $\tilde{a}_3 = 1$ and substitute it into (6.35) to obtain:

$$c_m[n+1] = c_m[n] - \mu q\left(\left(-\frac{E\left[|x[n]|^4\right]}{\sigma_x^2} \tilde{a}_3\right) z[n] + \tilde{a}_3 |z[n]|^2 z[n]\right) y^*[n-m] =$$

$$c_m[n] - \mu_{\text{new}}\left(\left(-\frac{E\left[|x[n]|^4\right]}{\sigma_x^2}\right) z[n] + |z[n]|^2 z[n]\right) y^*[n-m] \tag{6.36}$$

where $\mu_{\text{new}} = \mu q$. According to (6.36) we may see that even if $a_3 \neq 1$ the expression for a_1 given in (6.31) still corresponds to Godard's [5] algorithm.

Now, the question that may arise here is which expression for a_1 ((6.33) or (6.31)) will lead to a lower value for C_1. This question is important because it tells us which algorithm will lead to a lower residual ISI, the algorithm that uses (6.33) or (6.31). In order to answer on the question we substitute (6.31) into (6.29) and denote the outcome as C_{1_G}:

$$C_{1_G} = a_3^2 \left(E\left[|x[n]|^6\right] - \frac{\left(E\left[|x[n]|^4\right]\right)^2}{\sigma_x^2} \right) \qquad (6.37)$$

Next we substitute (6.33) into (6.29) and denote the outcome as C_{1_new}:

$$C_{1\_new} = a_3^2 \left(\sigma_x^2 \left(\frac{E\left[|x[n]|^6\right]}{E\left[|x[n]|^4\right]} \right)^2 - E\left[|x[n]|^6\right] \right) \qquad (6.38)$$

Now the big question is whether C_{1_new} is greater or less than C_{1_G}. We start our derivations by assuming that $C_{1_G} \leq C_{1\_new}$.

$$a_3^2 \left(E\left[|x[n]|^6\right] - \frac{\left(E\left[|x[n]|^4\right]\right)^2}{\sigma_x^2} \right) \leq a_3^2 \left(\sigma_x^2 \left(\frac{E\left[|x[n]|^6\right]}{E\left[|x[n]|^4\right]} \right)^2 - E\left[|x[n]|^6\right] \right)$$

$$(6.39)$$

Since $a_3^2 > 0$ we may write (6.39) as:

$$-\sigma_x^2 \left(\frac{E\left[|x[n]|^6\right]}{E\left[|x[n]|^4\right]} \right)^2 + 2E\left[|x[n]|^6\right] - \frac{\left(E\left[|x[n]|^4\right]\right)^2}{\sigma_x^2} \leq 0 \qquad (6.40)$$

which is a second order equation with respect to $E\left[|x[n]|^6\right]$. Solving the following equation

$$-\sigma_x^2 \left(\frac{E\left[|x[n]|^6\right]}{E\left[|x[n]|^4\right]} \right)^2 + 2E\left[|x[n]|^6\right] - \frac{\left(E\left[|x[n]|^4\right]\right)^2}{\sigma_x^2} = 0 \qquad (6.41)$$

for $E\left[|x[n]|^6\right]$ leads to a single solution:

$$E\left[|x[n]|^6\right] = \frac{\left(E\left[|x[n]|^4\right]\right)^2}{\sigma_x^2} \tag{6.42}$$

Since the expression $\left(E\left[|x[n]|^6\right]\right)^2$ is multiplied by a negative value in (6.40), we have according to (6.42) that (6.39) is always true.

SIMULATION

In this section, the new closed-form approximated expression for the achievable residual ISI (6.2) is tested via simulation. In addition, we use (6.28) in a simple test case in order to see how low the ISI has to be in order for (6.2) to be valid. In our simulation, we use Godard's [5] and Shalvi and Weinstein's [4] algorithm. Three different input constellations and two different channels are used. The equalizer taps for Godard's algorithm [5] are updated according to:

$$c_m\left[n+1\right] = c_m\left[n\right] - \mu_G \left(|z\left[n\right]|^2 - \frac{E\left[|x\left[n\right]|^4\right]}{E\left[|x\left[n\right]|^2\right]}\right) z\left[n\right] y^*\left[n-m\right] \tag{6.43}$$

where μ_G is the step-size. The values for a_1 and a_3 corresponding to Godards's [5] algorithm are defined as a_1^G and a_3^G respectively and are given by:

$$a_1^G = -\frac{E\left[|x\left[n\right]|^4\right]}{E\left[|x\left[n\right]|^2\right]}; \qquad a_3^G = 1 \tag{6.44}$$

The equalizer taps for algorithm [4] are updated according to:

$$c_m'\left(n+1\right) = c_m''\left(n\right) + \mu_{SW} \cdot \text{sgn}\Upsilon(x)\left|z\left(n\right)\right|^2 z\left(n\right) y^*\left(n-m\right)$$

$$c_m''\left(n\right) = \left(1 \bigg/ \sqrt{\sum_m |c_m'|^2}\right) c_m' \tag{6.45}$$

where $c_m''\left(n\right)$ is the vector of taps after iteration, $c_m''\left(0\right)$ is some reasonable initial guess, μ_{SW} is the step-size and $\Upsilon(x) = E\left[|x|^4\right] -$

$2E^2\left[|x|^2\right] - |E\left[x^2\right]|^2$ is the kurtosis associated to x where $x = x[n]$. In the following, we denote algorithm [4] as SW. The values for a_1, a_{12} and a_3 corresponding to Shalvi and Weinstein's algorithm [4] are defined as a_1^{SW}, a_{12}^{SW} and a_3^{SW} respectively and are given by:

$$a_1^{SW} = 0; \qquad a_3^{SW} = -\text{sgn}\Upsilon(x) \qquad (6.46)$$

Three input sources were considered: A **QPSK source** (a modulation using $\pm \{1\}$ levels for in-phase and quadrature components), a **V29 source** (Fig. (**6.1**)) and a **32QAM source** (Fig. (**6.2**)). As already was mentioned, two different channels were considered:

Channel1 (initial ISI = 0.44): The channel parameters were determined according to [4]:
$h_n = \{0 \quad \text{for} \quad n < 0; \quad -0.4 \quad \text{for} \quad n = 0; \quad 0.84 \cdot 0.4^{n-1} \quad \text{for} \quad n > 0\}$.

Channel2 (initial ISI = 0.88): The channel parameters were determined according to: $h_n = (0.4851, -0.72765, -0.4851)$.
The equalizer's taps are initialized by setting the center tap equal to one and all others to zero.
Fig. (**6.3**) to Fig. (**6.15**) show the simulated performance of Godard's equalization method for the 32QAM and V29 input case, namely the ISI as a function of iteration number for various step-size parameters, channel characteristics, signal to noise ratio (SNR) and equalizer's tap length, compared with the calculated residual ISI expression (6.2) proposed in this chapter. According to Fig. (**6.3**) to Fig. (**6.15**), the calculated outcome from the proposed expression for the residual ISI (6.2) is very close to the simulated result. Fig. (**6.16**) shows the simulated performance of Shalvi and Weinstein's algorithm [4] for the QPSK input case, namely the ISI as a function of iteration number for various step-size parameters and SNR = 30 [dB], compared with

the calculated residual ISI expression (6.2) proposed in this chapter. According to Fig. (**6.16**), the calculated outcome from the proposed expression for the residual ISI (6.2) is very close to the simulated result.

Next we turn to test the expression given in (6.28) which indicates how low the ISI has to be in order to make (6.2) valid. We use Godard's algorithm [5] for the 32QAM input case sent via **Channel2** with two different values for the step-size parameter and tap length set equal to nineteen. According to (6.28) and $\mu_G = 0.0002$, we obtain that the residual ISI has to be much lower than 1.45 [dB] ($ISI \ll 1.45$ [dB]) in order to make (6.2) valid. According to Fig. (**6.13**) we see that the residual ISI is approximately equal to -16 [dB] which is much lower than 1.45 [dB]. Thus it is not surprising that the calculated and simulated value for the residual ISI are very close. According to (6.28) and $\mu_G = 0.0003$, we obtain that the residual ISI has to be much lower than -3.044 [dB] ($ISI \ll -3.044$ [dB]) in order to make (6.2) valid. According to Fig. (**6.13**) we see that the residual ISI is approximately equal to -13.5 [dB] which is much lower than -3.044 [dB]. Thus again, it is not surprising that the calculated and simulated value for the residual ISI are very close.

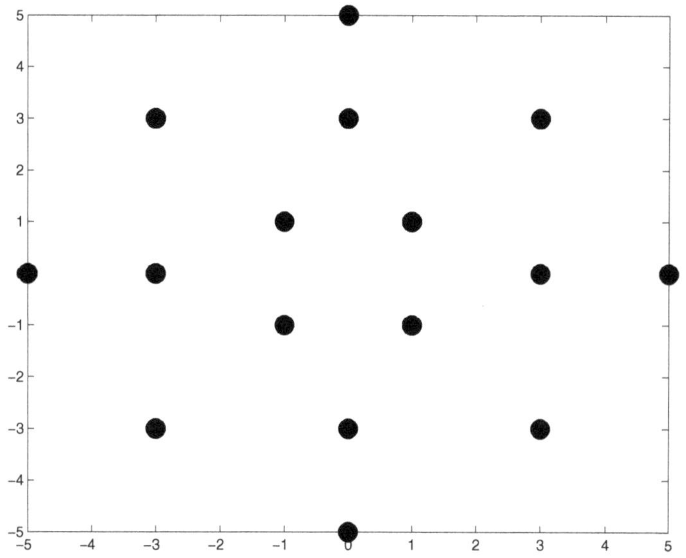

Figure 6.1: V29 source.

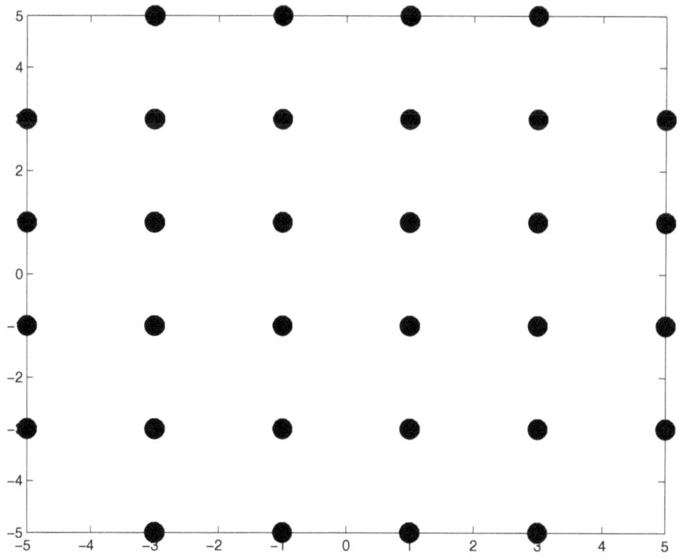

Figure 6.2: 32QAM source.

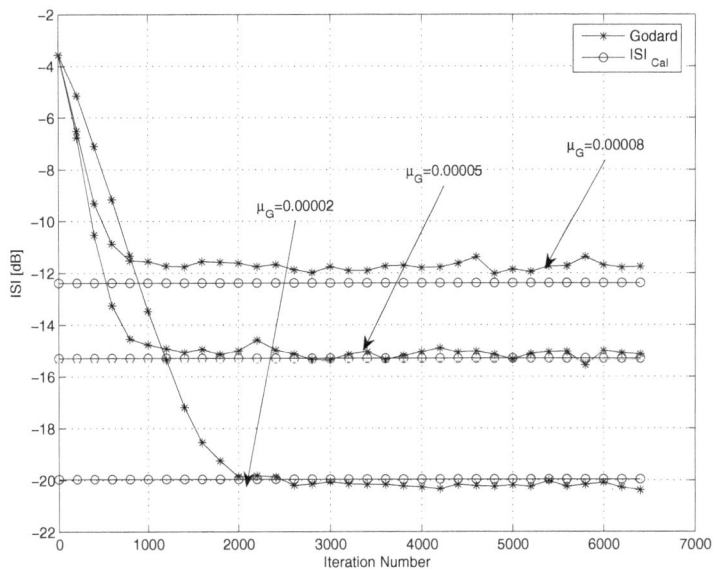

Figure 6.3: A comparison between the simulated (with Godard's algorithm) and calculated residual ISI for the V29 source input going through channel1. The averaged results were obtained in 100 Monte Carlo trials for SNR = 30 [dB]. The equalizer's length was set to 13.

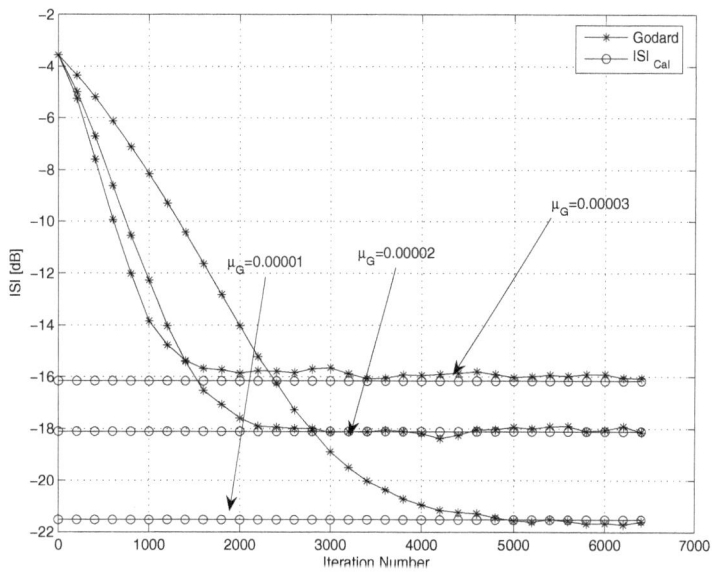

Figure 6.4: A comparison between the simulated (with Godard's algorithm) and calculated residual ISI for the V29 source input going through channel1. The averaged results were obtained in 100 Monte Carlo trials for SNR = 30 [dB]. The equalizer's length was set to 19.

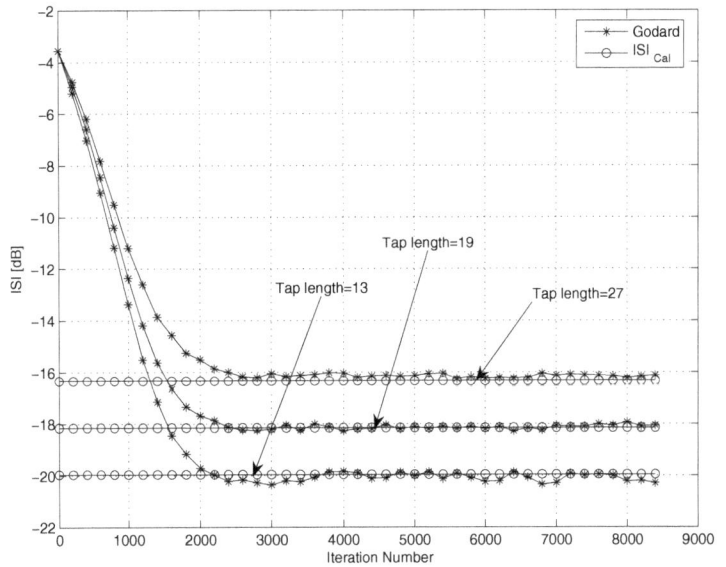

Figure 6.5: A comparison between the simulated (with Godard's algorithm) and calculated residual ISI for the V29 source input going through channel1. The averaged results were obtained in 100 Monte Carlo trials for SNR = 30 [dB]. The step-size parameter was set to $\mu_G = 0.00002$.

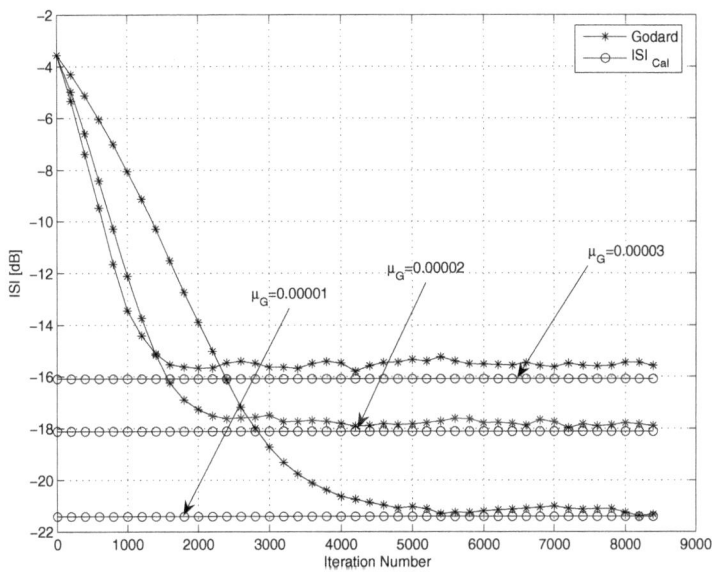

Figure 6.6: A comparison between the simulated (with Godard's algorithm) and calculated residual ISI for the V29 source input going through channel1. The averaged results were obtained in 100 Monte Carlo trials for SNR = 22 [dB]. The equalizer's length was set to 19.

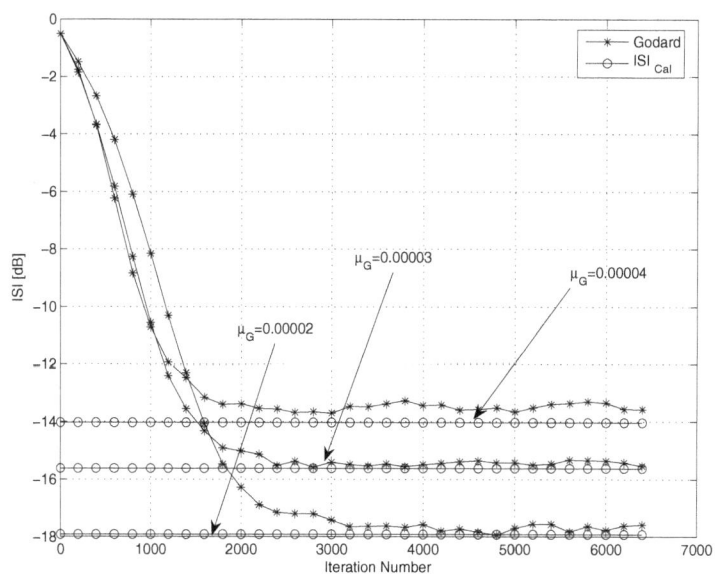

Figure 6.7: A comparison between the simulated (with Godard's algorithm) and calculated residual ISI for the V29 source input going through channel2. The averaged results were obtained in 100 Monte Carlo trials for SNR = 30 [dB]. The equalizer's length was set to 19.

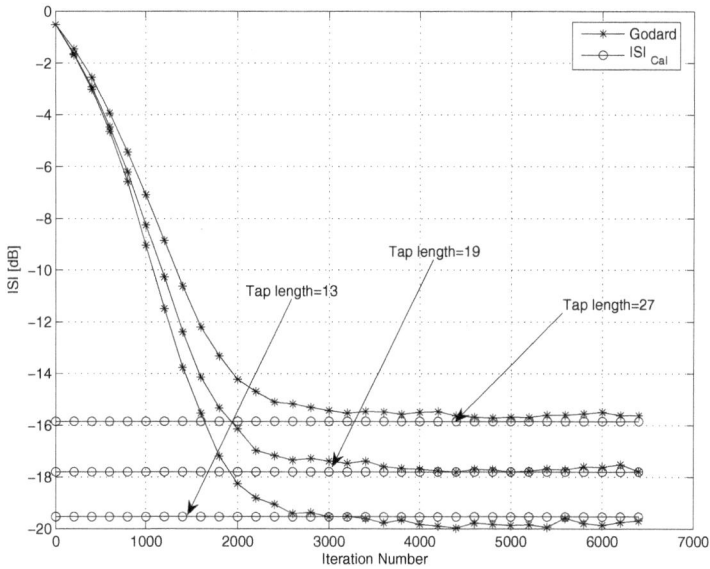

Figure 6.8: A comparison between the simulated (with Godard's algorithm) and calculated residual ISI for the V29 source input going through channel2. The averaged results were obtained in 100 Monte Carlo trials for SNR = 30 [dB]. The step-size parameter was set to $\mu_G = 0.00002$.

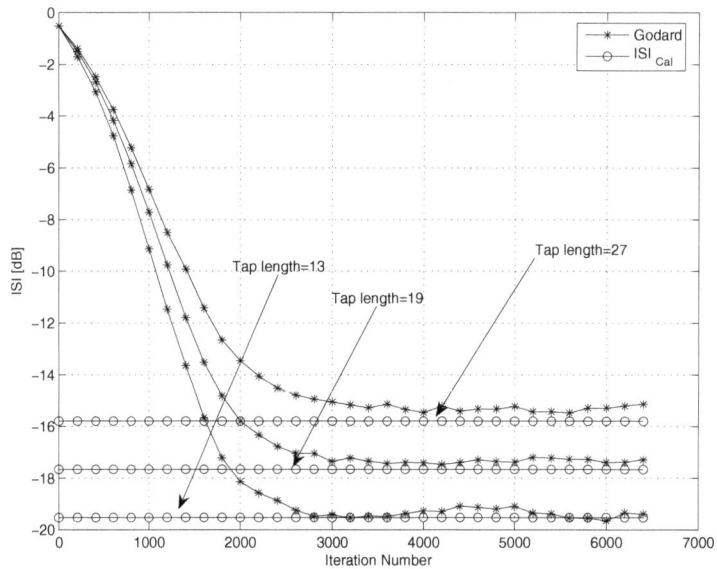

Figure 6.9: A comparison between the simulated (with Godard's algorithm) and calculated residual ISI for the V29 source input going through channel2. The averaged results were obtained in 100 Monte Carlo trials for SNR = 22 [dB]. The step-size parameter was set to $\mu_G = 0.00002$.

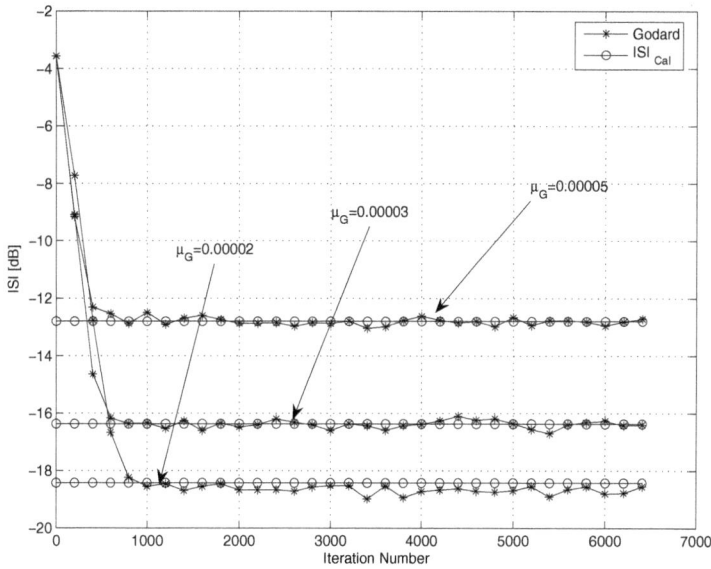

Figure 6.10: A comparison between the simulated (with Godard's algorithm) and calculated residual ISI for the 32QAM source input going through channel1. The averaged results were obtained in 100 Monte Carlo trials for SNR = 30 [dB]. The equalizer's length was set to 13.

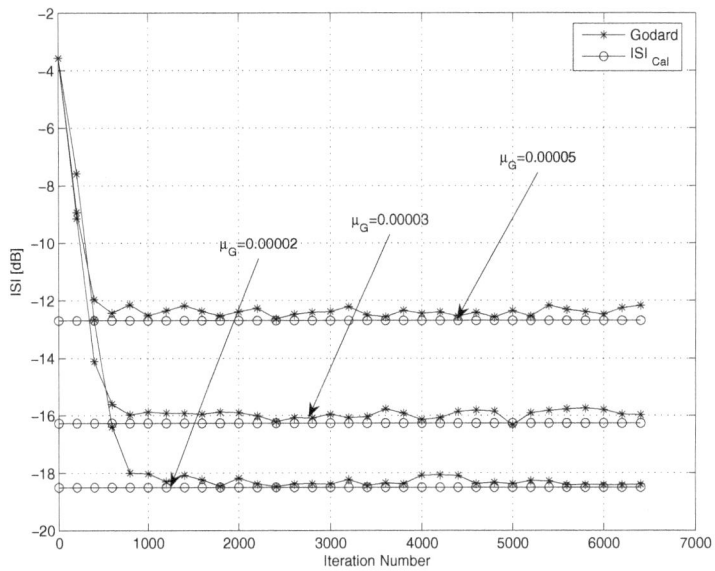

Figure 6.11: A comparison between the simulated (with Godard's algorithm) and calculated residual ISI for the 32QAM source input going through channel1. The averaged results were obtained in 100 Monte Carlo trials for SNR = 22 [dB]. The equalizer's length was set to 13.

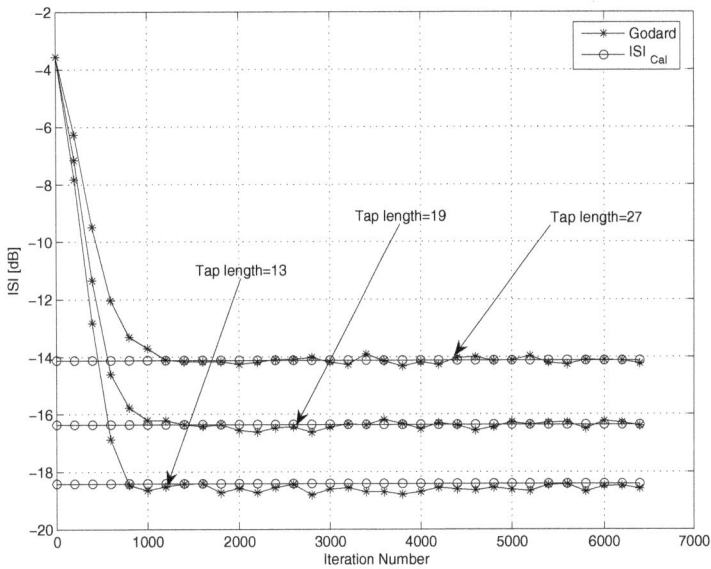

Figure 6.12: A comparison between the simulated (with Godard's algorithm) and calculated residual ISI for the 32QAM source input going through channel1. The averaged results were obtained in 100 Monte Carlo trials for SNR = 30 [dB]. The step-size parameter was set to $\mu_G = 0.00002$.

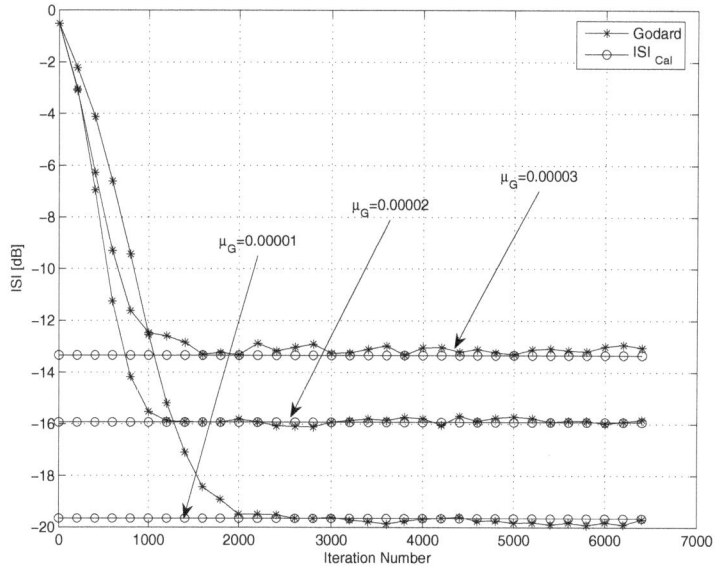

Figure 6.13: A comparison between the simulated (with Godard's algorithm) and calculated residual ISI for the 32QAM source input going through channel2. The averaged results were obtained in 100 Monte Carlo trials for SNR = 30 [dB]. The equalizer's length was set to 19.

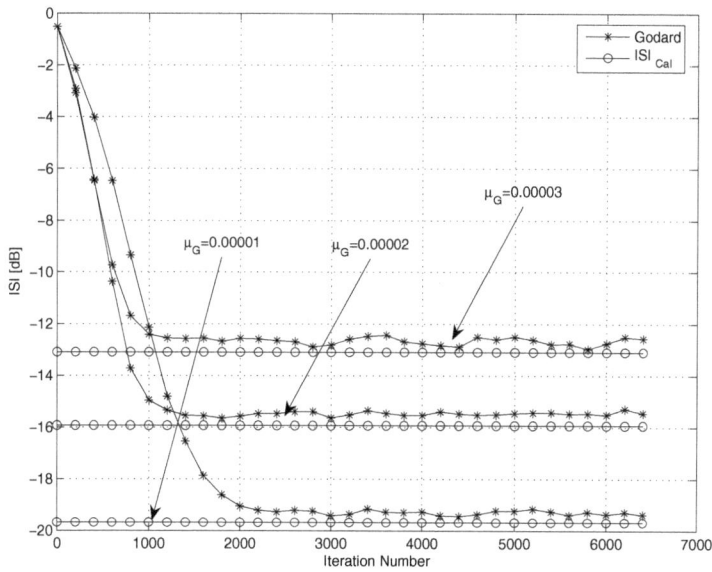

Figure 6.14: A comparison between the simulated (with Godard's algorithm) and calculated residual ISI for the 32QAM source input going through channel2. The averaged results were obtained in 100 Monte Carlo trials for SNR = 22 [dB]. The equalizer's length was set to 19.

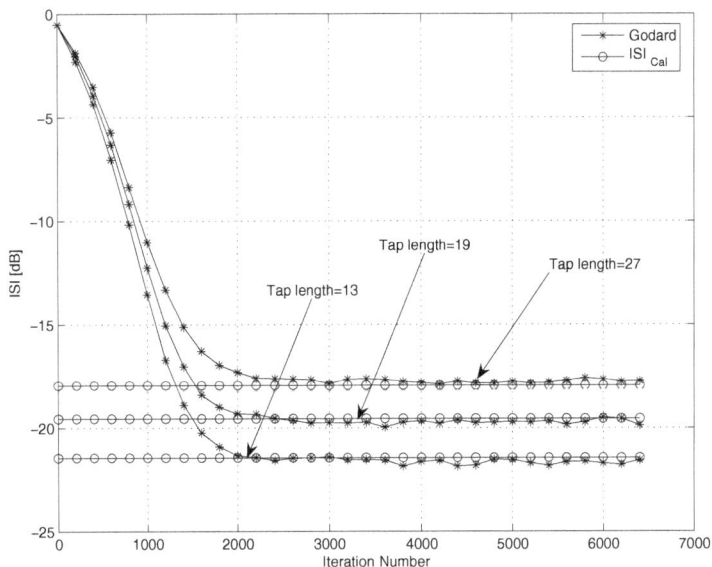

Figure 6.15: A comparison between the simulated (with Godard's algorithm) and calculated residual ISI for the 32QAM source input going through channel2. The averaged results were obtained in 100 Monte Carlo trials for SNR = 30 [dB]. The step-size parameter was set to $\mu_G = 0.00001$.

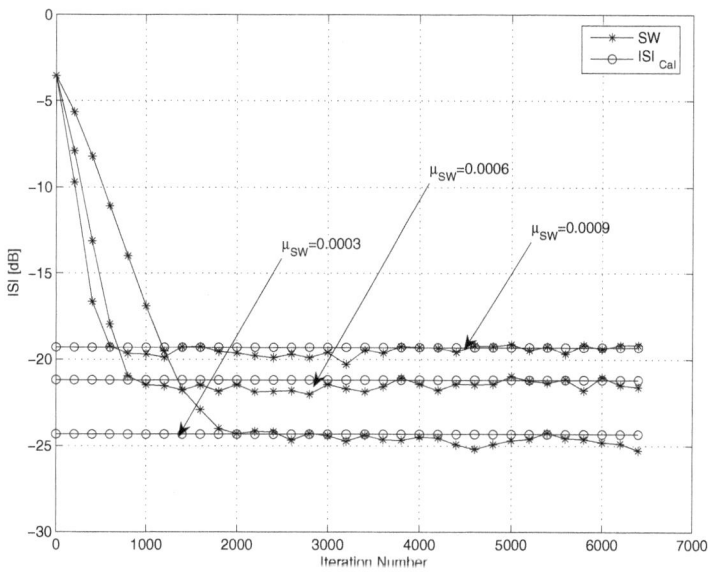

Figure 6.16: A comparison between the simulated (with SW algorithm) and calculated residual ISI for the QPSK source input going through channel1. The averaged results were obtained in 100 Monte Carlo trials for SNR = 30 [dB]. The equalizer's length was set to 13.

CONCLUSION

In this chapter a closed-form approximated expression for the residual ISI was derived for the noiseless, general complex input case (such as the 32QAM constellation) valid for type of blind adaptive equalizers where the error that is fed into the adaptive mechanism which updates the equalizer's taps can be expressed as a polynomial function of order three of the equalized output. The derived expression for the achievable residual ISI is dependent on the channel power (which is measurable or can be calculated if the channel coefficients are given), on the step-size parameter, equalizer's tap length and input signal statistics. Since the step-size parameter, the input signal statistics as well as the equalizer's tap length are known parameters by the system designer, there is no need anymore to carry out any simulation with various step-size parameters in order to obtain the optimal step-size parameter for a required residual ISI. Although the closed-form approximated expression for the achievable residual ISI was derived for the noiseless case, simulation results indicate that the new expression provides very good results also for the noisy case. In this chapter, a new closed-form expression was derived (named in the paper as C_1) that indicates if the chosen equalization method leads to perfect equalization performance from the residual ISI point of view. This new expression can be a useful tool for equalization performance comparison between blind equalizers. We have shown in this chapter for the general complex input case, a new technique for obtaining a closed-form approximated expression for the residual ISI for type of blind adaptive equalizers where the error that is fed into the adaptive mechanism which updates the equalizer's taps can be expressed as a polynomial function of order three of the equalized output. But, the same technique can be also applied for other type of blind adaptive equalizers where the error that is fed into the adaptive mechanism which updates the equalizer's taps is expressed as a polynomial function of the equalized output of order higher than three.

ABBREVIATIONS

ISI=   Intersymbol Interference

SNR=   Signal to Noise Ratio

QAM= Quadrature Amplitude Modulation

PAM=   Pulse Amplitude Modulation

QPSK=   Quadrature Phase Shift Keying

REFERENCES

[1] M. Pinchas, "A closed approximated formed expression for the achievable residual intersymbol interference obtained by blind equalizers," *Signal Processing Journal (Eurasip)*, vol. 90, pp. 1940–1962, 2010.

[2] ——, "A new closed approximated formed expression for the achievable residual isi obtained by adaptive blind equalizers for the noisy case," in *IEEE International Conference on Wireless Communications Networking and Information Security WCNIS2010*, 2010.

[3] M. R. Spiegel, Ed., *Mathematical Handbook, Schaum's Outline Series*. McGraw-Hill, 1997.

[4] O. Shalvi and E. Weinstein, "New criteria for blind deconvolution of non-minimum phase systems (channels)," *IEEE Transaction on Information Theory*, vol. IT-36, pp. 312–321, 1990.

[5] D. N. Godard, "Self recovering equalization and carrier tracking in two-dimensional data communication systems," *IEEE Transaction on Communications*, vol. COM-28, pp. 1867–1875, 1980.

Chapter 7

HOW DOES THE EQUALIZER'S PARAMETERS, CHANNEL CHARACTERISTICS OR INPUT SIGNAL CONSTELLATION AFFECT THE EQUALIZATION PERFORMANCE?

MONIKA PINCHAS

Department of Electrical and Electronic Engineering, Ariel University Center of Samaria, Ariel 40700, ISRAEL

ABSTRACT

It is very important for a system designer to understand the connection between the equalizer's performance (achievable residual intersymbol interference (ISI) and convergence speed) and the various parameters involved in the equalizer's design such as the equalizer's tap length, step-size parameter, channel power and input constellation in order to achieve optimal and expected equalization performance. In this chapter, we show the connection between the equalization performance (achievable residual ISI and convergence speed) and the step-size parameter, equalizer's tap length, channel power and input constellation statistics for type of equalizers where the error that is fed into the adaptive mechanism which updates the equalizer's taps can be expressed as a polynomial function of order three of the equalized output.

KEYWORDS

Blind deconvolution, intersymbol interference (ISI), convergence speed, convolutional noise power, residual ISI, step-size parameter, equalizer's tap-length, channel power, equalization performance, polynomial function of order three

THE CONNECTION BETWEEN THE EQUALIZER'S PARAMETERS AND THE EQUALIZATION PERFORMANCE

It is well known that an equalizer with insufficient tap length may lead to a high residual ISI. But does this mean that an equalizer with a high number of taps will always lead to improved equalization performance from the achievable residual ISI point of view and convergence speed? In this chapter we are going to answer on that question and show the connection between the equalization performance (achievable residual ISI and convergence speed) and the step-size parameter, equalizer's tap length, channel power and input constellation statistics for type of equalizers where the error that is fed into the adaptive mechanism which updates the equalizer's taps can be expressed as a polynomial function of order three of the equalized output. For that reason we recall from chapter 2 the equalized output signal for the noiseless case:

$$z\left[n\right] = x\left[n\right] + p\left[n\right] \tag{7.1}$$

where $p[n]$ is the convolutional noise and $x[n]$ is the source signal. Recently [1], a closed-form approximated expression was proposed for the achievable residual ISI valid for the real valued and two independent quadrature carrier case and for type of blind equalizers where the error that is fed into the adaptive mechanism which updates the equalizer's taps can be expressed as a polynomial function of order three of the equalized output. The recently proposed expression for the achievable residual ISI [1] can not be applied for input constellations such as the 32QAM (quadrature amplitude modulation) or V29 case. In the previous chapter, chapter 6, we proposed a new closed-form approximated expression for the achievable residual ISI that can be applied for various input cases including the 32QAM or V29 input constellation, under the following assumptions:

1. The convolutional noise $p[n]$, is a zero mean, white Gaussian process with variance $\sigma_p^2 = E[p[n]p^*[n]]$ where $E[\cdot]$ is the expectation operator and $()^*$ is the conjugate operation.

2. The source signal $x[n]$ is a signal with known variance ($\sigma_x^2 = E[x[n]x^*[n]]$) and higher moments.

3. The convolutional noise $p[n]$ and the source signal $x[n]$ are independent. Thus, $\sigma_z^2 = E[z[n]z^*[n]] = E[(x[n] + p[n])(x[n] + p[n])^*] = E[x[n]x^*[n]] + E[p[n]p^*[n]]$

4. No noise is added.

5. $\dfrac{\partial F(z[n])}{\partial z[n]}$ can be expressed as a polynomial function of the equalized output namely as $P[z[n]]$ of order three as defined in (7.5).

6. The gain between the source and equalized output signal is equal to one. Namely, $|\tilde{s}|_{max}^2 = 1$ where $\tilde{s}$ is given in (2.4).

The new closed-form approximated expression for the achievable residual ISI according to chapter 6 is given by:

$$ISI = 10\log_{10}(m_p) - 10\log_{10}(\sigma_x^2) \qquad (7.2)$$

where m_p is defined by:

$$m_p = \min\left[Sol_1^{mp_1}, Sol_2^{mp_1}\right] \quad \text{for} \quad Sol_1^{mp_1} > 0 \quad \text{and} \quad Sol_2^{mp_1} > 0$$

$$\text{or}$$

$$m_p = \max\left[Sol_1^{mp_1}, Sol_2^{mp_1}\right] \quad \text{for} \quad Sol_1^{mp_1} \cdot Sol_2^{mp_1} < 0$$

$$\text{where}$$

$$Sol_1^{mp_1} = \frac{-B_1 + \sqrt{B_1^2 - 4A_1C_1B}}{2A_1}; \quad Sol_2^{mp_1} = \frac{-B_1 - \sqrt{B_1^2 - 4A_1C_1B}}{2A_1}$$

$$A_1 = B\left(27\sigma_x^2 a_3^2 + 6a_1 a_3\right) - 6a_3$$
(7.3)

$$B_1 = B\left(8\sigma_x^2 a_1 a_3 + a_1^2 + 9E\left[|x[n]|^4\right] a_3^2\right) - 2\left(2a_3\sigma_x^2 + a_1\right)$$
(7.4)

$$C_1 = \sigma_x^2 a_1^2 + 2E\left[|x[n]|^4\right] a_1 a_3 + E\left[|x[n]|^6\right] a_3^2$$

$$B = \mu N \sigma_x^2 \sum_{k=0}^{k=R-1} |h(k)|^2$$

R is the channel length, N is the equalizer's tap length, μ is the step-size parameter, a_1 and a_3 are the property of the chosen blind adaptive equalizer via:

$$\frac{\partial F(z[n])}{\partial z[n]} = a_1 z[n] + a_3 |z[n]|^2 z[n]$$
(7.5)

It should be pointed out that the closed-form approximated expression for the residual ISI given in (7.2) is quite similar (but not equal) to the expression given in [1]. As a matter of fact the expression for "B" (7.4) $B = \mu N \sigma_x^2 \sum_{k=0}^{k=R-1} |h(k)|^2$ is the same as the one defined in [1]. Therefore, any conclusion we may get concerning the affect of the parameter "B" on the residual ISI may be also valid for the closed-form approximated expression for the residual ISI given in [1].

Fig. (**7.1**) to Fig. (**7.5**) show the ISI as a function of iteration number for Godard's [2] and Shalvi-Weinstein's [3] algorithm for various step-size parameters compared with the closed-form approximated expression for the residual ISI given in (7.2). The following channels were used:

Channel1 (initial ISI = 0.44): The channel parameters were determined according to [3]:

$$h_n = \{0 \quad \text{for} \quad n < 0; \quad -0.4 \quad \text{for} \quad n = 0; \quad 0.84 \cdot 0.4^{n-1} \quad \text{for} \quad n > 0\}.$$

Channel2 (initial ISI = 0.88): The channel parameters were determined according to: $h_n = (0.4851, -0.72765, -0.4851)$.

According to Fig. (**7.1**) to Fig. (**7.5**), the residual ISI is higher for higher values for the step-size parameter. This outcome can be also seen in [1]. Since the step-size parameter is connected with the parameter "B", we may say that if we enlarge "B", a higher residual ISI will be expected. But "B" may be also enlarged by choosing another constellation input with a higher input variance. The 32QAM input constellation has a higher variance compared with the V29 input case. Therefore, if the same channel, equalizer's tap length, same equalizer and the same step-size parameter are used for both the 32QAM and V29 input case, lower residual ISI is expected for the V29 input case compared with the 32QAM constellation. Indeed, according to Fig. (**7.2**) and Fig. (**7.4**), this outcome is seen very clearly when comparing both figures (Fig. (**7.2**) and Fig. (**7.4**)) for $\mu_G = 0.00002$ and $\mu_G = 0.00003$. Therefore, if we want to get for both cases (32QAM and V29 case) the same residual ISI for the same equalizer type, equalizer's tap length and channel, the step-size parameter for the 32QAM case should be set to:

$$\mu_{32\text{QAM}} = \frac{\sigma^2_{x\text{V29}}}{\sigma^2_{x32\text{QAM}}} \mu_{\text{V29}} \qquad (7.6)$$

where μ_{32QAM} and μ_{V29} are the step-size parameters for the 32QAM and V29 input case respectively. $\sigma^2_{x_{V29}}$ and $\sigma^2_{x_{32QAM}}$ are the input variances for the 32QAM and V29 input case respectively.

Next, we want to see how the equalizer's tap length affects the residual ISI. For that purpose, we first recall the expression for the convolution of the channel with the chosen equalizer:

$$\tilde{s}[n] = \tilde{c}\,[n] * h\,[n] = \delta\,[n] + \xi\,[n] \tag{7.7}$$

where $\xi[n]$ stands for the difference (error) between the ideal and non-ideal coefficients of $c[n]$ and $\tilde{c}[n]$ respectively and δ is the Kronecker delta function. According to (7.7), it is clear that the equalizer's tap length and the values of the coefficients are important here. Suppose for a moment that the ideal equalizer has 20 taps where all the taps have relative high values compared to the zero case and that we take an equalizer with only one tap. Obviously, the value for $\xi[n]$ will be high thus leading to a high residual ISI since the chosen equalizer is too short. Please note that in generally we do not know the length of the ideal equalizer since we do not know the channel we dealing with. So, the question that might arise here is why not using always an equalizer with a relative high number of taps? This question can be answered by looking on the parameter "B" given in (7.4). As it can be seen, "B" is a linear function of the equalizer's tap length. Therefore, if the equalizer's tap length is increased, a higher value for the residual ISI is expected. Fig. (**7.6**) to Fig. (**7.9**) show the residual ISI as a function of the iteration number for Godard's [2] algorithm for various equalizer's length compared with the closed-form approximated expression for the residual ISI given in (7.2). According to Fig. (**7.6**) to Fig. (**7.9**), increasing the equalizer's tap length causes an increase in the residual ISI. Please note that the closed-form approximated expression for the residual ISI (7.2) which uses the parameter "B" is valid for equalizers where the equalizer's tap length is considered to be sufficiently long in order to achieve the ideal case from the ISI point of view. It does not deal with the case where the equalizer's tap length is taken too short. Therefore, we may conclude that if the equalizer's tap length is taken too short or too high, a higher residual ISI is expected compared to the case where the equalizer's tap length is sufficiently long enough to reach the ideal case from the ISI point of view.

According to the parameter "B", the channel power $\sum_{k=0}^{k=R-1} |h(k)|^2$ plays also an important role in the achievable residual ISI. If we consider for a moment two channels having different channel powers, the channel with the higher channel power will lead to a higher residual ISI for the same step-size parameter, input constellation, equalizer's tap length and the same equalizer, compared to the channel with the lower channel power. Fig. (**7.10**) shows the residual ISI as a function of the iteration number for Godard's [2] algorithm for two different channel powers compared with the closed-form approximated expression for the residual ISI given in (7.2). It should be pointed out that **Channel3** was derived from **Channel1** where we just multiplied the gain by two. According to Fig. (**7.10**) we see that the channel with the higher channel power leads to a higher residual ISI compared to the channel with the lower channel power.

Up to now, we have seen that the step-size parameter is close related to the achievable equalization performance from the residual ISI point of view where a higher value for the step-size parameter will lead to a higher residual ISI. But a higher value for the step-size parameter will also lead to a faster convergence time as may be seen from Fig. (**7.1**), Fig. (**7.2**) and Fig. (**7.5**). The main task of a blind adaptive equalizer is to converge quickly and leave the system with a residual ISI which is low enough for the eye diagram to be considered as open. Therefore, we actually have here a trade off between the residual ISI and the convergence rate. The equalizer's tap length has also a great influence on the convergence rate. Increasing the equalizer's tap length will lead to a longer convergence time as may be seen from Fig. (**7.6**) to Fig. (**7.9**). According to [4], the best rate of convergence is dependent on the number of filter coefficients. The more coefficients (in the equalizer), the longer it takes for the coefficients to converge. The more coefficients there are, the more "noise" is introduced into the adaptation of each coefficient by the simultaneous adaptation of the other coefficients [4]. Therefore, we may conclude that if the equalizer's tap length is taken too high, a higher convergence time and residual ISI are expected.

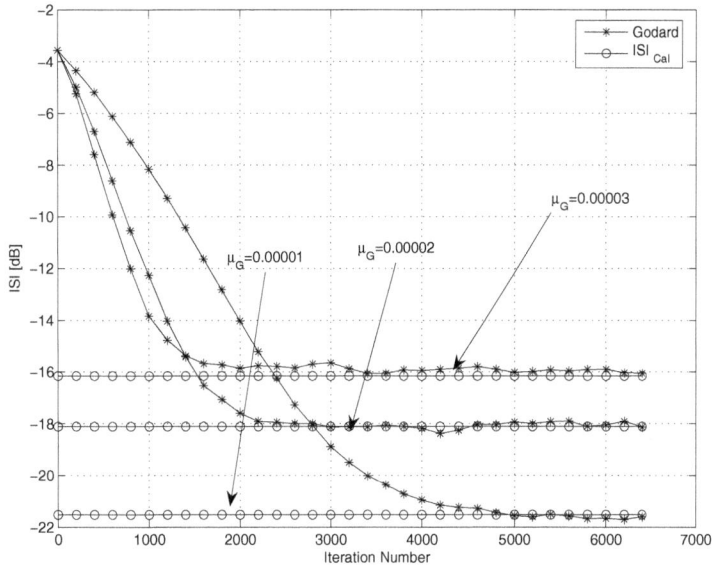

Figure 7.1: A comparison between the simulated (with Godard's algorithm) and calculated residual ISI for the V29 source input going through channel1. The averaged results were obtained in 100 Monte Carlo trials for SNR = 30 [dB]. The equalizer's length was set to 19.

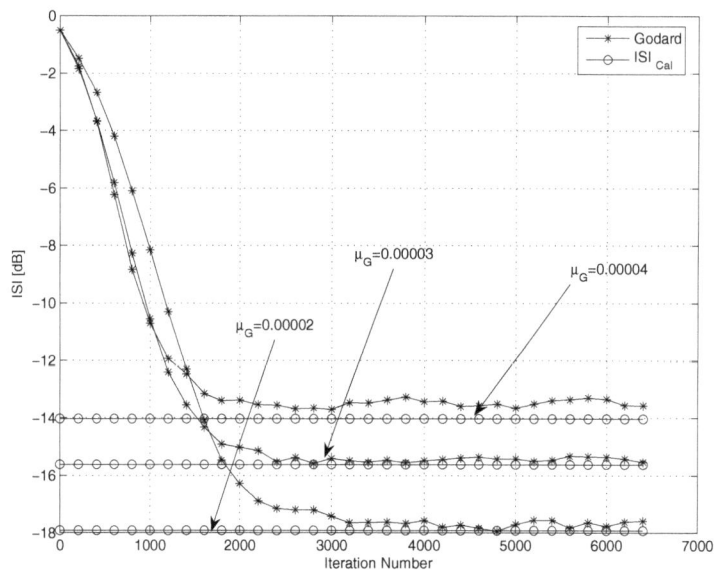

Figure 7.2: A comparison between the simulated (with Godard's algorithm) and calculated residual ISI for the V29 source input going through channel2. The averaged results were obtained in 100 Monte Carlo trials for SNR = 30 [dB]. The equalizer's length was set to 19.

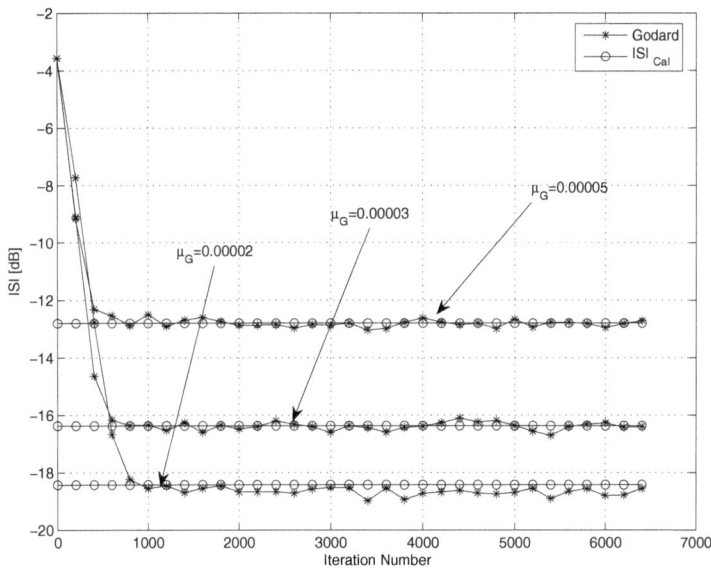

Figure 7.3: A comparison between the simulated (with Godard's algorithm) and calculated residual ISI for the 32QAM source input going through channel1. The averaged results were obtained in 100 Monte Carlo trials for SNR = 30 [dB]. The equalizer's length was set to 13.

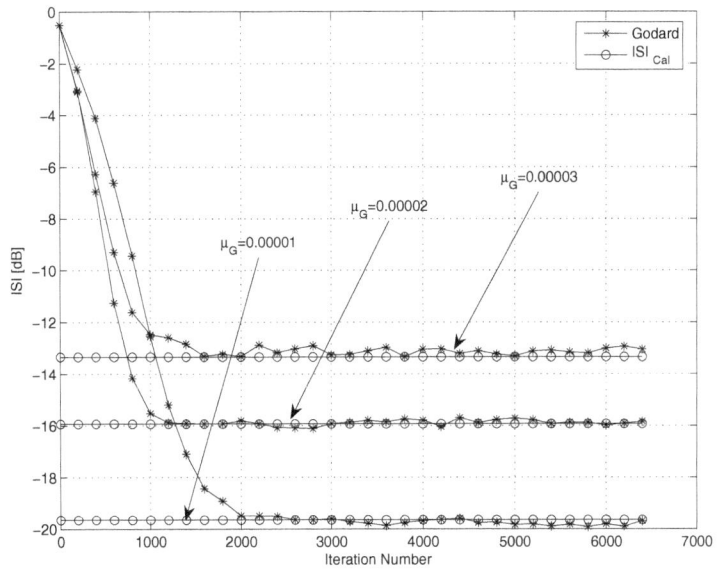

Figure 7.4: A comparison between the simulated (with Godard's algorithm) and calculated residual ISI for the 32QAM source input going through channel2. The averaged results were obtained in 100 Monte Carlo trials for SNR = 30 [dB]. The equalizer's length was set to 19.

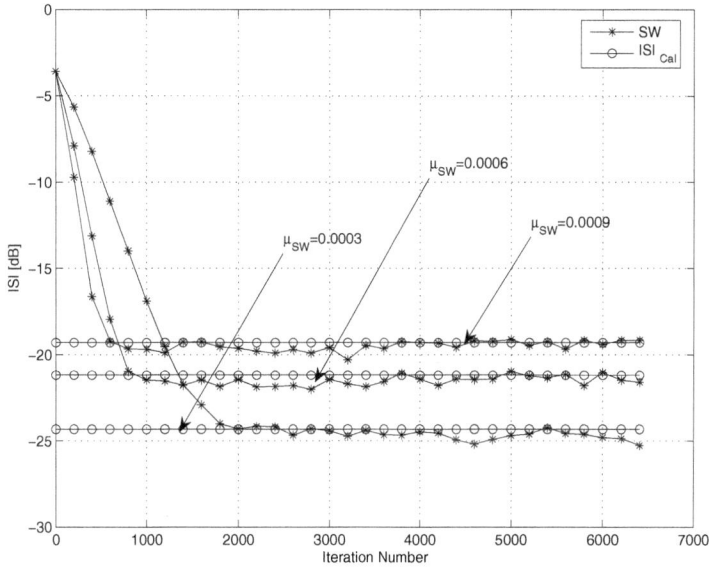

Figure 7.5: A comparison between the simulated (with SW algorithm) and calculated residual ISI for the QPSK (quadrature phase shift keying) source input going through channel1. The averaged results were obtained in 100 Monte Carlo trials for SNR = 30 [dB]. The equalizer's length was set to 13.

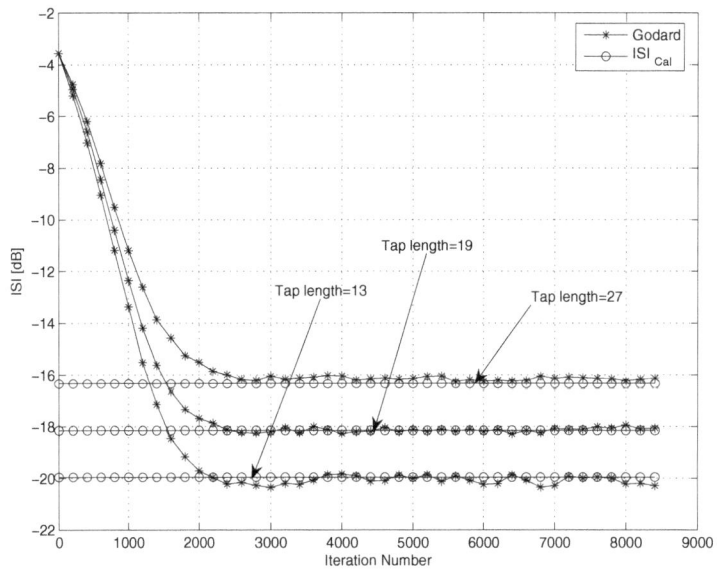

Figure 7.6: A comparison between the simulated (with Godard's algorithm) and calculated residual ISI for the V29 source input going through channel1. The averaged results were obtained in 100 Monte Carlo trials for SNR = 30 [dB]. The step-size parameter was set to $\mu_G = 0.00002$.

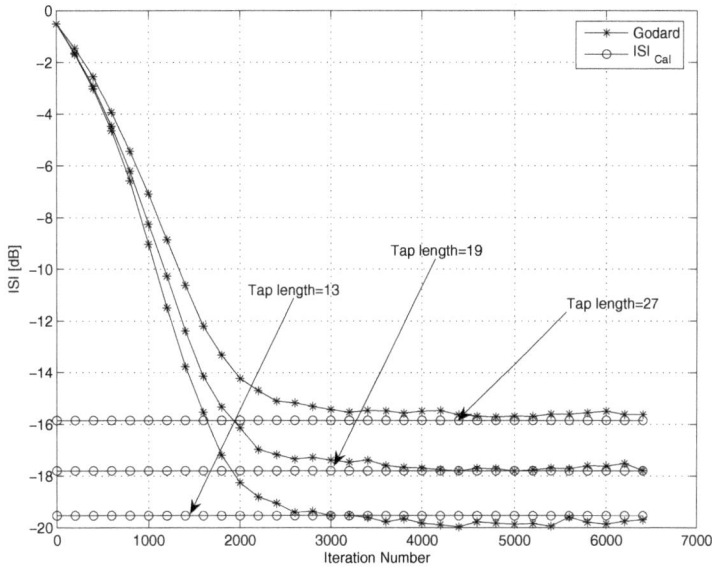

Figure 7.7: A comparison between the simulated (with Godard's algorithm) and calculated residual ISI for the V29 source input going through channel2. The averaged results were obtained in 100 Monte Carlo trials for SNR = 30 [dB]. The step-size parameter was set to $\mu_G = 0.00002$.

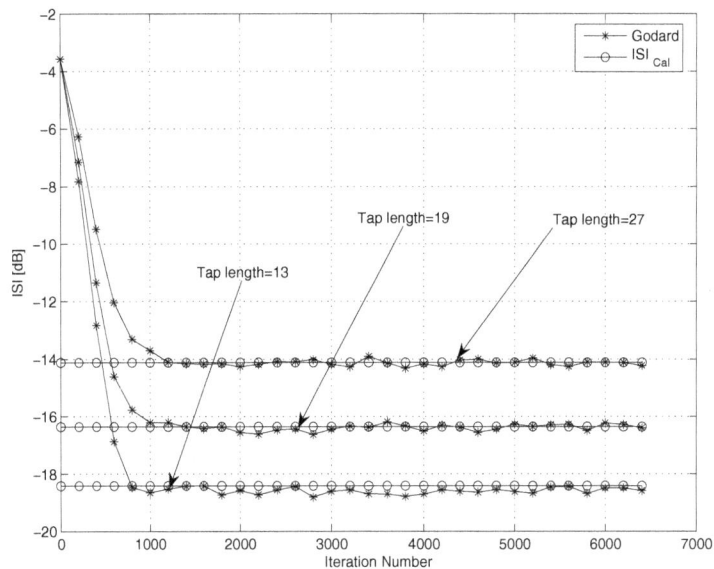

Figure 7.8: A comparison between the simulated (with Godard's algorithm) and calculated residual ISI for the 32QAM source input going through channel1. The averaged results were obtained in 100 Monte Carlo trials for SNR = 30 [dB]. The step-size parameter was set to $\mu_G = 0.00002$.

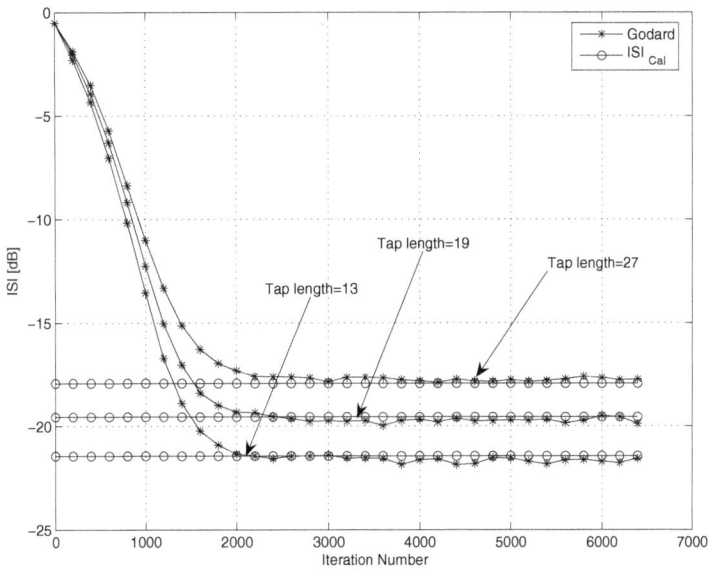

Figure 7.9: A comparison between the simulated (with Godard's algorithm) and calculated residual ISI for the 32QAM source input going through channel2. The averaged results were obtained in 100 Monte Carlo trials for SNR = 30 [dB]. The step-size parameter was set to $\mu_G = 0.00001$.

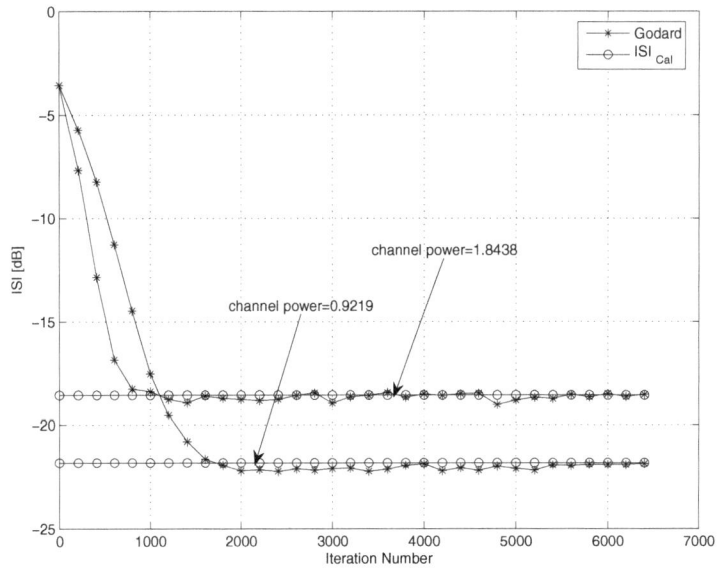

Figure 7.10: A comparison between the simulated (with Godard's algorithm) and calculated residual ISI for the 32QAM source input going through channel1 and channel3. The averaged results were obtained in 100 Monte Carlo trials for SNR = 30 [dB]. The equalizer's length was set to 13 and the step-size parameter was set to $\mu_G = 0.00001$.

ABBREVIATIONS

ISI=   Intersymbol Interference

SNR=   Signal to Noise Ratio

QAM= Quadrature Amplitude Modulation

QPSK=   Quadrature Phase Shift Keying

REFERENCES

[1] M. Pinchas, "A closed approximated formed expression for the achievable residual intersymbol interference obtained by blind equalizers," *Signal Processing Journal (Eurasip)*, vol. 90, pp. 1940–1962, 2010.

[2] D. N. Godard, "Self recovering equalization and carrier tracking in two-dimensional data communication systems," *IEEE Transaction on Communications*, vol. COM-28, pp. 1867–1875, 1980.

[3] O. Shalvi and E. Weinstein, "New criteria for blind deconvolution of non-minimum phase systems (channels)," *IEEE Transaction on Information Theory*, vol. IT-36, pp. 312–321, 1990.

[4] E. A. Lee and D. G. Messerschmitt, Eds., *Adaptive Equalization, in: E. A. Lee and D. G. Messerschmitt, Digital Communication*, 2nd ed. Kluwer Academic Publisher, third printing, 1997.

Chapter 8

DOES THE CHOSEN EQUALIZER LEAD TO OPTIMAL EQUALIZATION PERFORMANCE?

MONIKA PINCHAS

Department of Electrical and Electronic Engineering, Ariel University Center of Samaria, Ariel 40700, ISRAEL

ABSTRACT

By choosing a particular equalizer it is useful to know in advance if the chosen equalizer leads to perfect equalization performance. In this chapter, we explain how we can know without carrying out any simulation, if the chosen equalizer leads to perfect equalization performance for the real valued and two independent quadrature carrier case. We derive in this chapter for the real valued and two independent quadrature carrier case, some conditions on the input constellation statistics for which perfect equalization performance is obtained for type of blind equalizers where the error that is fed into the adaptive mechanism which updates the equalizer's taps is expressed as a polynomial function of order three. We show also that perfect equalization performance can not be obtained for type of blind equalizers where the error that is fed into the adaptive mechanism which updates the equalizer's taps is expressed as a polynomial function of order three, when dealing with the noiseless and 16QAM constellation input case.

KEYWORDS

Blind deconvolution, perfect equalization, intersymbol interference (ISI), convolutional noise, convolutional noise power, residual ISI, step-size parameter, equalizer's tap-length, mean square error (MSE), polynomial function of order three

CRITERIA FOR OPTIMAL EQUALIZATION PERFORMANCE

In this chapter we consider the single input single output (SISO) and single input multiple output (SIMO) system described in chapter 2 and adopt the various assumptions that were made there. According to chapter 2, the overall impulse response of the SIMO or SISO model is described by:

$$\tilde{s}[n] = \sum_{i=1}^{i=M} \tilde{c}^{(i)}[n] * h^{(i)}[n] \qquad \text{for SISO: M=1} \qquad \text{for SIMO: M > 1}$$
(8.1)

where "$*$" denotes the convolution operation, $h^{(i)}[n]$ and $\tilde{c}^{(i)}[n]$ are the channel and equalizer respectively used in the i-th path. Perfect equalization is described according to [1] by:

$$\tilde{s}[n] = \delta[n - \tau]e^{j\phi}$$
(8.2)

where δ is the Kronecker delta function. We assume here as was done in chapter 2 that $\tau = 0$ and $\phi = 0$. For the general case we can write (8.1) as:

$$\tilde{s}[n] = \delta[n] + \xi[n]$$
(8.3)

where $\xi[n]$ stands for the error not having perfect equalization. According to chapter 2, the equalized output signal may be defined for the noiseless, SISO or SIMO case as:

$$z[n] = x[n] + p[n]$$
(8.4)

where $x[n]$ is the sent signal via the channel and $p[n]$ is the convolutional noise arising from the difference between the non ideal and ideal equalizer $(p[n] = x[n] * \xi[n]$, see chapter 2). For the ideal case, $p[n] = 0$ or $p[n] \rightarrow 0$ for which the equalized output and input signal are approximately the same. The intersymbol interference (ISI) and the mean square error (MSE) criteria are both used as a measure of performance in equalizers' applications. The ISI is defined as:

$$ISI[n] = \frac{\sum_{\tilde{m}} |\tilde{s}[\tilde{m}]|^2 - |\tilde{s}|^2_{max}}{|\tilde{s}|^2_{max}} \tag{8.5}$$

where $|\tilde{s}|_{max}$ is the component of $\tilde{s}$, given in (8.3), having the maximal absolute value. For the ideal case where we have a single tap with non-zero value while the other taps are set to zero (8.2), the ISI given by (8.5) is zero since the single tap is also the tap of $\tilde{s}$ having the maximal absolute value. Thus an equalizer that achieves in the final stages of the deconvolutional process $ISI \cong 0$, has optimal equalization performance in the ISI sense or leads to perfect equalization. Let us turn now to the MSE criteria. According to chapter 2 and chapter 3, $T[z[n]]$ is the estimate of the input signal $x[n]$ and the following bound between the cost function $F[z[n]]$ and $T[z[n]]$ exists:

$$T[z[n]] = z[n] - \frac{\partial F[z[n]]}{\partial z[n]} \tag{8.6}$$

Thus choosing the cost function results in a corresponding choice of $T[z[n]]$. Therefore, we may say that an equalizer based on a cost function is implicitly based also on the corresponding function $T[z[n]]$. Since the function $T[z[n]]$ is the estimate of the input signal $x[n]$, the equalization performance will depend strongly on how well the function $T[z[n]]$ estimates the input signal $x[n]$. For that reason we define the MSE for the noiseless and real valued case as:

$$MSE = E[[T[z[n]] - x[n]]^2] \tag{8.7}$$

where $E[\cdot]$ and $(\cdot)^*$ are the expectation and conjugate operator respectively. Please note that the MSE for the two independent quadrature carrier case is twice the MSE obtained for the real valued situation. By using (8.4), the MSE (8.7) for the noiseless and real valued case may take the following shape:

$$MSE \cong \tilde{A} + \tilde{B}\sigma_p^2 \qquad (8.8)$$

where $\tilde{A}$ and $\tilde{B}$ are constants independent with σ_p^2 $(E[p^2[n]] = \sigma_p^2)$ which may depend on the input signal statistics. For example let us consider the Maximum Entropy [2] and Godard's algorithm [3]. For the noiseless and real valued case, the MSE for the Maximum Entropy algorithm [2] is given by [2]:

$$E\left[T[z[n]] - x\right]^2 \cong \frac{\sigma_p^2}{\left(1 + \sigma_p^2 \left(E\left[\frac{g''(x)}{2g(x)}\right]\right)\right)^2} \qquad (8.9)$$

where

$$E\left[\frac{g''(x)}{g(x)}\right] = \left(\sum_{k=2}^{N} \lambda_k m_{k-2}\left(k-1\right)k\right) + \sum_{L=2}^{N}\sum_{k=2}^{N} \lambda_k \lambda_L m_{k+L-2} kL \qquad (8.10)$$

$m_k = E[x^k]$ $(x = x[n])$ and λ_k are the lagrange multipliers. For the very low ISI case (σ_p^2 is very low) we may rewrite (8.9) as:

$$E\left[T[z[n]] - x\right]^2 \cong \sigma_p^2 \qquad (8.11)$$

Thus for the Maximum Entropy algorithm [2] we have $\tilde{A} = 0$ and $\tilde{B} = 1$. Next we turn to Godard's algorithm [3]. For the noiseless and real valued case, the MSE for Godard's algorithm [3] is given by [2]:

$$\text{MSE} = E[x^6] - 2E[x^4]\left(\frac{E[x^4]}{\sigma_x^2}\right) + \sigma_x^2 \left(\frac{E[x^4]}{\sigma_x^2}\right)^2 +$$
$$\left(15E[x^4] + \left(\frac{E[x^4]}{\sigma_x^2}\right)^2 - 12E[x^4] + 1 - 6\sigma_x^2 + 2\left(\frac{E[x^4]}{\sigma_x^2}\right)\right)\sigma_p^2$$
$$+O(\sigma_p^4)$$

$$(8.12)$$

where $\sigma_x^2 = E[x^2[n]]$. According to (8.12):

$$\tilde{A} = E[x^6] - 2E[x^4]\left(\frac{E[x^4]}{\sigma_x^2}\right) + \sigma_x^2\left(\frac{E[x^4]}{\sigma_x^2}\right)^2$$

$$\tilde{B} = \left(15E[x^4] + \left(\frac{E[x^4]}{\sigma_x^2}\right)^2 - 12E[x^4] + 1 - 6\sigma_x^2 + 2\left(\frac{E[x^4]}{\sigma_x^2}\right)\right)\sigma_p^2$$

$$(8.13)$$

An equalizer is said to have optimal equalization performance according to the MSE criteria if $\tilde{A}$ in the MSE expression (8.8) is zero. This means that when the convolutional noise power (σ_p^2) tends to zero, the MSE will also go to zero. Namely, perfect equalization is obtained. The Maximum Entropy algorithm [2] achieves optimum equalization performance while Godard's algorithm [3] depends on the constellation input. For the MPSK (M-phase shift keying modulation) input case for example, the MSE for Godard's algorithm [3] will tend to zero $(\tilde{A} = 0)$ when $\sigma_p^2 \to 0$. But for the 16QAM input case the MSE for Godard's algorithm [3] will stay with a constant value $(\tilde{A} \neq 0)$ even when we put $\sigma_p^2 \to 0$. It should be pointed out that $ISI = 0$ leads to $MSE = 0$ and vice versa. The MSE can be calculated in most cases thus no simulation is needed in those cases in order to see if the chosen equalizer reaches perfect equalization performance. But, this is not true when considering the ISI. Up to recently [4], the ISI could be obtained only via simulation. In [4], a closed-form approximated expression was derived for the achievable ISI valid for the noiseless, real and two independent quadrature carrier case that depends on the step-size parameter, equalizer's tap length, input signal statistics and channel power and is given by:

$$ISI = 10\log_{10}(m_p) - 10\log_{10}\left(\sigma_{x_r}^2\right) \qquad (8.14)$$

where $\sigma^2_{x_r}$ is the variance of the real part of the input sequence $x[n]$ and m_p is defined by:

$$m_p = \min\left[Sol_1^{mp_1}, Sol_2^{mp_1}\right] \quad \text{for} \quad Sol_1^{mp_1} > 0 \quad \text{and} \quad Sol_2^{mp_1} > 0$$

or

$$m_p = \max\left[Sol_1^{mp_1}, Sol_2^{mp_1}\right] \quad \text{for} \quad Sol_1^{mp_1} \cdot Sol_2^{mp_1} < 0$$

where

$$Sol_1^{mp_1} = \frac{-B_1 + \sqrt{B_1^2 - 4A_1C_1B}}{2A_1}; \quad Sol_2^{mp_1} = \frac{-B_1 - \sqrt{B_1^2 - 4A_1C_1B}}{2A_1}$$
(8.15)

$$A_1 = \left(B\left(45o^2_{x_r}a_3^2 + 18\sigma^2_{x_r}a_3a_{12} + 6a_1a_3 + 9\sigma^2_{x_r}a_{12}^2 + 2a_1a_{12}\right) - 2\left(3a_3 + a_{12}\right)\right)$$

$$B_1 = \left(B\left(12\left(\sigma^2_{x_r}\right)^2 a_3a_{12} + 6\left(\sigma^2_{x_r}\right)^2 a_{12}^2 + 12\sigma^2_{x_r}a_1a_3 + 4\sigma^2_{x_r}a_1a_{12} + \right.\right.$$
$$\left.\left. a_1^2 + 15E\left[x_r^4\right]a_3^2 + 2E\left[x_r^4\right]a_3a_{12} + E\left[x_r^4\right]a_{12}^2\right) - 2\left(a_1 + 3\sigma^2_{x_r}a_3 + \sigma^2_{x_r}a_{12}\right)\right)$$

$$C_1 = 2\left(\sigma^2_{x_r}\right)^2 a_1a_{12} + \sigma^2_{x_r}a_1^2 + 2E\left[x_r^4\right]\sigma^2_{x_r}a_3a_{12} + E\left[x_r^4\right]\sigma^2_{x_r}a_{12}^2 + $$
$$2E\left[x_r^4\right]a_1a_3 + E\left[x_r^6\right]a_3^2$$

$$B = \mu N\sigma_x^2 \sum_{k=0}^{k=R-1}\left|h\left(k\right)\right|^2$$
(8.16)

$m_p = E[(Re(p[n]))^2]$, $Re(\cdot)$ is the real part of $(\cdot)$, $x_r = Re(x[n])$, R is the channel length, N is the equalizer's tap length and a_1, a_{12}, a_3 are properties of the chosen equalizer and found by [4]:

$$Re\left(\frac{\partial F\left[z[n]\right]}{\partial z\left[n\right]}\right) = \left(a_1\left(z_r\right) + a_3\left(z_r\right)^3 + a_{12}\left(z_r\right)\left(z_i\right)^2\right) \qquad (8.17)$$

where z_r and z_i are the real and imaginary parts of the equalized output $z[n]$ respectively.

The closed-form approximated expression for the achievable ISI [4] was obtained for the following assumptions:

1. The convolutional noise $p[n]$, is a zero mean, white Gaussian process with variance $\sigma_p^2 = E[p[n]p[n]^*]$ where $()^*$ is the conjugate operation.

2. The source signal $x[n]$ is an independent non-Gaussian signal with known variance and higher moments. $x[n]$ is a real or two independent quadrature carrier signal

3. The convolutional noise $p[n]$ and the source signal are independent. Thus,

$\sigma_z^2 = E[z[n]z[n]^*] = E[(x[n] + p[n])(x[n] + p[n])^*] = E[x[n]x[n]^*] + E[p[n]p[n]^*]$

4. No noise is added.

5. $\frac{\partial F[n]}{\partial z[n]}$ can be expressed as a polynomial function of the equalized output namely as $P[z[n]]$ of order three.

6. $|\tilde{s}|_{max}^2 = 1$

According to [4], if C_1 is not equal to zero then the chosen equalizer will never reach perfect equalization performance. According to (8.16), the expression for C_1 depends on the input constellation statistics and equalizer's properties (a_1, a_{12}, a_3) only. This implies as was already mentioned in [4], that there might be some kind of equalization methods for which we will be left with a residual ISI dependent on the input constellation statistics and on the other hand, there might be other algorithms that might reach perfect equalization performance, namely will leave the system with no residual ISI. Thus for the noiseless, real and two independent quadrature carrier case and for type of blind equalizers where the error that is fed into the adaptive mechanism which updates the equalizer's taps can be expressed as a polynomial function of order three we can just use the expression for C_1 (8.16) in order to see if the chosen equalizer is capable of reaching perfect equalization performance.

DO WE GET PERFECT EQUALIZATION FOR EQUALIZERS BASED ON A POLYNOMIAL FUNCTION OF ORDER THREE?

In this section, we consider the real valued and two independent quadrature carrier case and a type of blind adaptive equalizer where the error that is fed into the adaptive mechanism which updates the equalizer's taps is expressed as a polynomial function of order three. For this type of equalizer, we derive some conditions on the input constellation statistics for which perfect equalization performance is obtained and show that perfect equalization performance can not be obtained when dealing with the noiseless and 16QAM constellation input case.

We start with showing that perfect equalization performance can not be obtained when dealing with the noiseless and 16QAM constellation input case for type of blind adaptive equalizers defined in this section.

Let us first recall the expression for C_1:

$$C_1 = 2 \left(\sigma_{x_r}^2\right)^2 a_1 a_{12} + \sigma_{x_r}^2 a_1^2 + 2E\left[x_r^4\right]\sigma_{x_r}^2 a_3 a_{12} + E\left[x_r^4\right]\sigma_{x_r}^2 a_{12}^2 + \\ 2E\left[x_r^4\right]a_1 a_3 + E\left[x_r^6\right]a_3^2$$

$$(8.18)$$

According to (8.18), the expression for C_1 is a function of the input signal statistics and a function of the property of the chosen equalizer via a_1, a_{12} and a_3. In order to get perfect equalization performance, we wish to find those values for a_1, a_{12} and a_3 that lead the function C_1 to zero. In order to find those values for a_1, a_{12} and a_3 we use the following equations:

$$\frac{\partial C_1}{\partial a_1} = 2 \left(\sigma_{x_r}^2\right)^2 a_{12} + 2\sigma_{x_r}^2 a_1 + 2E\left[x_r^4\right] a_3 = 0$$

$$\frac{\partial C_1}{\partial a_3} = 2E\left[x_r^4\right] \sigma_{x_r}^2 a_{12} + 2E\left[x_r^4\right] a_1 + 2E\left[x_r^6\right] a_3 = 0 \qquad (8.19)$$

$$\frac{\partial C_1}{\partial a_{12}} = 2 \left(\sigma_{x_r}^2\right)^2 a_1 + 2E\left[x_r^4\right] \sigma_{x_r}^2 a_3 + 2E\left[x_r^4\right] \sigma_{x_r}^2 a_{12} = 0$$

From the first and third equation of (8.19) we obtain:

$$a_{12} \left(\sigma_{x_r}^4 - E[x_r^4]\right) = 0 \qquad (8.20)$$

Now, (8.20) will be zero if $\sigma_{x_r}^4 = E[x_r^4]$ or $a_{12} = 0$. For the 16QAM case, $\sigma_{x_r}^4 \neq E[x_r^4]$. Thus, a_{12} has to be zero in order to comply with (8.20). Next we substitute $a_{12} = 0$ into (8.19) and obtain:

$$E[x_r^4]a_1 + E[x_r^6]a_3 = 0$$
$$\sigma_{x_r}^2 a_1 + E[x_r^4]a_3 = 0 \qquad (8.21)$$

From (8.21) we obtain:

$$a_3 \left(E[x_r^6] - \frac{(E[x_r^4])^2}{\sigma_{x_r}^2}\right) = 0 \qquad (8.22)$$

For the 16QAM case, $E[x_r^6] \neq (E[x_r^4])^2/\sigma_{x_r}^2$. Thus, a_3 has to be zero in order to comply with (8.22). Since $E[x_r^4] \neq 0$, $\sigma_{x_r}^2 \neq 0$, $a_{12} = 0$ and $a_3 = 0$, a_1 has to be zero. Therefore, the only solution for (8.19) to be true is $a_1 = a_3 = a_{12} = 0$. This outcome implies that no perfect equalization can be obtained for the 16QAM constellation input with a blind adaptive equalizer where the error that is fed into the adaptive mechanism which updates the equalizer's taps is expressed as a polynomial function of order three.

Next, we turn to derive some conditions on the input constellation statistics for which perfect equalization performance is obtained. We consider here two different cases based on (8.20). In the first case, we assume $\sigma_{x_r}^4 = E[x_r^4]$ for which (8.20) is fulfilled. Substituting $\sigma_{x_r}^4 = E[x_r^4]$ back into (8.19) we get:

$$\sigma_{x_r}^4 a_1 + \sigma_{x_r}^4 \sigma_{x_r}^2 a_{12} + E[x_r^6]a_3 = 0$$
$$a_1 + \sigma_{x_r}^2 a_{12} + \sigma_{x_r}^2 a_3 = 0$$

$$(8.23)$$

From (8.23) we obtain the following equation:

$$a_3 \left(E[x_r^6] - \sigma_{x_r}^6 \right) = 0 \qquad (8.24)$$

Thus, we may conclude that for the real valued and two independent quadrature carrier case and $a_{12} \neq 0$, perfect equalization may be obtained when:

For the two independent quadrature carrier case:

$$E[x_r^6] = \sigma_{x_r}^6; \qquad E[x_j^6] = \sigma_{x_j}^6; \qquad E[x_r^4] = \sigma_{x_r}^4; \qquad E[x_j^4] = \sigma_{x_j}^4$$

$$a_1 + \sigma_{x_r}^2 a_{12} + \sigma_{x_r}^2 a_3 = 0; \qquad a_1 + \sigma_{x_j}^2 a_{12} + \sigma_{x_j}^2 a_3 = 0$$

For the real valued case:

$$E[x_r^6] = \sigma_{x_r}^6; \qquad E[x_r^4] = \sigma_{x_r}^4$$

$$a_1 + \sigma_{x_r}^2 a_{12} + \sigma_{x_r}^2 a_3 = 0$$

$$(8.25)$$

where x_j is the imaginary part of the input signal $x[n]$. The QPSK (quadrature phase shift keying) constellation input is an example for holding the conditions described in (8.25). Let us consider for a moment Godard's algorithm [3] where

$$\frac{\partial F\left[z[n]\right]}{\partial z\left[n\right]} = \left(\left|z\left[n\right]\right|^2 - \frac{E\left[\left|x\left[n\right]\right|^4\right]}{E\left[\left|x\left[n\right]\right|^2\right]}\right) z\left[n\right] \qquad (8.26)$$

The values for a_1, a_{12} and a_3 corresponding to Godards's [3] algorithm are defined as a_1^G, a_{12}^G and a_3^G respectively and are given by:

$$a_1^G = -\frac{E\left[\left|x\left[n\right]\right|^4\right]}{E\left[\left|x\left[n\right]\right|^2\right]}; \qquad a_{12}^G = 1; \qquad a_3^G = 1 \qquad (8.27)$$

For the QPSK case we have

$$E\left[\left|x\left[n\right]\right|^4\right] = 4; \qquad E\left[\left|x\left[n\right]\right|^2\right] = 2; \qquad \sigma_{x_r}^2 = \sigma_{x_j}^2 = 1 \qquad (8.28)$$

By using (8.28) we have for the QPSK case:

$$a_1^G = -2; \qquad a_{12}^G = 1; \qquad a_3^G = 1 \qquad (8.29)$$

Now, according to (8.29), a_1^G, a_{12}^G and a_3^G comply for the QPSK case with the condition on a_1, a_{12} and a_3 described in (8.25). Thus, Godard's algorithm [3] reaches perfect equalization performance for the QPSK case. It is already known in the literature [5] that Godard's algorithm [3] reaches perfect equalization performance for the QPSK input case. We have brought the algorithm here to show that our proposed conditions on a_1, a_{12} and a_3 (8.25) for perfect equalization performance, are achievable and hold.

Next, we turn to the second case, $a_{12} = 0$, for which (8.20) is fulfilled. Substituting $a_{12} = 0$ back into (8.19) leads to (8.21) and (8.22). Thus, we may conclude that for the real valued and two independent quadrature carrier case and $a_{12} = 0$, perfect equalization may be obtained when:

For the two independent quadrature carrier case:

$$E[x_r^6] = \frac{\left(E[x_r^4]\right)^2}{\sigma_{x_r}^2}; \qquad E[x_j^6] = \frac{\left(E[x_j^4]\right)^2}{\sigma_{x_j}^2};$$

$$\sigma_{x_r}^2 a_1 + E[x_r^4]a_3 = 0; \qquad \sigma_{x_j}^2 a_1 + E[x_j^4]a_3 = 0$$

$$(8.30)$$

For the real valued case:

$$E[x_r^6] = \frac{\left(E[x_r^4]\right)^2}{\sigma_{x_r}^2}$$

$$\sigma_{x_r}^2 a_1 + E[x_r^4]a_3 = 0$$

The QPSK constellation input is an example for holding the condition described in (8.30). Let us consider for a moment the NEW algorithm [6] where

$$\frac{\partial F\left[z[n]\right]}{\partial z\left[n\right]} = \left(\frac{E\left[(x_r\left[n\right])^4\right]}{E\left[(x_r\left[n\right])^6\right]} z_r^3\left[n\right] + j\frac{E\left[(x_r\left[n\right])^4\right]}{E\left[(x_r\left[n\right])^6\right]} z_i^3\left[n\right] - z\left[n\right]\right)$$

$$(8.31)$$

$x_r\left[n\right]$ and $z_r\left[n\right]$ are the real parts of $x[n]$ and $z[n]$ respectively. $x_j\left[n\right]$ and $z_j\left[n\right]$ are the imaginary parts of $x[n]$ and $z[n]$ respectively. The values for a_1, a_{12} and a_3 corresponding to the NEW algorithm [6] are defined as a_1^N, a_{12}^N and a_3^N respectively and are given by:

$$a_1^N = -1; \qquad a_{12}^N = 0; \qquad a_3^N = \frac{E\left[(x_r\left[n\right])^4\right]}{E\left[(x_r\left[n\right])^6\right]} \qquad (8.32)$$

For the QPSK case we have:

$$E\left[(x_r\left[n\right])^4\right] = 1; \qquad E\left[(x_r\left[n\right])^6\right] = 1 \qquad (8.33)$$

By using (8.33) we have for the QPSK case:

$$a_1^N = -1; \qquad a_{12}^N = 0; \qquad a_3^N = 1 \qquad (8.34)$$

Now, according to (8.34), a_1^N, a_{12}^N and a_3^N comply for the QPSK case with the condition on a_1, a_{12} and a_3 described in (8.30). Thus, the NEW algorithm [6] reaches perfect equalization performance for the QPSK case.

ABBREVIATIONS

ISI=	Intersymbol Interference
SNR=	Signal to Noise Ratio
QAM=	Quadrature Amplitude Modulation
QPSK=	Quadrature Phase Shift Keying
SISO=	Single Input Single Output
SIMO=	Single Input Multiple Output
MSE=	Mean Square Error
MPSK=	M- Phase Shift Keying

REFERENCES

[1] A. K. Nandi, Ed., *Blind estimation using higher-order statistics.*   Boston: Kluwer Academic, 1999.

[2] M. Pinchas and B. Z. Bobrovsky, "A maximum entropy approach for blind deconvolution," *Signal Processing Journal (Eurasip)*, vol. 86, pp. 2913–2931, 2006.

[3] D. N. Godard, "Self recovering equalization and carrier tracking in two-dimensional data communication systems," *IEEE Transaction on Communications*, vol. COM-28, pp. 1867–1875, 1980.

[4] M. Pinchas, "A closed approximated formed expression for the achievable residual intersymbol interference obtained by blind equalizers," *Signal Processing Journal (Eurasip)*, vol. 90, pp. 1940–1962, 2010.

[5] E. A. Lee and D. G. Messerschmitt, Eds., *Adaptive Equalization, in: E. A. Lee and D. G. Messerschmitt, Digital Communication*, 2nd ed.   Kluwer Academic Publisher, third printing, 1997.

[6] M. Pinchas, "What are the analytical conditions for which a blind equalizer will loose the converge state," *Signal, Image and Video Processing*, DOI: 10.1007/s11760-011-0221-0.

Chapter 9

THE CONVERGENCE TIME OF A BLIND ADAPTIVE EQUALIZER

MONIKA PINCHAS

Department of Electrical and Electronic Engineering, Ariel University Center of Samaria, Ariel 40700, ISRAEL

ABSTRACT

Closed-form approximated expressions were recently proposed for the convergence time (or number of iterations required for convergence) and for the Intersymbol Interference (ISI) as a function of time valid during the stages of the iterative deconvolution process. The new derivations are valid for the noiseless, real valued and two independent quadrature carrier case and for type of blind equalizers where the error that is fed into the adaptive mechanism which updates the equalizer's taps can be expressed as a polynomial function of order three of the equalized output like in Godard's algorithm. The derivations are based on the knowledge of the initial ISI and channel power (which is measurable) and do not need the knowledge of the channel coefficients. However, they are not based on strong mathematical foundations and were tested via simulation with Godard's algorithm for two different channels only. Thus, it could be argued that they hold only for a particular equalization method and for special cases. In this chapter, we present by simulation the usefulness of the recently derived expressions for more channel types and step-size parameters as well as for another type of equalizer. Thus, the question if the obtained expressions hold also for other types of equalization methods and other channels is answered.

Based on the recently proposed expression for the ISI as a function of time, we derive in this chapter a closed-form approximated expression for the mean square error (MSE) as a function of time (or as a function of iteration number).

KEYWORDS

Blind deconvolution, intersymbol interference (ISI), convolutional noise power, residual ISI, step-size parameter, equalizer's tap-length, mean square error (MSE), polynomial function of order three, convergence speed, channel power

INTRODUCTION

It is well known that fast convergence speed and reaching a residual ISI where the eye diagram is considered to be open are the main requirements from a blind equalizer. Fast convergence speed may be obtained by increasing the step-size parameter. But increasing the step-size parameter will lead to a higher residual ISI which might not meet any more the system's requirements. Up to recently [1], there was no closed-form expression for the ISI as a function of time, nor a closed-form expression for the convergence time (or number of iterations required for convergence) even when all the channel coefficients were known. Therefore, the system designer had to carry out many simulations for a given channel and type of equalizer, in order to obtain the optimal step-size parameter and equalizer's tap length for a required convergence speed and residual ISI.

The recently proposed expression [1] for the ISI as a function of time is a closed-form approximated expression valid during the stages of the iterative deconvolution process for type of blind equalizers where the error that is fed into the adaptive mechanism which updates the equalizer's taps can be expressed as a polynomial function of order three of the equalized output. It should be pointed out that Godard's algorithm [2] for example, belongs to the mentioned type of blind equalizers. Based on the closed-form approximated expression for the ISI as a function of time, a closed-form approximated expression for the convergence speed (or number of iterations required for convergence) as a function of initial ISI, step-size parameter, equalizer's tap length, input signal statistics and channel power was derived [1]. The new obtained expressions [1] are applicable for the noiseless, low ISI condition, real valued and two independent quadrature carrier case.

The new proposed expression for the ISI as a function of time [1] was tested via simulation using Godard's algorithm [2] with two different channels only. It was not shown to be applicable for different types of algorithms nor for more than two different channels. In this chapter, we will show via simulation that the proposed expression for the ISI as a function of time [1] is also applicable for another type of equalizer, type of channel and works well also with different step-size parameters for a chosen channel. In addition, we will explain and show via simulation for which type of channel the new proposed expression for the ISI as a function of time [1] does not hold.

In this chapter we show the connection between the MSE and ISI and propose a closed-form approximated expression for the MSE as a function of time that is valid for the noiseless, real valued and two independent quadrature carrier case and for type of blind equalizers where the error that is fed into the adaptive mechanism which updates the equalizer's taps can be expressed as a polynomial function of order three.

ISI AS A FUNCTION OF TIME

In this section, we consider the system described in 2 where we recall the update equation for the equalizer's coefficients:

$$\underline{c}_{eq}[n+1] = \underline{c}_{eq}[n] - \mu \frac{\partial F[n]}{\partial z[n]} \underline{y}^*[n] =$$

(9.1)

$$\underline{c}_{eq}[n] - \mu P(z[n]) \underline{y}^*[n]$$

where μ is the step-size parameter, $\underline{c}_{eq}[n]$ is the equalizer vector where the input vector is $\underline{y}[n] = [y[n] \ldots y[n-L+1]]^T$ and L is the equalizer's tap length. The operator $()^T$ denotes for transpose of the function $()$, $()^*$ is the conjugate operation, $F[n]$ is the cost function and $P(z[n])$ is expressed as a polynomial function of the equalized output. In this chapter, we assume that the order of $P(z[n])$ is three. The real part of $P(z[n])$ may be expressed according to [3] as:

$$P_r(z[n]) = \left(a_1 (z_1[n]) + a_3 (z_1[n])^3 + a_{12} (z_1[n]) (z_2[n])^2 \right) =$$

$$\left(a_1 (x_r + p_r[n]) + a_3 (x_r + p_r[n])^3 + a_{12} (x_r + p_r[n]) (x_i + p_i[n])^2 \right)$$

(9.2)

where $z_1[n]$ and $z_2[n]$ are the real and imaginary parts of the equalized output signal $z[n]$ respectively $(z[n] = x[n] + p[n])$, $p[n]$ is the convolutional noise, $x_r = x_1[n]$, $x_i = x_2[n]$ ($x_1[n]$ and $x_2[n]$ are the real and imaginary parts of the input signal $x[n]$ respectively), $p_r[n]$ and $p_i[n]$ are the real and imaginary parts of $p[n]$ respectively and a_1, a_{12}, a_3 are parameters of the chosen equalizer. According to [1], the closed-form approximated expression for the ISI as a function of time is given by:

$$\widetilde{ISI}(t) \cong \left(\widetilde{ISI}(0) - 10^{\frac{ISI_r}{10}} \right) e^{\frac{\gamma B t B_1}{\Delta t}} + 10^{\frac{ISI_r}{10}}$$

(9.3)

where Δt is defined as $t_{i+1} - t_i$, $\widetilde{ISI}(0)$ is the ISI at $t = 0$ and γ is given according to [1] by:

$$\gamma = \left(\min \left[\frac{1}{\sigma_{x_r}^2} \sqrt{\left| \frac{B_1}{BD} \right|}, \frac{1}{\sigma_{x_r}^2} \left| \frac{B_1}{A_1} \right| \right] \right) \frac{1}{\widetilde{ISI}(0)L} \qquad (9.4)$$

$$A_1 = \left(B \left(45\sigma_{x_r}^2 a_3^2 + 18\sigma_{x_r}^2 a_3 a_{12} + 6a_1 a_3 + 9\sigma_{x_r}^2 a_{12}^2 + 2a_1 a_{12} \right) \right.$$
$$\left. -2 \left(3a_3 + a_{12} \right) \right)$$

$$B_1 = \left(B \left(12 \left(\sigma_{x_r}^2 \right)^2 a_3 a_{12} + 6 \left(\sigma_{x_r}^2 \right)^2 a_{12}^2 + 12\sigma_{x_r}^2 a_1 a_3 + 4\sigma_{x_r}^2 a_1 a_{12} + \right.\right.$$
$$a_1^2 + 15 E \left[x_r^4 \right] a_3^2 + 2 E \left[x_r^4 \right] a_3 a_{12} + E \left[x_r^4 \right] a_{12}^2 \right) -$$
$$\left. 2 \left(a_1 + 3\sigma_{x_r}^2 a_3 + \sigma_{x_r}^2 a_{12} \right) \right)$$

$$C_1 = \left(2 \left(\sigma_{x_r}^2 \right)^2 a_1 a_{12} + \sigma_{x_r}^2 a_1^2 + 2 E \left[x_r^4 \right] \sigma_{x_r}^2 a_3 a_{12} + \right.$$
$$\left. E \left[x_r^4 \right] \sigma_{x_r}^2 a_{12}^2 + 2 E \left[x_r^4 \right] a_1 a_3 + E \left[x_r^6 \right] a_3^2 \right)$$

$$B = \mu L \sigma_x^2 \sum_{k=0}^{k=R-1} |h\left[k \right]|^2 ; \qquad D = 15a_3^2 + 6a_3 a_{12} + 3a_{12}^2$$
$$(9.5)$$

where R is the channel length, σ_x^2 is the source variance defined by $E[x[n]x^*[n]]$ where $E[\cdot]$ is the expectation operator and $()^*$ is the conjugate operation, $\sigma_{x_r}^2 = E[x_r^2]$ and ISI_r is the residual ISI expressed in dB units and is defined for $|\tilde{s}|_{max}^2 = 1$ ($\tilde{s}$ is given in (2.4)) according to [3] as:

$$ISI_r = 10 \log_{10} \left(\hat{m}_p \right) - 10 \log_{10} \left(\sigma_{x_r}^2 \right) \qquad (9.6)$$

where $\hat{m}_p$ is defined according to [3] as:

$$\hat{m}_p = \min \left[Sol_1^{mp1}, Sol_2^{mp1}\right] \quad \text{for} \quad Sol_1^{mp1} > 0 \quad \text{and} \quad Sol_2^{mp1} > 0$$

or

$$\hat{m}_p = \max \left[Sol_1^{mp1}, Sol_2^{mp1}\right] \quad \text{for} \quad Sol_1^{mp1} \cdot Sol_2^{mp1} < 0$$

where

$$Sol_1^{mp1} = \frac{-B_1 + \sqrt{B_1^2 - 4A_1 C_1 B}}{2A_1}; \quad Sol_2^{mp1} = \frac{-B_1 - \sqrt{B_1^2 - 4A_1 C_1 B}}{2A_1}$$

(9.7)

Based on (9.3), the total number of iteration required for convergence was obtained [1]:

$$n = 8\left|\frac{1}{\gamma BB_1}\right| = \frac{\widetilde{ISI}(0)L}{\min\left[\frac{1}{\sigma_{xr}^2}\sqrt{\left|\frac{B_1}{BD}\right|}, \frac{1}{\sigma_{xr}^2}\left|\frac{B_1}{A_1}\right|\right]} \frac{8}{B}\left|\frac{1}{B_1}\right| \quad (9.8)$$

where $|(\cdot)|$ is the absolute operation of $(\cdot)$. It should be pointed out that the closed-form approximated expression (9.3) was obtained for the following assumptions:

1. The convolutional noise $p[n]$, is a zero mean, white Gaussian process with variance $\sigma_p^2[n] = E[p[n]p^*[n]]$.

2. The source signal $x[n]$ is an independent non-Gaussian signal with known variance and higher moments. The source signal $x[n]$ may belong to a real valued or two independent quadrature carrier case.

3. The convolutional noise $p[n]$ and the source signal are independent. Thus,

$\sigma_z^2[n] = E[z[n]z^*[n]] = E[(x[n] + p[n])(x[n] + p[n])^*] = E[x[n]x^*[n]] + E[p[n]p^*[n]]$

4. Low ISI case

5. $|\tilde{s}|^2_{max} = 1$

6. No noise is added

7. $\partial F[n]/\partial z[n]$ can be expressed as a polynomial function of order three of the equalized output namely as $P(z[n])$ of order three.

Comments:

Assumptions 1 and 3 were also made in [4], [5], [6] and in [7]. It should be noted that the described model for the convolutional noise $p[n]$ is applicable during the latter stages of the process where the process is close to optimality [7]. According to [7], in the early stages of the iterative deconvolution process, the ISI is typically large with the result that the data sequence and the convolutional noise are strongly correlated and the convolutional noise sequence is more uniform than Gaussian [8]. However, satisfying equalization performance were obtained by [6] and others [9] in spite of the fact that the described model for the convolutional noise $p[n]$ was used. These results [6], [9] may indicate that the described model for the convolutional noise $p[n]$ can be used (maybe not in the optimum way) in the early stages where the "eye diagram" is still closed.

Although the proposed expression for the ISI as a function of time (9.3) was derived for the noiseless case, it was shown via simulation [1] that it is also applicable for signal to noise ratio (SNR) values down to 22 $[dB]$ for Godard's algorithm [2]. In this chapter we will show via simulation that the proposed expression for the ISI as a function of time (9.3) is also applicable for SNR values down to 22 $[dB]$ for another type of equalizer. Thus, it can not be claimed any more that the proposed expression holds only for a particular algorithm and for the high SNR case only.

MSE AS A FUNCTION OF TIME

Recently [10], a closed-form approximated expression was obtained for the achievable MSE for the real valued and two independent quadrature carrier case and for type of blind equalizers where the error that is fed into the adaptive mechanism which updates the equalizer's taps can be expressed as a polynomial function of order three. This expression [10], is valid only in the convergence state and cannot be applied in the earlier stages of the deconvolution process. In this section, we derive a closed-form approximated expression for the MSE as a function of time based on the previously proposed expression for the ISI as a function of time. The obtained expression is valid for the noiseless, real valued and two independent quadrature carrier case and for type of blind equalizers where the error that is fcd into the adaptive mechanism which updates the equalizer's taps can be expressed as a polynomial function of order three.

We start our derivations by first obtaining the relationship between the ISI and the convolutional noise power for the noiseless case. Then, we will show the relationship between the MSE and the convolutional noise power and at last we will make the mathematical connection between the ISI and MSE. Once we will have the mathematical relationship between the ISI and MSE, we will use the recently proposed expression for the ISI as a function of time to obtain the MSE as a function of time. We start our derivations for the real valued case and then turn to the two independent quadrature carrier one.

Based on the system description introduced in chapter 2 we may write the equalized output signal for the noiseless case as:

$$
\begin{gathered}
z[n] = \tilde{c}[n] * h[n] * x[n] = \tilde{s}[n] * x[n] \\
\Downarrow \\
E[z[n]^2] = \sigma_x^2 \sum_{\tilde{m}} |\tilde{s}[\tilde{m}]|^2
\end{gathered}
\tag{9.9}
$$

where "$*$" denotes the convolution operation, $h[n]$ is the channel, $\tilde{c}[n]$ is the equalizer used in the system and $\tilde{s}[n] = \tilde{c}[n] * h[n]$. Now we use assumption 3 from the previous section, namely the relation of $\sigma_p^2[n] = \sigma_z^2[n] - \sigma_x^2$ (where $\sigma_z^2[n] = E[z[n]^2]$) together with (9.9) to obtain:

$$\sigma_p^2[n] = \sigma_z^2[n] - \sigma_x^2 = \sigma_x^2 \left[\sum_{\tilde{m}} |\tilde{s}[\tilde{m}]|^2 - 1 \right] \qquad (9.10)$$

which by the help of (2.7) may be expressed for $|\tilde{s}|^2_{max} = 1$ as:

$$\sigma_p^2[n] = \sigma_x^2 \cdot ISI[n] \qquad \text{for} \quad |\tilde{s}|^2_{max} = 1 \qquad (9.11)$$

Note that for the real valued and two independent quadrature carrier case we may write:

$$m_p[n] = \sigma_{x_r}^2 \cdot ISI[n] \qquad \text{for} \quad |\tilde{s}|^2_{max} = 1 \qquad (9.12)$$

where $E[p_r[n]^2] = m_p[n]$. Now we turn to derive the MSE. The equalized output signal $z[n]$ can be described for the noiseless case as:

$$z[n] = x[n] + p[n] \qquad (9.13)$$

By using (9.13) we may write the MSE as:

$$MSE = E[[z[n] - x[n]]^2] = E[p^2[n]] = \sigma_p^2[n] \qquad (9.14)$$

Now by using (9.11) and (9.14) we obtain:

$$MSE = \sigma_x^2 \cdot ISI[n] \qquad \text{for} \quad |\tilde{s}|^2_{max} = 1 \qquad (9.15)$$

Next we turn to the two independent quadrature carrier case. The MSE value for the two independent quadrature carrier case is twice of that obtained for the real valued situation and $\sigma_x^2 = 2\sigma_{x_r}^2$. Thus we may write:

For the real valued case :

$$MSE = \sigma_x^2 \cdot ISI[n] \qquad \text{for} \quad |\tilde{s}|_{max}^2 = 1 \quad \text{and} \quad \sigma_x^2 = \sigma_{x_r}^2$$

$$(9.16)$$

For the two independent quadrature carrier case :

$$MSE = \sigma_x^2 \cdot ISI[n] \qquad \text{for} \quad |\tilde{s}|_{max}^2 = 1$$

Based on (9.16), we obtain by multiplying (9.3) with σ_x^2 the MSE as a function of time:

$$MSE(t) = \sigma_x^2 \widetilde{ISI}(t) \cong \sigma_x^2 \left(\widetilde{ISI}(0) - 10^{\frac{ISI_r}{10}} \right) e^{\frac{\gamma BtB_1}{\Delta t}} + \sigma_x^2 10^{\frac{ISI_r}{10}} \quad (9.17)$$

By substituting $t = n\Delta t$ in (9.17) we obtain the MSE as a function of iteration number:

$$MSE[n] \cong \sigma_x^2 \left(\widetilde{ISI}(0) - 10^{\frac{ISI_r}{10}} \right) e^{\gamma BnB_1} + \sigma_x^2 10^{\frac{ISI_r}{10}} \qquad (9.18)$$

SIMULATION

In this section, the closed-form approximated expression for the convergence time (or number of iteration required for convergence) and the expressions for the ISI and MSE as a function of time are tested via simulation. In the following we use Godard's equalizer [2] and the 16QAM constellation (a modulation using $\pm \{1,3\}$ levels for in-phase and quadrature components) as the source. The equalizer taps for Godard's equalizer [2] are updated according to:

$$c_l[n+1] = c_l[n] - \mu_G \left(|z[n]|^2 - \frac{E\left[|x[n]|^4\right]}{E\left[|x[n]|^2\right]} \right) z[n] y^*[n-l] \quad (9.19)$$

where μ_G is the step-size parameter and $l = 0, 1, 2, 3, ..., L - 1$. The values for a_1, a_{12} and a_3 corresponding to Godards's [2] algorithm are defined as a_1^G, a_{12}^G and a_3^G respectively and are given by:

$$a_1^G = -\frac{E\left[|x[n]|^4\right]}{E\left[|x[n]|^2\right]}; \qquad a_{12}^G = 1; \qquad a_3^G = 1 \qquad (9.20)$$

In order to show that the derivations for the convergence time (or number of iteration required for convergence) and ISI as a function of time are valid not only for Godard's algorithm [2], another blind algorithm with the following properties is needed:

1. The algorithm leaves the system in the steady state with $|\tilde{s}|_{max}^2 = 1$ (meaning that the gain between the equalized output and input sequence is equal to one).

2. The error that is fed into the adaptive mechanism which updates the equalizer's taps can be expressed as a polynomial function of the equalized output of order three.

For that purpose we use the blind algorithm derived in (3.23) which was also used in [11]:

$$c_m[n+1] = c_m[n] -$$

$$\mu_N \left(\frac{E\left[(x_r[n])^4\right]}{E\left[(x_r[n])^6\right]} z_r^3[n] + j\frac{E\left[(x_r[n])^4\right]}{E\left[(x_r[n])^6\right]} z_i^3[n] - z[n] \right) y^*[n-m]$$

$$(9.21)$$

where μ_N is the step-size parameter, $m = 0, 1, 2, 3, ..., L - 1$, $z_r[n]$ and $z_i[n]$ are the real and imaginary parts of $z[n]$ respectively. The values for a_1, a_{12} and a_3 corresponding to the new algorithm are defined as a_1^N, a_{12}^N and a_3^N respectively and are given by:

$$a_1^N = -1; \qquad a_{12}^N = 0; \qquad a_3^N = \frac{E\left[(x_r[n])^4\right]}{E\left[(x_r[n])^6\right]} \qquad (9.22)$$

In the following we denote algorithm (9.21) as "New". Three different channels are considered.

Channel1 (initial ISI = 0.44): The channel parameters are determined according to [12]:

$h_n = 0$ for $n < 0$; -0.4 for $n = 0$ $0.84 \cdot 0.4^{n-1}$ for $n > 0$.

Channel2 (initial ISI = 0.5): The channel parameters are determined according to [6]:

$h_n = (-0.0144, 0.0006, 0.0427, 0.0090, -0.4842, -0.0376, 0.8163, 0.0247,$
$0.2976, 0.0122, 0.0764, 0.0111, 0.0162, 0.0063)$

Channel3 (initial ISI = 0.598): The channel parameters are determined according to: $h_n = (0.2851, -0.72765, -0.4851)$

For Channel1, Channel2 and Channel3 an equalizer with 13, 21 and 27 taps is used respectively. In the simulation, the equalizer is initialized by setting the center tap equal to one and all others to zero. Fig. (**9.1**) to Fig. (**9.11**) show the simulated performance of the "New" algorithm and Godard's equalization method [2] for the 16QAM input case, namely the ISI as a function of iteration number for various step-size parameters, channel characteristics and equalizer's tap length, compared with the calculated ISI as a function of iteration number (9.3). According to Fig. (**9.1**) to Fig. (**9.11**), the closed-form approximated expression for the ISI as a function of time (or iteration number) (9.3), fits very well the simulated results.

The approximated expression for the ISI as a function of time (9.3) is an exponentially decaying function which will have similar performance compared with the simulated ISI only when the initial ISI is not too far away from the very low ISI case. Note that for the very low ISI case, the ISI can be described as an exponential function with a single time constant in the exponent [1]. An example where the new proposed expression for the ISI as a function of time (9.3) does not hold is the case where the channel (denoted in the following as Channel4) causes an initial ISI of 0.888 which is sufficiently high.

Channel4 (initial ISI = 0.88): The channel parameters were determined according to: $h_n = (0.4851, -0.72765, -0.4851)$.

Fig. (**9.12**) shows the simulated performance of Godard's equalization method [2] for the 16QAM input case, namely the simulated ISI as a function of iteration number compared with the calculated ISI as a function of iteration number (9.3) proposed in this paper. According to Fig. (**9.12**), the simulated ISI can not be described as an exponential decaying function having only a single time constant in the exponent. Thus, the proposed expression for the ISI as a function of time (9.3) does not fit the simulated ISI. In generally, the approximated expression for the ISI as a function of iteration number (9.3) is not expected to be a good approximation for the high ISI condition.

Next, the expression for the total number of iterations required for convergence (9.8) was calculated for each simulation:

Case I – described in Fig. (9.1):
The calculated number of iterations required for convergence according to (9.8) is 2413.

Case II – described in Fig. (9.2):
The calculated number of iterations required for convergence according to (9.8) is 4266.

Case III – described in Fig. (9.3):
The calculated number of iterations required for convergence according to (9.8) is 4457.

Case IV – described in Fig. (9.4):
The calculated number of iterations required for convergence according to (9.8) is 6492.

Case V – described in Fig. (9.5):
The calculated number of iterations required for convergence according

to (9.8) is 3535.

Case VI – described in Fig. (9.6):

The calculated number of iterations required for convergence according to (9.8) is 1334.

Case VII – described in Fig. (9.7):

The calculated number of iterations required for convergence according to (9.8) is 1334.

Case VIII – described in Fig. (9.8):

The calculated number of iterations required for convergence according to (9.8) is 3404.

Case IX – described in Fig. (9.9):

The calculated number of iterations required for convergence according to (9.8) is 932.

Case X – described in Fig. (9.10):

The calculated number of iterations required for convergence according to (9.8) is 1389.

Case XI – described in Fig. (9.11):

The calculated number of iterations required for convergence according to (9.8) is 3343.

According to Fig. (**9.1**) to Fig. (**9.11**), there is a high correlation between the simulated and calculated (9.8) results for the number of iterations required for convergence. The reader may find a summary of these results in Table (**9.1**), Table (**9.2**) and Table (**9.3**).

Next we turn to the noisy case. Fig. (**9.13**) shows the simulated performance of the "New" algorithm for the 16QAM input case, namely the ISI as a function of iteration number for various SNR values, compared with the calculated ISI as a function of iteration number (9.3). According to Fig. (**9.13**), the approximated expression for the ISI as a function of iteration number (9.3) is valid also for the noisy case.

Fig. (**9.14**) and Fig. (**9.15**) show the simulated performance of Godard's equalization method [2] for the 16QAM input case, namely the MSE as a function of iteration number for various step-size parameters, channel characteristics and equalizer's tap length, compared with the calculated MSE as a function of iteration number (9.18) proposed in this chapter. According to Fig. (**9.14**) and Fig. (**9.15**), the closed-form approximated expression for the MSE as a function of time (or iteration number) (9.18), fits very well the simulated results.

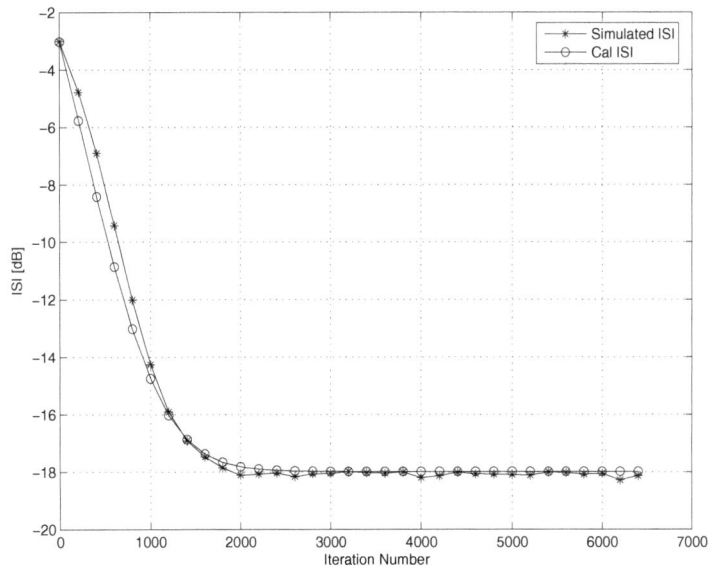

Figure 9.1: A comparison between the simulated (with Godard's algorithm) and calculated ISI as a function of time for the 16QAM source input going through channel2. The averaged results were obtained in 100 Monte Carlo trials for the noiseless case. The equalizer's length was set to 21 and $\mu_G = 0.00004$.

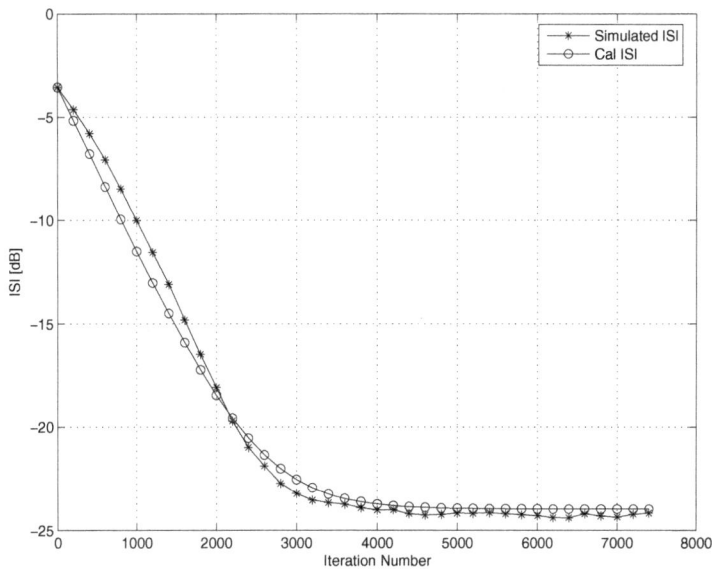

Figure 9.2: A comparison between the simulated (with Godard's algorithm) and calculated ISI as a function of time for the 16QAM source input going through channel1. The averaged results were obtained in 100 Monte Carlo trials for the noiseless case. The equalizer's length was set to 13 and $\mu_G = 0.00002$.

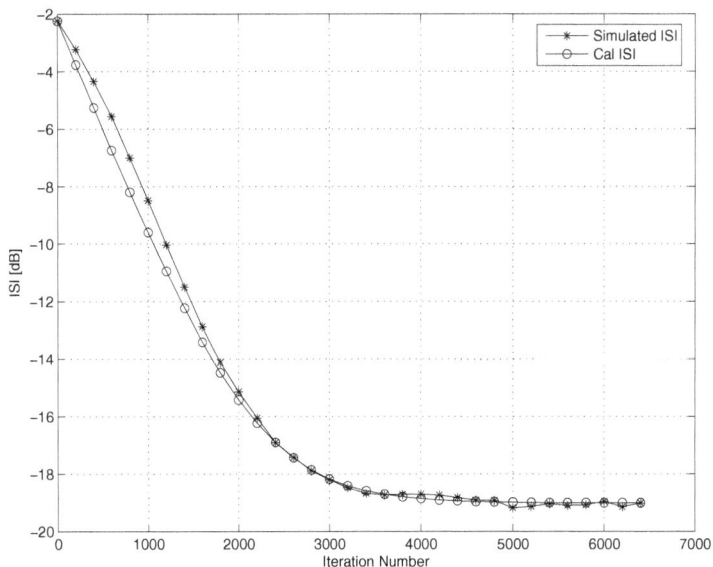

Figure 9.3: A comparison between the simulated (with Godard's algorithm) and calculated ISI as a function of time for the 16QAM source input going through channel3. The averaged results were obtained in 100 Monte Carlo trials for the noiseless case. The equalizer's length was set to 27 and $\mu_G = 0.00003$.

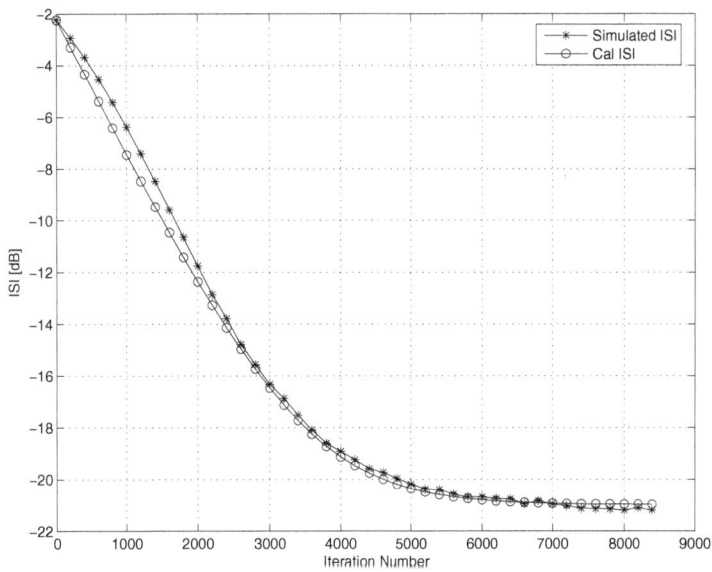

Figure 9.4: A comparison between the simulated (with Godard's algorithm) and calculated ISI as a function of time for the 16QAM source input going through channel3. The averaged results were obtained in 100 Monte Carlo trials for the noiseless case. The equalizer's length was set to 27 and $\mu_G = 0.00002$.

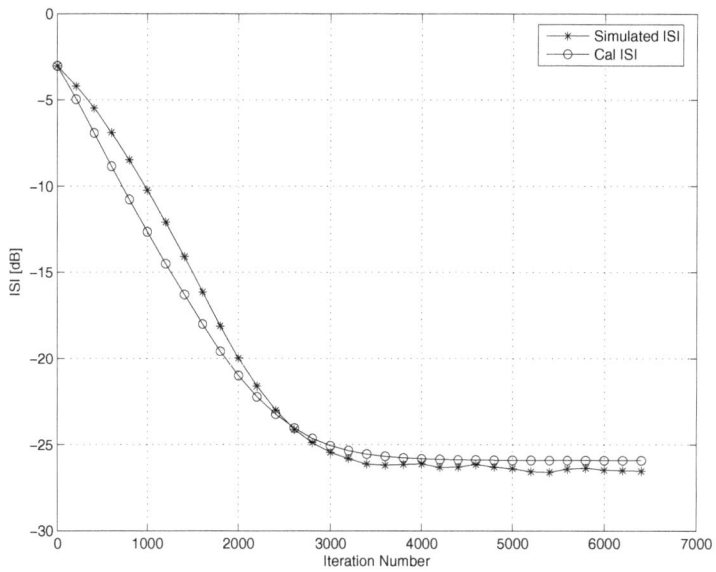

Figure 9.5: A comparison between the simulated (with the "New" algorithm) and calculated ISI as a function of time for the 16QAM source input going through channel2. The averaged results were obtained in 100 Monte Carlo trials for the noiseless case. The equalizer's length was set to 21 and $\mu_N = 0.0002$.

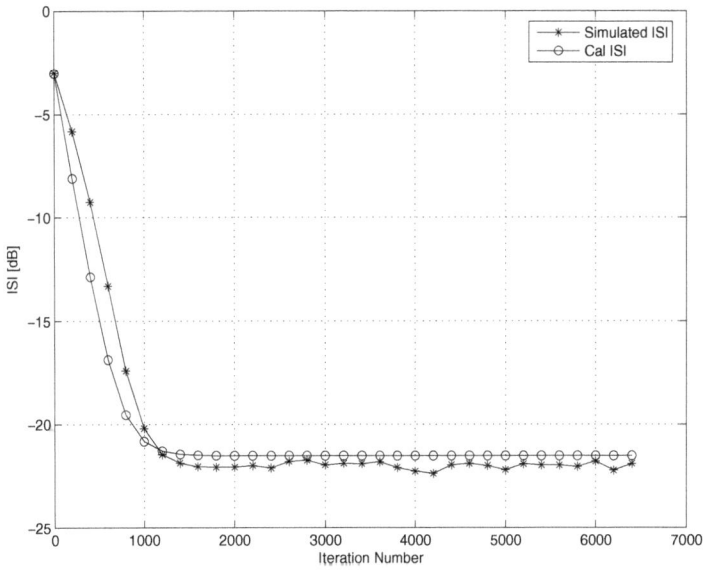

Figure 9.6: A comparison between the simulated (with the "New" algorithm) and calculated ISI as a function of time for the 16QAM source input going through channel2. The averaged results were obtained in 100 Monte Carlo trials for the noiseless case. The equalizer's length was set to 21 and $\mu_N = 0.0005$.

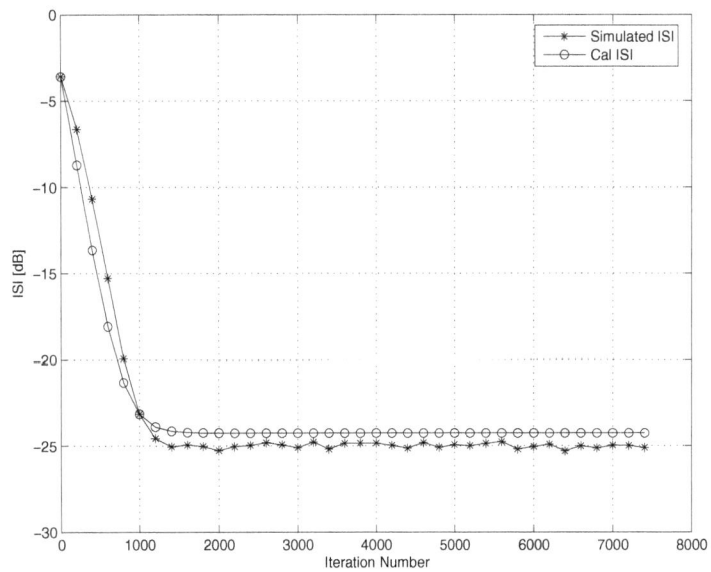

Figure 9.7: A comparison between the simulated (with the "New" algorithm) and calculated ISI as a function of time for the 16QAM source input going through channel1. The averaged results were obtained in 100 Monte Carlo trials for the noiseless case. The equalizer's length was set to 13 and $\mu_N = 0.0005$.

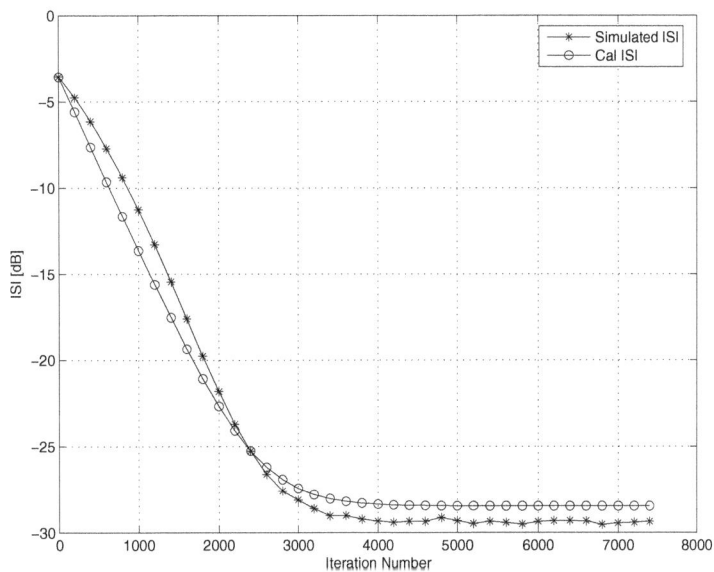

Figure 9.8: A comparison between the simulated (with the "New" algorithm) and calculated ISI as a function of time for the 16QAM source input going through channel1. The averaged results were obtained in 100 Monte Carlo trials for the noiseless case. The equalizer's length was set to 13 and $\mu_N = 0.0002$.

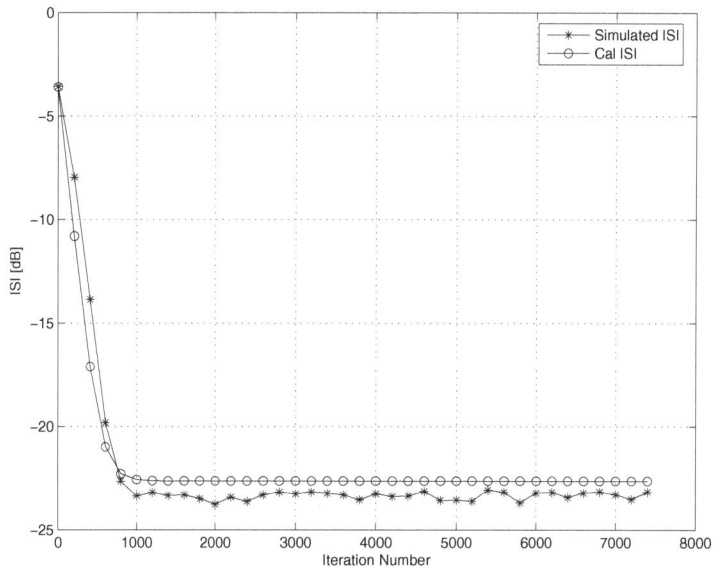

Figure 9.9: A comparison between the simulated (with the "New" algorithm) and calculated ISI as a function of time for the 16QAM source input going through channel1. The averaged results were obtained in 100 Monte Carlo trials for the noiseless case. The equalizer's length was set to 13 and $\mu_N = 0.0007$.

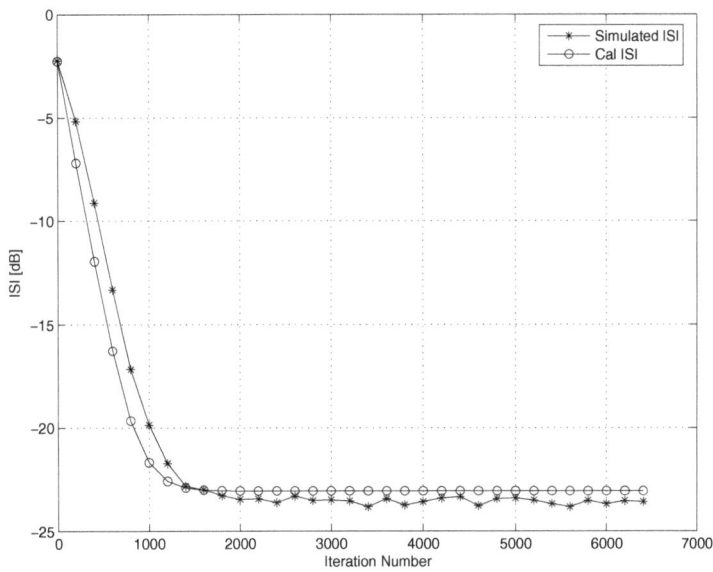

Figure 9.10: A comparison between the simulated (with the "New" algorithm) and calculated ISI as a function of time for the 16QAM source input going through channel3. The averaged results were obtained in 100 Monte Carlo trials for the noiseless case. The equalizer's length was set to 13 and $\mu_N = 0.0007$.

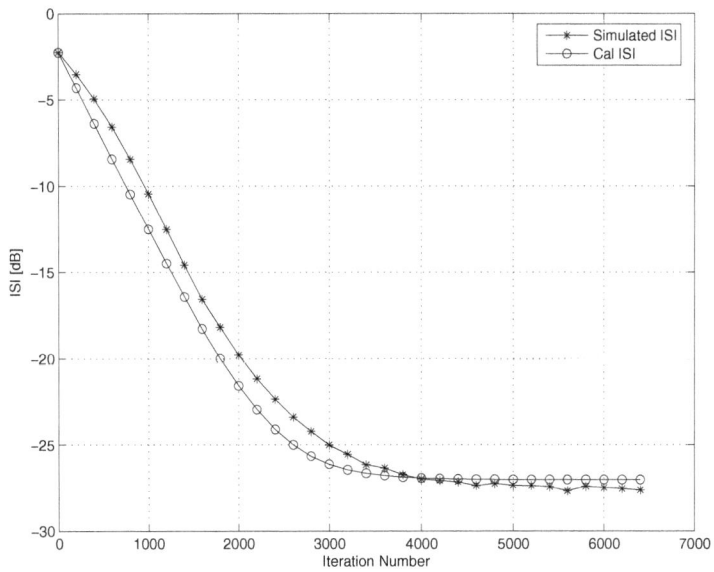

Figure 9.11: A comparison between the simulated (with the "New" algorithm) and calculated ISI as a function of time for the 16QAM source input going through channel3. The averaged results were obtained in 100 Monte Carlo trials for the noiseless case. The equalizer's length was set to 13 and $\mu_N = 0.0003$.

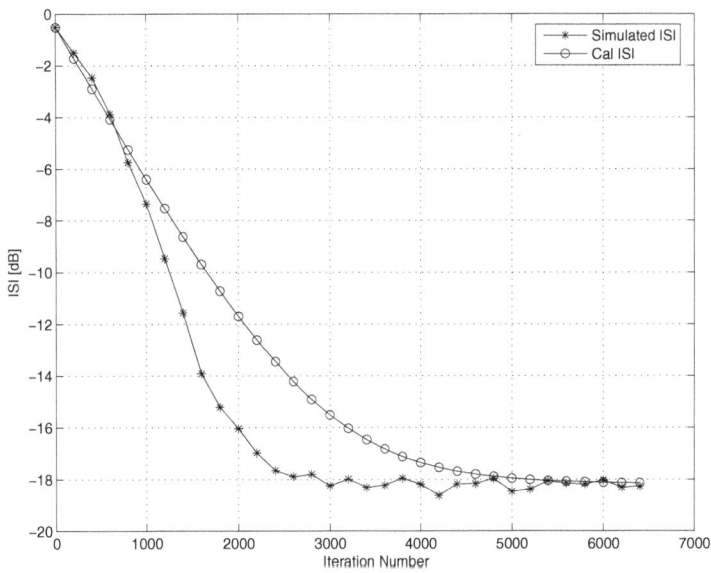

Figure 9.12: A comparison between the simulated (with Godard's algorithm) and calculated ISI as a function of time for the 16QAM source input going through channel4. The averaged results were obtained in 10 Monte Carlo trials for the noiseless case. The equalizer's length was set to 27 and $\mu_G = 0.00003$.

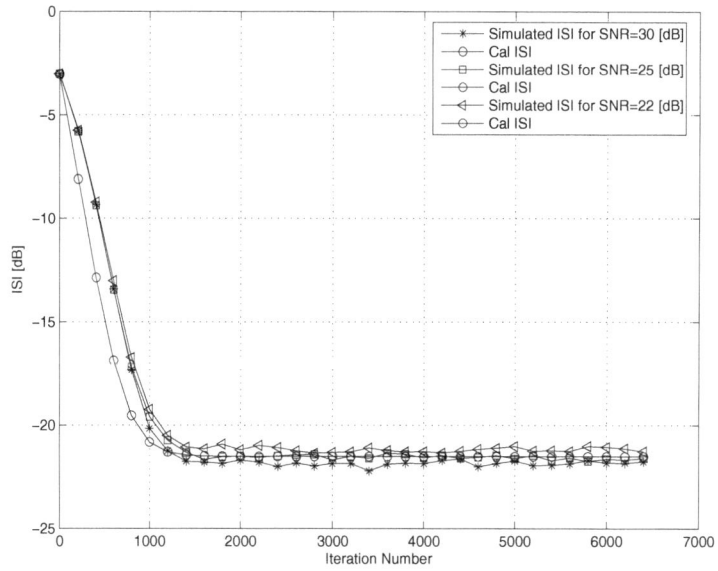

Figure 9.13: A comparison between the simulated (with the "New" algorithm) and calculated ISI as a function of time for the 16QAM source input going through channel2. The averaged results were obtained in 100 Monte Carlo trials for the noisy case. The equalizer's length was set to 21 and $\mu_N = 0.0005$.

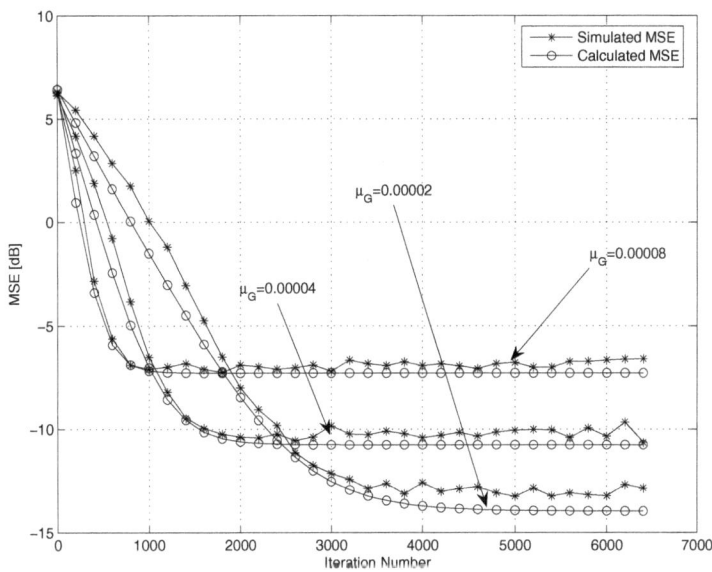

Figure 9.14: A comparison between the simulated (with Godard's algorithm) and calculated ISI as a function of time for the 16QAM source input going through channel1. The averaged results were obtained in 500 Monte Carlo trials for $SNR = 30\ [dB]$. The equalizer's length was set to 13.

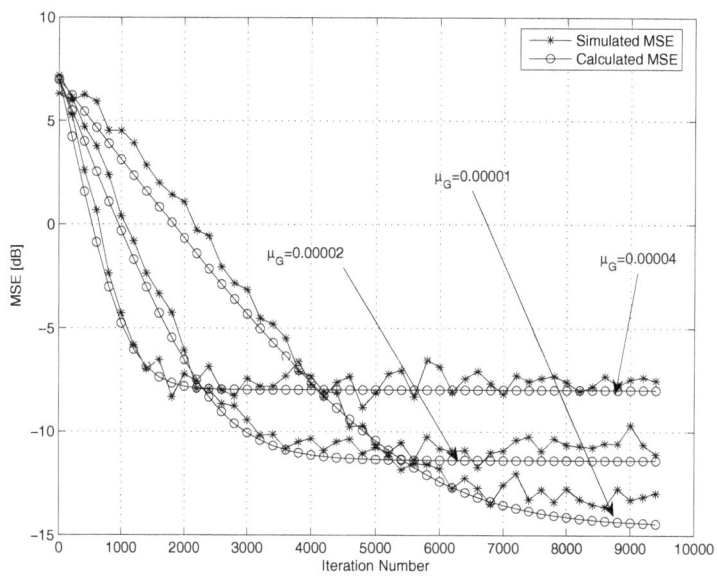

Figure 9.15: A comparison between the simulated (with Godard's algorithm) and calculated ISI as a function of time for the 16QAM source input going through channel1. The averaged results were obtained in 100 Monte Carlo trials for $SNR = 30\ [dB]$. The equalizer's length was set to 21.

Case	Simulation Results	Calculated Results
Type of Alg. : Godard [2] Source input: 16QAM Equalizer's length: 21 Step-size: $\mu_G = 0.00004$ Channel : 2; No Noise	2500	2413
Type of Alg. : Godard [2] Source input: 16QAM Equalizer's length: 13 Step-size: $\mu_G = 0.00002$ Channel : 1; No Noise	4500	4266
Type of Alg. : Godard [2] Source input: 16QAM Equalizer's length: 27 Step-size: $\mu_G = 0.00003$ Channel : 3; No Noise	4600	4457

Table 9.1: Number of iteration required for convergence.

Case	Simulation Results	Calculated Results
Type of Alg. : Godard [2] Source input: 16QAM Equalizer's length: 27 Step-size: $\mu_G = 0.00002$ Channel : 3; No Noise	6600	6492
Type of Alg. : "New" [11] Source input: 16QAM Equalizer's length: 21 Step-size: $\mu_N = 0.0002$ Channel : 2; No Noise	3500	3535
Type of Alg. : "New" [11] Source input: 16QAM Equalizer's length: 21 Step-size: $\mu_N = 0.0005$ Channel : 2; No Noise	1500	1334
Type of Alg. : "New" [11] Source input: 16QAM Equalizer's length: 13 Step-size: $\mu_N = 0.0007$ Channel : 3; No Noise	1800	1389

Table 9.2: Number of iteration required for convergence.

Case	Simulation Results	Calculated Results
Type of Alg. : "New" [11] Source input: 16QAM Equalizer's length: 13 Step-size: $\mu_N = 0.0005$ Channel : 1; No Noise	1400	1334
Type of Alg. : "New" [11] Source input: 16QAM Equalizer's length: 13 Step-size: $\mu_N = 0.0002$ Channel : 1; No Noise	3700	3404
Type of Alg. : "New" [11] Source input: 16QAM Equalizer's length: 13 Step-size: $\mu_N = 0.0007$ Channel : 1; No Noise	1000	932
Type of Alg. : "New" [11] Source input: 16QAM Equalizer's length: 13 Step-size: $\mu_N = 0.0003$ Channel : 3; No Noise	4000	3343

Table 9.3: Number of iteration required for convergence.

CONCLUSION

In this chapter we have shown via simulation that the recently proposed expression for the ISI as a function of time is useful not only for Godard's algorithm but also for another type of equalization method, for different channel types and works well for SNR values down to 22 $[dB]$. Thus the question if the recently proposed expression for the ISI as a function of time is useful only for a particular equalization method is answered. In addition, we have proposed in this chapter a closed-form approximated expression for the MSE as a function of time that is valid for the noiseless, real and two independent quadrature carrier case and for type of blind equalizers where the error that is fed into the adaptive mechanism which updates the equalizer's taps can be expressed as a polynomial function of order three. Simulation results have shown that the simulated and calculated expression for the MSE as a function of time (or iteration number) are very close.

ABBREVIATIONS

ISI=   Intersymbol Interference

SNR=   Signal to Noise Ratio

QAM= Quadrature Amplitude Modulation

QPSK=   Quadrature Phase Shift Keying

MSE=   Mean Square Error

REFERENCES

[1] M. Pinchas, "An analytical expression for the convergence time of adaptive blind equalizers," in *ICINCO-8th International Conference on Informatics in Control, Automation and Robotics*, 2011.

[2] D. N. Godard, "Self recovering equalization and carrier tracking in two-dimensional data communication systems," *IEEE Transaction on Communications*, vol. COM-28, pp. 1867–1875, 1980.

[3] M. Pinchas, "A closed approximated formed expression for the achievable residual intersymbol interference obtained by blind equalizers," *Signal Processing Journal (Eurasip)*, vol. 90, pp. 1940–1962, 2010.

[4] C. L. Nikias and A. P. Petropulu, Eds., *Higher-Order Spectra Analysis A Nonlinear Signal Processing Framework*. Prentice-Hall, 1993.

[5] S. Bellini, "Bussgang techniques for blind equalization," in *IEEE Global Telecommunication Conference Records*, 1986, pp. 1634–1640.

[6] S. Fiori, "A contribution to (neuromorphic) blind deconvolution by flexible approximated bayesian estimation," *Signal Processing*, vol. 81, pp. 2131–2153, 2001.

[7] S. Haykin, Ed., *Adaptive Filter Theory*. Prentice-Hall, Englewood cliffs,NJ, 1991.

[8] R. Godfrey and F. Rocca, "Zero memory non-linear deconvolution," *Geophys. Prospect.*, vol. 29, pp. 189–228, 1981.

[9] M. Pinchas and B. Z. Bobrovsky, "A maximum entropy approach for blind deconvolution," *Signal Processing Journal (Eurasip)*, vol. 86, pp. 2913–2931, 2006.

[10] M. Pinchas, "A novel expression for the achievable MSE performance obtained by blind adaptive equalizers, " *Signal, Image and Video Processing*, DOI: 10.1007/s11760-011-0208-x.

[11] M. Pinchas, "What are the analytical conditions for which a blind equalizer will loose the converge state," *Signal, Image and Video Processing*, DOI: 10.1007/s11760-011-0221-0.

[12] O. Shalvi and E. Weinstein, "New criteria for blind deconvolution of non-minimum phase systems (channels)," *IEEE Transaction on Information Theory*, vol. IT-36, pp. 312–321, 1990.

Chapter 10

ADVANTAGES AND DISADVANTAGES OF BLIND ADAPTIVE EQUALIZERS COMPARED WITH THE NON ADAPTIVE AND NON BLIND APPROACH

MONIKA PINCHAS

Department of Electrical and Electronic Engineering, Ariel University Center of Samaria, Ariel 40700, ISRAEL

ABSTRACT

In this chapter we present the advantages and disadvantages of the blind serially adaptive equalizer compared with the non-blind adaptive version as well as with the blind but non adaptive approach.

KEYWORDS

Blind adaptive equalizer, non-blind adaptive equalizer, non adaptive blind equalizer, training sequences, nonlinear transformation, convergence speed, computational cost, higher order statistics (HOS), stochastic gradient method, LMS (least mean squares) algorithm

ADVANTAGES AND DISADVANTAGES OF THE APPROACHES

In the non-blind adaptive equalization approach, the equalizer's taps are updated according to:

$$\underline{c}_{eq}[n+1] = \underline{c}_{eq}[n] + \mu \left(x\left[n\right] - z\left[n\right] \right) \underline{y}^*\left[n\right] \tag{10.1}$$

where $()^*$ is the conjugate operation, μ is the step-size parameter, $\underline{c}_{eq}[n]$ is the equalizer vector where the input vector is $\underline{y}[n] = [y[n] \ldots y[n - N + 1]]^T$ and N is the equalizer's tap length. The operator $()^T$ denotes for transpose of the function $()$. The equalized output signal is $z\left[n\right]$ and the transmitted signal is given by $x\left[n\right]$. In the non-blind adaptive equalization version, training sequences are thus needed to generate the error that is fed into the adaptive mechanism which updates the equalizer's taps. But those training sequences sacrifice bandwidth. In the blind version, the algorithm is designed in such a way that we do not require the external supply of a desired response to generate the error signal in the output of the adaptive equalization filter [1]. The algorithm itself generates an estimate of the desired response by applying a nonlinear transformation to sequences involved in the adaptation process [1]. Since blind adaptive equalizers do not need any training sequence, they conserve bandwidth and are useful for point-to-multipoint network applications, such as the fiber to the curb (FTTC) systems [2]. The blind adaptive approach is used for ease of interoperability between different manufactures and just offers lot of simplicity in multipoint communication. When a communication link is reset, equalizer adjustment from scratch is necessary. In this case, using a training sequence is inefficient, since the transmitter has to retransmit the training sequence specifically for each receiver which is reset. But, the non-blind adaptive approach yields in most cases to better equalization performance considering convergence speed and equalization quality compared with the blind adaptive version. In addition, the blind adaptive version has a higher computational cost compared to its non blind version.

In the non-adaptive blind approach ([3] for an example), a batch of received data is collected of some length. The algorithm that derives the equalizer's taps is usually based on higher order statistics (HOS) and converges within few iterations only. On the other hand, the blind serially adaptive approach based on the stochastic gradient method which is considered as a "blind" counterpart to the classic LMS (least mean squares) algorithms is well known to have relatively slow convergence speed.

The length of the collected data in the non-adaptive blind approach, plays a major role in the equalization quality. Increasing the collected data length will usually decrease the residual ISI and the number of iterations for convergence for non time varying channels. But when the channel is time-varying even slow time-varying, a high value for the collected data length may cause to unsatisfying residual ISI. On the other hand, the blind serially adaptive equalizer is able to track changes in a time-varying and non predictable channel, if not limited by convergence speed.

ABBREVIATIONS

ISI= Intersymbol Interference

LMS=   Least Mean Square

HOS=Higher Order Statistics

FTTC=   Fiber To The Curb

REFERENCES

[1] C. L. Nikias and A. P. Petropulu, Eds., *Higher-Order Spectra Analysis A Non-linear Signal Processing Framework.* Prentice-Hall, 1993.

[2] G. H. Im and H. C. Won, "Rf interference suppression for vdsl system," *IEEE Trans. on Consumer Electronics*, vol. 47, pp. 715–722, 2001.

[3] O. Shalvi and E. Weinstein, "Super-exponential methods for blind deconvolution," *IEEE Trans. on Information Theory*, vol. 39, pp. 504–519, 1993.

APPENDIX

MONIKA PINCHAS

Department of Electrical and Electronic Engineering, Ariel University Center of Samaria, Ariel 40700, ISRAEL

ABSTRACT

In this section, we bring a simple Matlab program for a serially blind adaptive equalizer.

KEYWORDS

Blind adaptive equalizer, serially blind adaptive equalizer, intersymbol interference (ISI), step-size parameter, equalizer's tap-length, channel coefficients, simulation code, equalized constellation output, Godard's algorithm, SISO case

Matlab Program

Appendix   *The Whole Story Behind Blind Adaptive Equalizers/ Blind Deconvolution* **187**

26/11/11 15:22   C:\temp_mat_files\Godard.m   1 of 3

```matlab
% ***********************************************************************%
%    This Matlab program simulates Godard's algorithm for the     %
%    16QAM input case by considering an easy channel              %
%    (FIR channel with static coefficients), SISO and             %
%    noiseless case. The code is written with minimum use         %
%    of inbuilt Matlab functions in order that also a very        %
%    beginner with Matlab can easily understand the code.         %
%    In other words, the code is not written in a sophisticated   %
%    way or in a way that reduces the simulation runtime.         %
%    This program is a code for a serially adaptive equalizer     %
% ***********************************************************************%

clear;
num_iter=100;                                    % The number of Monte Carlo trials
for bbb=1:num_iter
zz1=0;
zz2=0;
delta=0.00004;                                   %step-size
tap_length=13;                                   % Equalizer's tap length
c=[zeros(1,(tap_length-1)/2) 1 zeros(1,(tap_length-1)/2)];   % Initialize the equalizer
%***********************************************%
%      Channel preparation        %
%***********************************************%
for i=1:13
  if i>1
    h(i)=0.8*0.4^(i-2);
  else
    h(i)=-0.4;
  end
end
%***********************************************************************************%
symbol_length=7000;                              % Length of the simulation
y=zeros(1,symbol_length);                        % Output channel initialization
z=zeros(1,symbol_length);                        % Initialization of the equalized output

for i=1:symbol_length
%*************************************%
%      Data Preparation       %
%*************************************%
xx=4*(1-2*rand(1));
if abs(xx)>=2
  d_a(i)=(sign(xx))*3;
else
  d_a(i)=(sign(xx))*1;
end

xxx=4*(1-2*rand(1));
if abs(xxx)>=2
  d_b(i)=(sign(xxx))*3;
else
  d_b(i)=(sign(xxx))*1;
end
d_c(i)=d_a(i)+j*d_b(i);
%*************************************************************************%
    zz1=zz1+(abs(d_c(i)))^4;
    zz2=zz2+(abs(d_c(i)))^2;
```

```matlab
%******************************************************%
% Generating the data from the channel output %
%******************************************************%
if i>length(h)-1
   for k=1:length(h)
      y(i)=y(i)+h(k)*d_c(i-(k-1));
   end
 end

end
%******************************************************%
%    Generating the equalized output      %
%******************************************************%
p=0;
for f=415:length(y)  % enables the use of comparison with other type of equalizers that need an
                     % initialization phase. Those equalizers that need the initialization
                     % phase should be intitialized until the time of f=415.

   for k=1:tap_length

     z(f)=z(f)+c(k)*y(f-(k-1));                       % equalized output
   end
%******************************************************%
    p=p+1;
    iteration(p)=p;
    s=conv(c,h);                                 % Convolving the channel with the equalizer
    %******************************************************%
    %        Generating the ISI          %
    %******************************************************%
    isi(bbb,p)=(sum((abs(s(1:length(s) ))).^2)-max(((abs(s(1:length(s) ))).^2)))/max(((abs(s(1:length(s) ))).^2));

    %******************************************************%
    %    Generating the equalized output        %
    %******************************************************%
      for k=1:tap_length

      %******************************************************%
      %         Godard                     %
      %******************************************************%
        c_a(k)=c(k)-delta*((abs(z(f)))^2-(zz1/symbol_length)/(zz2/symbol_length))*z(f)*conj(y(f-(k-1)));
      %******************************************************************************************%

      end

      c=c_a;

 end
end
isi_new=(sum(isi,1))./num_iter;                       % Averaged ISI
isi_new_db=10*log10(isi_new);                         % Generating the ISI in dB units
plot(iteration(1:200:length(iteration)),isi_new_db(1:200:length(iteration)),'k*-')
xlabel('Iteration Number')
ylabel('ISI [dB]')
```

```
grid
legend('Godard')
figure(2); plot(real(z(2000:end)), imag(z(2000:end)),'+')   % shows the equalized constellation
                                              % output in the steady state.
grid
figure(3); plot( real(z),'+')  % should show 4 levels (-3, -1, 1, 3) in the steady state when
                    % there is no gain between the source and equalized output

grid
```

BIBLIOGRAPHY

[1] M. G. A. Benveniste and G. Ruget, "Robust identification of a non-minimum phase system: Blind adjustment of a linear equalizer in data communications," *IEEE Transactions on Automatic Control,* vol. 25, pp. 385–399, 1980.

[2] S. A. Assaf and L. D. Zirkle, "Approximate analysis of nonlinear stochastic systems," *International Journal of Control,* vol. 23, pp. 477–492, 1976.

[3] S. Bellini, "Bussgang techniques for blind equalization," in *IEEE Global Telecommunication Conference Records,* 1986, pp. 1634–1640.

[4] ——, "Blind equalization," *Alta Freq.,* vol. 57, pp. 445–450, 1988.

[5] S. Bellini and F. Rocca, Eds., *Blind deconvolution: Polyspectra or Bussgang techniques?, in Digital Communications.* Elsevier Science Publishers B.V., 1986.

[6] A. Benveniste and M. Goursat, "Blind equalizers," *IEEE Transaction on Communications,* vol. COM-32, pp. 871–883, 1984.

[7] D. C. C. Bover, "Moment equation methods for nonlinear stochastic systems," *Journal of Mathematical Analysis and Applications,* vol. 65, pp. 306–320, 1978.

[8] J. A. Cadzow, "Blind deconvolution via cumulant extrema," *IEEE Signal Processing Mag.*, vol. 13, pp. 24–42, 1996.

[9] C. Y. Chi and M. C. Wu, "Inverse filter criteria for blind deconvolution and equalization using two cumulants," *Signal Processing*, vol. 43, pp. 55–63, 1995.

[10] ——, "A unified class of inverse filter criteria using two cumulants for blind deconvolution and equalization," in *Proc. IEEE Int. Conf. Acoustics, Speech, Signal Processing*, Detroit, MI, 1995, pp. 1960–1963.

[11] C.-Y. Chi, C.-Y. Chen, C.-H. Chen, and C.-C. Feng, "Batch processing algorithms for blind equalization using higher-order statistics," *IEEE Signal Processing Magazine*, pp. 25–49, 2003.

[12] H. Cramer, Ed., *Mathematical Methods of Statistics.* Princeton University Press, 1951.

[13] D. L. Donoho, *On minimum entropy deconvolution*, ser. In Applied Time Series Analysis II, D. F. Findly, Ed. New York: Academic, 1981.

[14] C. Feng and C. Chi, "Performance of cumulant based inverse filters for blind deconvolution," *IEEE Transaction on Signal Processing*, vol. 47, pp. 1922–1935, 1999.

[15] S. Fiori, "A contribution to (neuromorphic) blind deconvolution by flexible approximated bayesian estimation," *Signal Processing*, vol. 81, pp. 2131–2153, 2001.

[16] ——, "A fast fixed-point neural blind deconvolution algorithm," *IEEE Transaction on Neural Networks*, vol. 15, pp. 455–459, 2004.

[17] D. N. Godard, "Self recovering equalization and carrier tracking in two-dimensional data communication systems," *IEEE Transaction on Communications*, vol. COM-28, pp. 1867–1875, 1980.

[18] R. Godfrey and F. Rocca, "Zero memory non-linear deconvolution," *Geophys. Prospect.*, vol. 29, pp. 189–228, 1981.

[19] D. Hatzinakos and C. L. Nikias, "Blind equalization using a tricespectrum based algorithm," *IEEE Transaction on Comm.*, vol. 39, pp. 669–682, 1991.

[20] D. Hatzinakos and C. L. Nikias, Eds., *Blind equalization based on higher-order statistics (H.O.S.), in Blind Deconvolution (S. Haykin,ed.).* Prentice Hall, 1994.

[21] S. Haykin, Ed., *Adaptive Filter Theory.* Prentice-Hall, Englewood cliffs,NJ, 1991.

[22] ——, *Bussgang Techniques for blind deconvolution and equalization, in: S. Haykin (Ed.), Blind Deconvolution.* Prentice Hall, 1994.

[23] G. H. Im and H. C. Won, "Rf interference suppression for vdsl system," *IEEE Trans. on Consumer Electronics*, vol. 47, pp. 715–722, 2001.

[24] G.-H. Im, C. J. Park, and H. C. Won, "A blind equalization with the sign algorithm for broadband access," *IEEE Comm. Letters*, vol. 5, pp. 70–72, 2001.

[25] L. T. J. K. Tugnait and Z. Ding, "Single-user channel estimation and equalization," *IEEE Signal Processing Magazine*, vol. 17, pp. 16–28, 2000.

[26] B. Jellonek, D. Boss, and K. D. Kammeyer, "Generalized eigenvector algorithm for blind equalization," *EURASIP Signal Processing*, pp. 237–264, 1997.

[27] B. Jellonek and K. D. Kammeyer, "A closed-form solution to blind equalization," *EURASIP Signal Processing*, vol. 36, pp. 251–259, 1994.

[28] G. Jumarie, "Nonlinear filtering. a weighted mean squares approach and a bayesian one via the maximum entropy principle," *Signal Processing*, vol. 21, pp. 323–338, 1990.

[29] M. Lazaro *et al.*, "Stochastic blind equalization based on pdf fitting using parzen estimator," *IEEE Trans. on Signal Processing*, vol. 53, pp. 696–704, 2005.

[30] E. A. Lee and D. G. Messerschmitt, Eds., *Adaptive Equalization, in: E. A. Lee and D. G. Messerschmitt, Digital Communication*, 2nd ed. Kluwer Academic Publisher, third printing, 1997.

[31] A. K. Nandi, Ed., *Blind estimation using higher-order statistics*. Boston: Kluwer Academic, 1999.

[32] C. L. Nikias and A. P. Petropulu, Eds., *Higher-Order Spectra Analysis A Nonlinear Signal Processing Framework*. Prentice-Hall, 1993.

[33] S. A. Orszag and C. M. Bender, Eds., *Advanced Mathematical Methods for Scientist Engineers, International Series in Pure and Applied Mathematics*. McDraw-Hill, 1978.

[34] A. Papoulis, Ed., *Probability, Random Variables, and Stochastic Processes*, second international ed. Kogakusha: McGraw-Hill, 1984.

[35] G. Picchi and G. Prati, "Blind equalization and carrier recovery using a 'stop-and-go' decision-directed algorithm," *IEEE Transaction on Communications*, vol. COM-35, pp. 877–887, 1987.

[36] M. Pinchas, Ed., *Blind Equalizers By Techniques Of Optimal Non-Linear Filtering Theory.* ISBN 978-3-639-15530-3, VDM Verlagsservice gesellschaft mbH, 2009.

[37] M. Pinchas, "A closed approximated formed expression for the achievable residual intersymbol interference obtained by blind equalizers," *Signal Processing Journal (Eurasip)*, vol. 90, pp. 1940–1962, 2010.

[38] ——, "A new closed approximated formed expression for the achievable residual isi obtained by adaptive blind equalizers for the noisy case," in *IEEE International Conference on Wireless Communications Networking and Information Security WCNIS2010*, 2010.

[39] ——, "An analytical expression for the convergence time of adaptive blind equalizers," in *ICINCO-8th International Conference on Informatics in Control, Automation and Robotics*, 2011.

[40] ——, "A mse optimized polynomial equalizer for 16qam and 64qam constellation," *Signal, Image and Video Processing*, vol. 5 , No. 1, pp. 29–37, 2011.

[41] ——, "What are the analytical conditions for which a blind equalizer will loose the converge state," *Signal, Image and Video Processing*, DOI: 10.1007/s11760-011-0221-0.

[42] ——, "A novel expression for the achievable MSE performance obtained by blind adaptive equalizers, " *Signal, Image and Video Processing*, DOI: 10.1007/s11760-011-0208-x.

[43] M. Pinchas and B. Z. Bobrovsky, "A maximum entropy approach for blind deconvolution," *Signal Processing Journal (Eurasip)*, vol. 86, pp. 2913–2931, 2006.

[44] ——, "A novel hos approach for blind channel equalization," *IEEE Wireless Communication Journal*, vol. 6, 2007.

[45] J. G. Proakis and C. L. Nikias, "Blind equalization (overview paper)," in *SPIE The International Society for Optical Engineering*, 1991, pp. 1565:76–87.

[46] Y. Sato, "A method of self-recovering equalization for multilevel amplitude-modulation systems," *IEEE Transaction on Communications*, vol. COM-23, pp. 679–682, 1975.

[47] O. Shalvi and E. Weinstein, "New criteria for blind deconvolution of non-minimum phase systems (channels)," *IEEE Transaction on Information Theory*, vol. IT-36, pp. 312–321, 1990.

[48] ——, "Super-exponential methods for blind deconvolution," *IEEE Trans. on Information Theory*, vol. 39, pp. 504–519, 1993.

[49] M. R. Spiegel, Ed., *Mathematical Handbook, Schaum's Outline Series*. McGraw-Hill, 1997.

[50] Z. Xu and M. Tsatsanis, "Adaptive minimum variance methods for direct blind multichannel equalization," *Signal Processing Journal (Eurasip)*, vol. 73, no. 1-2, pp. 125–138, Feb. 1999.

[51] Z. Xu and M. K. Tsatsanis, "Blind adaptive algorithms for minimum variance CDMA receivers," *IEEE Transaction on Communications*, vol. 49, no. 1, Jan. 2001.

Index

[52] C. Xu, G. Feng and K. S. Kwak, "A Modified Constrained Constant Modulus Approach to Blind Adaptive Multiuser Detection," *IEEE Transaction on Communications*, vol. 49, no. 9, 2001.

[53] J. K. Tugnait and T. Li, "Blind detection of asynchronous CDMA signals in multipath channels using code-constrained inverse filter criterion," *IEEE Transaction on Signal Processing*, vol. 49, pp. 1300–1309, July 2001.

[54] Z. Xu and P. Liu, "Code-Constrained Blind Detection of CDMA Signals in Multipath Channels," *IEEE Signal Processing Letters*, vol. 9, no. 12, Dec. 2002.

[55] R. C. de Lamare and R. Sampaio-Neto, "Blind Adaptive Code-Constrained Modulus Algorithms for CDMA Interference Suppression in Multipath," *IEEE Communications Letters*, vol. 9, no. 4, pp. 334–336, Apr. 2005.

[56] R. C. de Lamare and R. Sampaio-Neto, "Blind adaptive MIMO receivers for space-time block-coded DS-CDMA systems in multipath channels using the constant modulus criterion," *IEEE Transaction on Communications*, vol. 58, no. 1, pp. 21–27, Jan. 2010.

Liver Cirrhosis and Related Diseases: Pathophysiology, Prognosis and Treatment

Liver Cirrhosis and Related Diseases: Pathophysiology, Prognosis and Treatment

Edited by Dallas Bowman

hayle
medical

New York

Hayle Medical,
750 Third Avenue, 9th Floor,
New York, NY 10017, USA

Visit us on the World Wide Web at:
www.haylemedical.com

ISBN: 978-1-63241-637-7

Cataloging-in-Publication Data

Liver cirrhosis and related diseases : pathophysiology, prognosis and treatment / edited by Dallas Bowman.
    p. cm.
Includes bibliographical references and index.
ISBN 978-1-63241-637-7
1. Liver--Cirrhosis. 2. Liver--Diseases. 3. Liver--Cirrhosis--Pathophysiology.
4. Liver--Cirrhosis--Prognosis. 5. Liver--Cirrhosis--Treatment. I. Bowman, Dallas.
RC848.C5 L58 2019
616.362 4--dc23

Table of Contents

Preface

In my initial years as a student, I used to run to the library at every possible instance to grab a book and learn something new. Books were my primary source of knowledge and I would not have come such a long way without all that I learnt from them. Thus, when I was approached to edit this book; I became understandably nostalgic. It was an absolute honor to be considered worthy of guiding the current generation as well as those to come. I put all my knowledge and hard work into making this book most beneficial for its readers.

Liver cirrhosis is a liver condition in which the liver fails to function properly, due to irreversible liver damage. This is caused by the replacement of normal liver tissue with scar tissue. Some of the consequences of abnormal functioning of liver cells include spider angiomata, gynecomastia, hypogonadism, etc. The scar tissue in liver cirrhosis increases the resistance to blood flow resulting in portal hypertension. Damage to the hepatic parenchyma triggers the activation of stellate cells, which obstructs hepatic blood flow and increases fibrosis. As the disease progresses, hepatic encephalopathy, hepatorenal syndrome, and bruising and bleeding may occur. Liver biopsy, blood test for platelet count, ultrasound and various other lab tests can help diagnose liver cirrhosis. Out of all liver cirrhosis cases, 57% may be attributed to hepatitis B and C. 20% of such cases are due to alcohol consumption. Abstaining from alcohol, prescription of corticosteroids and interferon, chelation therapy, etc. slows the progression of the disease and also helps to manage its symptoms. This book contains some path-breaking studies in the area of liver cirrhosis and its related diseases. It provides significant information of the pathophysiology, prognosis and treatment of liver cirrhosis. Those with an interest in hepatology and gastroenterology would find this book helpful.

I wish to thank my publisher for supporting me at every step. I would also like to thank all the authors who have contributed their researches in this book. I hope this book will be a valuable contribution to the progress of the field.

Editor

Case Report: Long-Term Survival in Patient with Cirrhosis of the Liver and Colon Cancer K-ras Wild-Type

Emiddio Barletta*, Lucia Cannella, Vincenza Tinessa, Domenico Germano, Bruno Daniele

Department of Medical Oncology, G. Rummo Hospital, Benevento, Italy
Email: *emiddiobarletta@libero.it

Abstract

K-ras wild-type carcinoma is a tumour that is sensitive to treatment with anti-cancer and anti-EGFR drugs: the combination of Cetuximab and Panitumumab with chemotherapy (Cetuximab) or as a single therapy (Panitumumab). Case Report: The clinical case presented here refers to a 68-year-old patient who had been diagnosed with adenocarcinoma of the recto sigmoid with pelvic recurrence three years after surgery. The patient had a severe co-morbidity: correlated B-type liver cirrhosis. First-line chemotherapy was begun with Oxaliplatin plus Capecitabine (CAPOXI) following a relapse, and this continued for six months (six cycles), when the treatment was interrupted because of the disease's progression and hematological and gastrointestinal toxicity. Following an assessment of the K-ras, diagnosed as wild type, the patient was excluded from second-line chemotherapy treatment because of decompensated cirrhosis and the persistence of thrombocytopenia and leukopenia. The patient was put forward for biological treatment with an anti-EGFR monoclonal antibody (Panitumumab). Panitumumab was administered at a dosage of 6 mg/kg every 2 weeks for 17 months; the treatment was well tolerated, despite the cirrhosis, and the main toxicity was the skin rash. Conclusion: In patients with severe comorbidities such as cirrhosis of the liver and K-ras wild-type carcinomas, therapy with a monoclonal antibody such as Panitumumab is a treatment that is well tolerated, with few serious toxic side-effects; it also offers advantages in terms of survival and clinical benefits.

Keywords

K-ras Wild-Type Carcinoma, Metastatic Colorectal Cancer, Panitumumab, Anti-EGFR Treatment, Cirrhosis

---

*Corresponding author.

1. Introduction

Colorectal cancer represents one of the major causes of morbidity and mortality from neoplasias in all western and highly-technologically developed countries. Every year 678,000 new cases are reported worldwide, with 150,000 cases in Europe and 30,000 in Italy. At present, despite these neoplasias showing high levels of curability when compared to those in other areas of the digestive system, the five-year survival rate stands at around 40% - 50%, possibly reaching 80% - 90% in the early forms. The treatment of carcinoma of the colon will depend on the stage of the disease; in stages I - III surgery is of primary importance, in stage IV surgery is only an option in selected cases. In stage IV the role of medical oncology treatment is of primary importance; chemotherapy with fluorinated pyrimidines (Fluorouracil and Capecitabine) with Oxaliplatin or Irinotecan (FOLFOX, FOLFIRI, or XELIRI or XELOX). [1] [2] are schemes that have resulted in an increase in average survival rates ranging from 6 months to more than 20. The use of anti-EGFR antibodies has demonstrated how it has been possible to achieve limited benefits, in terms of survival, in patients that were resistant to previous lines of chemotherapy including Oxaliplatin and Irinotecan. This has been demonstrated for both Cetuximab (hybrid antibody) and Panitumumab (humanized antibody) [3], but appears limited to patients who do not present mutations of the KRAS. This mutation should therefore be looked for before using these drugs [4]. Panitumumab in particular is only registered for use in patients who have already been treated with previous chemotherapies and that are EGFR positive and KRAS wild-type [5]. The K-ras oncogene is a central component of the system of signal transduction downstream of EGFR and plays a critical role in the regulation of cell growth. E was shown that the mutation status of the KRAS gene in tumor cells affects the response to cetuximab (and panitumumab), and that to benefit from treatment with these monoclonal antibodies are only the carriers of the KRAS protein not mutated (wild type) [6]. In cases where the KRAS is mutated the drug is ineffective [7]; in fact, in the presence of mutations in KRAS, which determine the constitutive activation, the proliferative signal is independent from the stimulus EGF. In these cases the block EGFR is not able to inhibit cell proliferation, making compartment the therapeutic effect of the antibody. These mutations were found in 40% of cases of cancer of the colon and rectum [8].

2. Case Presentation

A 68-year-old man suffering from HBV-related cirrhosis of the liver underwent a segmental resection of the sigmoid as a result of a stage pT3 pN2 M0 G2 adenocarcinoma (stage C and Koller Astler sec.). Family histori: parents died at age 70 from cardiovascular disease. In 2005 he received additional chemotherapy for 6 months with Fluorouracile + folic acid as per the Roswell Park scheme. After 3 years of follow-ups the patient presented a neoplastic recurrence at a pelvic and abdominal level, with a 35mm lesion with other satellite lesions and periaortic lymphadenopathies. Tumour markers: CEA and CA19.9 normal. Not being open to new surgical treatment he was started on a course of treatment with CAPOXI, given the comorbidity (cirrhosis of the liver), the patient began a treatment with Oxaliplatin 85 mg/m^2 every 3 weeks + Capecitabine 1800 mg/m²/day d1 −14 every 3 weeks. During the treatment he showed signs of significant hematological toxicity, in the main grade 3 thrombocytopenia (NCI) and grade 2 leukopenia (NCI). Six cycles of CAPOXI were administered and these stabilised the disease. In light of the disease stabilising, it was decided to continue treatment with Capecitabine alone at 1800 mg/mq/day, interrupting the Oxaliplatin because the patient had a peripheral grade 2 (NCI) neuropathy. Treatment continued until July of 2009 for a further 3 cycles. The instrumental restaging showed a slight radiological increase of the lesion, but one that could not be considered as a progression, according to the RECIST criteria. Treatment was interrupted because of the appearance of gastrointestinal toxicity, grade 2 diarrhoea and grade 2 stomatomycosis (NCI) as well as palmar-plantar erythrodysesthesia, and it was decided to proceed with a follow-up at 3 months. In October 2009 a whole body restaging CT scan (**Figure 1**) was performed that revealed the clear progression of the disease in the pelvic area (max. diameter of the lesion l 92mm) with the presence of intense pain in the area of the pelvis, radiating to the ipsilateral gluteus and the left leg; for this reason analgesic treatment was started with Fentanyl transdermal 25 - 50 - 100 - 125 ug. In order to programme a new line of therapy the determination of the EFGR expression and the mutational status of K-ras on the tumorous tissue of the surgical specimen was begun. The result was as follows: expression of EGFR and KRAS wild-type, and therefore predictors of a response to an anti-EGFR treatment with monoclonal Cetuximab and Panitumumab-type antibodies. Our evaluation on the choice of the best treatment for our patient was made in the wake of these data (laboratory) and given the more important comorbidity: cirrhosis of the liver—we excluded treatment with Irinotecan +/− with 5-FU (a drug with a predominant liver and intestinal toxicity) and

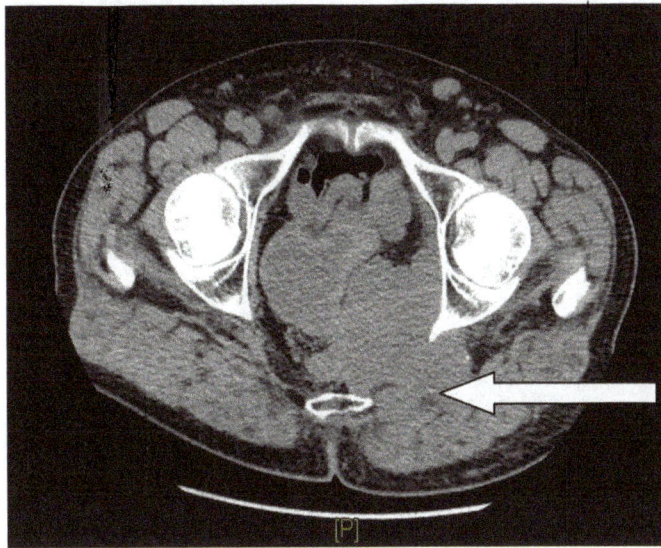

Figure 1. CT scan baseline.

taking into account the patient's blood-biochemical readings showing an increase in transaminase levels: GOT 70 U/l, ALT = 80 U/l; Leukopenia: GB = 3200; Neutrophils = 1700: Platelets = 80,000. We therefore excluded chemotherapy treatment with Irinotecan because it was hepatotoxic and opted for Panitumumab. As a consequence treatment began with Panitumumab 6 mg/kg IV every 14 days; after 2 doses a grade 3 rash (NCI) appeared, and so treatment was temporarily interrupted, postponing for another 2 weeks, and a symptomatic treatment with corticosteroids and salicylic sulphur based creams and antibiotics was ordered. At the end of the two week postponement period the rash reduced to a grade 1 - 2 (NCI). In addition to the rash, the patient also had grade 1 - 2 diarrhoea (NCI). Treatment was resumed with a grade 1 cutaneous rash, and 6 cycles were administered without the need for any interruptions and with the rash stabilising at grade 1. After the sixth cycle a whole-body restaging CT scan with medium contrast revealed a stabilising (with a slight radiological reduction) of the pelvic mass. Treatment with Panitumumab was continued without interruption, by carrying out CT restagings every 3 months with the resultant stabilising of the disease (the pelvic lesion measured around 90 mm); the treatment continued for 17 months (**Figure 2**). The cutaneous rash stabilised at grade 1 (NCI) in the absence of other significant toxicities.

3. Discussion

Currently the strategy for the treatment of metastatic colorectal carcinoma cannot take into account the evaluation of the mutational status of K-ras ref. [9]-[11], which may be present in two forms: mutated and wild-type. This evaluation allows us to identify those patients that are eligible for treatment with anti-EGFR. Approximately 40% of patients have mutated K-ras; in this case there is no inhibitory effect with anti-EGFR and therefore no benefit. 60% of patients have wild-type K-ras obtaining a clinical benefit from the use of an anti-EGFR. Panitumumab is a monoclonal humanised anti-EGFR antibody indicated as a single therapy after the failure of chemotherapies containing Oxaliplatin, Fluoropyridines and Irinotecanin in patients with K-ras wild-type metastatic colorectal carcinoma [12]. In the pivotal study, 463 patients who received multiple treatments with different lines of chemotherapy, 57% of which were wild-type, the arm treated with Vectibix + BSC with K-ras wild-type the PFS was greater compared to BSC alone with a ($p < 0.0001$), [13] no clinical benefit in patients with mutated K-ras. Treatment with Panitumumab was usually well tolerated with a grade 3 - 4 (NCI) cutaneous toxicity of 10%, a grade 3 - 4 (NCI) diarrhoea of 2% and grade 3 - 4 (NCI) Hypomagnesemia of 3% [14] [15]. The clinical case we examined was a K-ras wild-type and was treated with Panitumumab for 17 months; this produced a moderate clinical benefit (reduced pain, improved appetite, weight recovery, etc.) [16]. The predominant toxicity was cutaneous with a grade 3 - 4 rash after the first 2 doses, which then reduced and stabilised at grade 1 - 2 (NCI). Diarrhoea, on the other hand, established itself at grade 1, with short-term periods of worsening, passing to a grade 2. Hematological toxicity was nonexistent, but periodically there was a hypomagnesemia with

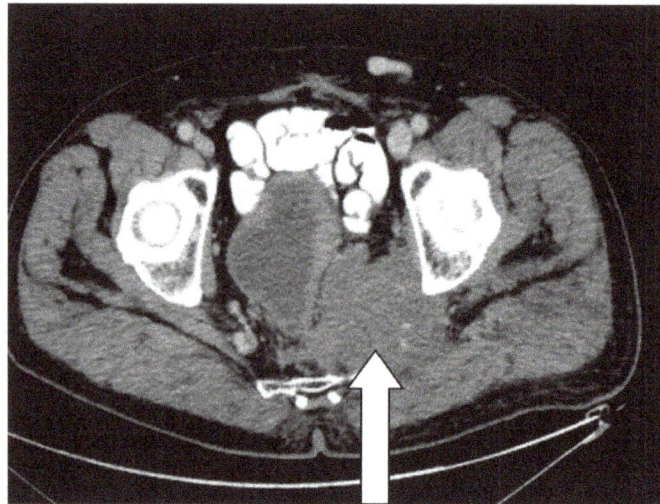

Figure 2. CT scan after 17 months.

hypocalcemia, both corrected with magnesium and calcium by os. Our patient had a PS (ECOG) = 1, however, given that the comorbidities include cirrhosis of the liver (even if compensated) and type II diabetes mellitus, thrombocytopenia and leukopenia (hepatopathy) and the persistence of toxicities from chemotherapy (peripheral neurotoxicity from Oxaliplatin), it was decided not to treat with chemotherapy + anti-EGFR (Cetuximab) [17]-[20], but only with an anti-EGFR: Panitumumab, which in the years from 2009 to 2011 had a therapeutic indication in single therapy after the failure of different chemotherapy regimens . The patient died in July 2013.

4. Conclusion

In the clinical case examined the patient in question was cirrhotic (cirrhosis is an important comorbidity that significantly limits chemotherapy treatments) and with a recurrent inoperable carcinoma of the colon (negative prognosis), maintained a stability of the disease for 17 months in succession, and overall survival of 8 years, with a humanised anti-EGFR antibody (Panitumumab) which, among other things, was well-tolerated and had a toxicity that was in the main cutaneous [21]. The patient also experienced a reduction in pain in the pelvic region, with a clear-cut reduction of analgesics, having begun at the start of the treatment with 125 µg of transdermal Fentanyl which was subsequently reduced to 25 µg every 72 hours and therefore a specific control over the symptoms associated with disease. In bygone times, we would probably have stopped after a first metastatic line because the toxicity induced by chemotherapy would have not allowed treatment to be continued. However, in an age of "target therapies" a Panitumumab-type monoclonal antibody allows us to have good control over the disease, with improvements in the quality of life and the survival rate of the patient; these represent two funda-mental end-points in oncology. Obviously all of this is only possible in patients who are K-ras wild-type but not in mutated K-ras cases; undoubtedly, in light of recent data, there are other molecular markers such as: B-raf, N-ras that can identify the patient best suited for anti-EGFR treatment.

Consent

For the clinical case was called informed consent.

References

[1]   Cassidy, J., Tabernero, J., Twelves, C., *et al.* (2004) XELOX (Capecitabine plus Oxaliplatin): Active First-Line Ther-apy for Patients with Metastatic Colorectal Cancer. *Journal of Clinical Oncology*, **22**, 2084-2091.
       http://dx.doi.org/10.1200/JCO.2004.11.069

[2]   Oh, S.C., Sur, H.Y. and Sung, H.J. (2007) A Phase II Study of Biweekly Dose-Intensified Oral Capecitabine plus Iri-notecan (bXELIRI) for Patients with Advanced or Metastatic Gastric Cancer. *British Journal of Cancer*, **96**, 1514-1519.
       http://dx.doi.org/10.1038/sj.bjc.6603752

[3]   Wisinski, K.B., Mulcahy, M.F. and Benson, A.B. (2007) Panitumumab in Metastatic Colorectal Cancer. *Clinical Ad-*

vances in Hematology and Oncology, **5**, 10-11.

[4] Allegra, C.J., Jessup, J.M. and Somerfield, M.R. (2009) American Society of Clinical Oncology Provisional Clinical Opinion: Testing for KRAS Gene Mutations in Patients with Metastatic Colorectal Carcinoma to Predict Response to Anti-Epidermal Growth Factor Receptor Monoclonal Antibody Therapy. *Journal of Clinical Oncology*, **27**, 2091-2096. http://dx.doi.org/10.1200/JCO.2009.21.9170

[5] Di Fiore, F., Charbonnier, F. and Lefebure, B. (2008) Clinical Interest of KRAS Mutation Detection in Blood for Anti-EGFR Terapie in Metastatic Colorectal Cancer. *British Journal of Cancer*, **99**, 551-552. http://dx.doi.org/10.1038/sj.bjc.6604451

[6] Karapetis, C.S., Khambata-Ford, S. and Jonker, D.J. (2008) KRAS Mutations and Benefit for Cetuximab in Advanced Colorectal Cancer. *New England Journal of Medicine*, **359**, 1747-1765. http://dx.doi.org/10.1056/NEJMoa0804385

[7] Siddiqui, A.D. and Piperdi, B. (2010) KRAS Mutation in Colon Cancer: A Marker of Resistance to EGFR-I Therapy. *The Annals of Surgical Oncology*, **17**, 1168-1176.

[8] McNeil, C. (2008) KRAS Mutations Are Changing Practice in Advanced Colorectal Cancer. *Journal of the National Cancer Institute*, **100**, 1667-1669. http://dx.doi.org/10.1093/jnci/djn429

[9] Amado, R.G., Wolf, M. and Peeters, M. (2008) Wild-Type KRAS Is Required for Panitumumab Efficacy in Patients with Metastatic Colorectal Cancer. *Journal of Clinical Oncology*, **26**, 626-634. http://dx.doi.org/10.1200/JCO.2007.14.7116

[10] Lievre, A., Bachet, J.B. and Le Corre, D. (2008) KRAS Mutations as an Independent Prognostic Factor in Patients with Advanced Colorectal Cancer Treated with Cetuximab. *Journal of Clinical Oncology*, **26**, 374-379. http://dx.doi.org/10.1200/JCO.2007.12.5906

[11] Di Fiore, F., Charbonnier, F. and Lefebure, B. (2008) Clinical Interest of KRAS Mutation Detection in Blood for Anti-EGFR Terapie in Metastatic Colorectal Cancer. *British Journal of Cancer*, **99**, 551-552. http://dx.doi.org/10.1038/sj.bjc.6604451

[12] Seront, E., Marot, L. and Coche, E. (2010) Successful Long-Term Management of a Patient with Late-Stage Metastatic Colorectal Cancer Treated with Panitumumab. *Cancer Treatment Reviews*, **36**, S11-S14. http://dx.doi.org/10.1016/S0305-7372(10)70002-5

[13] Van Cutsem, E., Peeters, M. and Siena, S. (2007) Open-Label Phase III Trial of Panitumumab plus Best Supportive Care Compared with Best Supportive Care Alone in Patients with Chemotherapy-Refractory Metastatic Colorectal Cancer. *Journal of Clinical Oncology*, **25**, 1658-1664. http://dx.doi.org/10.1200/JCO.2006.08.1620

[14] Lacouture, M.E., Mitchell, E.P. and Piperdi, B. (2010) Skin Toxicity Evaluation Protocol with Panitumumab (STEEP), a Phase II, Open-Label, Randomized Trial Evaluating the Impact of Pre-Emptive Skin Treatment Regimen on Skin Toxicties and Quality of Life in Patients with Metastatic Colorectal Cancer. *Journal of Clinical Oncology*, **28**, 1351-1357. http://dx.doi.org/10.1200/JCO.2008.21.7828

[15] Pinto, C., Barone, C.A., Girolomoni, G., *et al.* (2011) Managemente of Skin Toxcity Associated with Cetuximab Treatment in Combination with Chemotherapy or Radiotherapy. *Oncologist*, **16**, 228-238. http://dx.doi.org/10.1634/theoncologist.2010-0298

[16] Ishiyama, Y., Kotake, M. and Matsunaga, M. (2012) Two Cases of Metastatic Colorectal Cancer in Wild-Type KRAS Effectively Treated by Panitumumab. *Gan To Kagaku Ryoho*, **39**, 1567-1570.

[17] Peeters, M., Price, T.J. and Cervantes, A. (2010) Randomized Phase III Study of Panitumumab with Fluorouracil, Leucovorin an Irinotecan (FOLFIRI) Compared with FOLFIRI Alone as Second-Line Treatment in Patients with Metastatic Colorectal Cancer. *Journal of Clinical Oncology*, **28**, 4706-4713. http://dx.doi.org/10.1200/JCO.2009.27.6055

[18] Raymond, E., Boige, V. and Faivre, S. (2002) Dosage Adjustment and Pharmacockinetic Profile of Irinotecan in Cancer Patients with Hepatic Dysfunction. *Journal of Clinical Oncology*, **20**, 4303-4312. http://dx.doi.org/10.1200/JCO.2002.03.123

[19] Duillard, J.Y., Peeters, M. and Siena, S. (2008) Phase III Study (PRIME) of Panitumumab with FOLFOX-4 Compared to FOLFOX4 Alone in Patients with Previously Untreated Metastatic Colorectal Cancer. Preliminary Safety Data. *Journal of Clinical Oncology*, **28**, 153.

[20] Cunningham, D., Humblet, Y. and Siena, S. (2004) Cetuximab Monotherapy a Cetuximab plus Irinotecan in Irinotecan Refractory Metastatic Colorectal Cancer. *New England Journal of Medicine*, **351**, 337-345. http://dx.doi.org/10.1056/NEJMoa033025

[21] Lacouture, M.E. and Melosky, B.L. (2007) Cutaneous Reactions to Anticancer Agents Targeting the Epidermal Growth Factor Receptor: A Dermatology-Oncology Persepective. *Skin Therapy Letters*, **12**, 1-5.

Causes of Peripheral Blood Cytopenias in Patients with Liver Cirrhosis Portal Hypertension and Clinical Significances

Yunfu Lv

Department of General Surgery, People's Hospital of Hainan Province, Haikou, China
Email: yunfu_lv@126.com

Abstract

Liver cirrhosis portal hypertension patients to reduce the number of blood cells are common in clinical, and often affect the prognosis. This paper discusses cirrhotic portal hypertension patients complicated by the reason of the decrease in the number of peripheral blood cells and what is the clinical significance of these reasons so as to provide theoretical support for the choice of treatment. Splenomegaly and hypersplenism caused should be the main reason for reducing the number of blood cells, but not all, other reasons are alcohol and virus inhibition of bone marrow, liver function impairment, autoimmune damage and loss of blood, etc. If it is a function of the spleen hyperfunction caused by blood cells decreases, blood should rise to normal after splenectomy, or consider other reason or there are other reasons at the same time.

Keywords

Liver Cirrhosis Portal Hypertension, Peripheral Blood Cytopenias, Causes, Cilinical Significances

1. Introduction

There are approximately 350 million carriers of hepatitis B virus (HBV) worldwide, and more than half of them are in the Asia-Pacific region. China has a high carrier rate of HBV, with 9.8% of the population being HBV positive; the rate is as high as 16.4% in Hainan Province. Overall, 20% of HBV infections develop into chronic hepatitis. The incidence of the resulting nonalcoholic cirrhotic portal hypertension is thus very high and most patients are complicated by monolineage or multilineage cytopenias [1]. Cytopenias indicate that a leukocyte (WBC) counts of $<4.0 \times 10^9$/L, a erythrocyte (RBC) counts of $<4.0 \times 10^{12}$/L and/or a platelet (PLT) counts of $<100 \times 10^9$/L. People usually put the cirrhotic portal hypertension patients to reduce the number of blood cells are attributed to the splenic function, actually otherwise, the splenic function must have blood cells decreases, but are not necessarily blood cells caused by the splenic function. There are numerous causes for cytopenias in patients with hepatocirrhotic portal hypertension, including the toxic effects of hepatic viruses and alcohol on the bone marrow, hypofunctioning of the liver [2], splenomegaly, hypersplenism, gastrointestinal bleeding, and hematopoietic dysfunction caused by malnutrition. In most cases, cytopenias are caused by multiple factors.

2. Causes

2.1. Toxic Effects of Hepatic Virus

1) Hepatic viruses can directly suppress the differentiation and proliferation of hemopoietic stem cells and progenitor cells [3]. 2) Hepatic virus can cause disorders of cellular immunity and humoral immunity *in vivo*, to compromise the body's capacity to eliminate the viruses. The constant presence of viruses damages the hemopoietic functioning of the bone marrow [4]. 3) Viruses can impair the activity of bone marrow stromal cells to reduce the secretion of cytokines and to affect the proliferation of hemopoietic cells. 4) During pathogenesis caused by cytokines, the increase in the γ-interferon level and decrease in the interleukin-6 and erythropoietin levels, can affect the proliferation of hemopoietic cells [5]. The hepatitis B virus (HBV) and hepatitis C virus (HCV) can suppress the bone marrow, and affect the growth of all karyocytes in the bone marrow. This may lead to hypoplastic anemia, and patients must undergo a bone marrow transplant to survive.

The liver and bone marrow are target tissues of HBV. This virus can kill or injure hemopoietic cells directly, causing myelosuppression, and leading to leukopenia and reduction in the detoxification ability of the liver. This renders the body more sensitive to certain medicines, toxins and environmental pollutants, and cause hypofunctioning of bone marrow hematopoiesis. Leukopenia further damages immunity to cause the active replication of HBV, forming a vicious cycle. Currently, antiviral therapy is the first choice for chronic hepatitis B patients; however, antiviral medications also lead to myelosuppression. Therefore, monitoring leukocytes in the peripheral blood is conducive to the regulation of antiviral therapy. If the leukocyte count is lower than 2×10^9/L, antiviral therapy should be discontinued. Both HBV and HCV can induce suppression of the precursor cells of the bone marrow, and affect the lymph cells, causing lymphopenia and hypofunctioning of the bone marrow.

2.2. Toxic Effects of Alcohol

In the 1980s, studies of patients with alcoholic liver disease reported that neutrophil granulocytes demonstrated retarded growth and delayed release in the bone marrow. Later studies showed increased apoptosis of neutrophil granulocytes. Patients with end-stage cirrhosis complicated with neutropenia underwent Granulocyte-Macrophage Colony Stimulating Factor (GM-CSF) therapy for 7 days, and the leukocyte count increased more than 100%. However, the increased leukocytes could not be destroyed in the spleen, for no leukocyte fragments were found in the spleen. Ethanol can suppress or stimulate cellular proliferation, but in most cases, it suppresses cellular growth and increases cytotoxic effects. Its mechanism includes retarded cellular proliferation and induced apoptosis and necrosis [6]-[8]. A foreign study reported [9] that long-term alcoholism could cause abnormalities in the bone marrow and peripheral blood. In that study, 91% patients manifested changes in the peripheral blood including granulopenia, thrombopenia, etc., and changes in bone marrow included highly-differentiated hemopoietic tissue and myelofibrosis. Long-term alcohol consumption can reduce the absorption of folic acid and vitamin B_{12}, which impairs the synthesis of erythrocytes. Djordjevic *et al.* [10] believed both that hepatic viruses and alcoholism were able to cause cytopenias.

2.3. Hypofunctioning of Liver

Hypofunctioning of the liver reduces degradation of toxic metabolites by liver cells; in this case, the liver cannot detoxify the toxins that suppress the bone marrow, thus affecting hemopoietic function. The incidence of liver disease combined with thrombopenia is 15% - 70%. It is usually at mild or moderate level, and its severity is a prognostic indicator. In liver diseases, thrombopenia is closely related to hepatocirrhosis, anti-platelet autoantibodies [11], bone marrow suppression caused by HBV and HCV, and toxic effects from excessive alcohol consumption [12]. The discovery of thrombopoietin (TPO) in 1994 ushered in a new era in the study of cirrhotic thrombopenia. TPO is almost exclusively produced in liver cells; a small proportion of TPO is produced in the kidneys, bone marrow stromal cells and muscle. The production of TPO depends on the function and amount of liver cells. In cirrhosis, functional liver cells become less able to decrease the secretion of TPO. A study by Wolber *et al.* [13], of cirrhotic patients developing from the compensation to decompensation stage, demonstrated that the expression or serum level of TPO changed from an increase to a decrease, and that the platelet count decreased gradually. The decrease in liver function, to some extent, was related to hemocytopenia and bone marrow dysfunction. Forbes *et al.* [14] suggested that hepatic exogenous myofibroblasts played an important role in hepatic fibrosis. In hepatic fibrosis, bone marrow stem cells differentiate into hepatic endothelial parenchymal cells but not into myofibroblasts. This indicates that the change in hemopoietic function and inner environment of the bone marrow might be somehow related to or interactive with the occurrence and development of hepatic fibrosis or even hepatic cirrhosis. These observations suggest that changes in the bone marrow of cirrhotic patients do not result from one single factor but a combination of multiple factors, with a complicated regulation mechanism. The changes in bone marrow might be directly or indirectly related to the severity of hepatic cirrhosis and changes in liver or spleen function. Their relationship and the detailed mechanism remain to be further explored. Solving this puzzle will be of significant importance to clinical practice.

2.4. Splenomegaly and Hypersplenism

Hypersplenism is secondary to splenomegaly. Two mechanisms for splenomegaly caused by liver diseases exist. The first mechanism is expansionary splenomegaly, including congestive splenomegaly caused by increased venous pressure and hyperemic splenomegaly caused by increased splenic arterial flow; the former is the main cause. The second mechanism is hypertrophic splenomegaly, including: 1) Hepatic virus antigen and exogenous antigens unprocessed by the liver due to a shunting procedure, can stimulate the spleen and lead to hypertrophy of the immune tissue in the spleen (splenic corpuscle, periarterial lymphatic sheath, marginal zone). 2) In hepatic cirrhosis, increased necrotic cells and hypofunctioning of the hepatic reticuloendothelial system promote compensatory hypertrophy and lead to hyperfunctioning of the splenic reticuloendothelial system. 3) Increased intrasplenic pressure, stasis of blood circulation, change in the metabolic environment and other factors can cause fibroplastic proliferation. Generally speaking, intrasplenic immune tissues show obvious hypertrophy during hepatitis, and middle or end stage cirrhotic patients mainly manifest splenic sinus dilation, hypertrophy of reticuloendothelial system and fibrous tissues.

Currently, there are several hypotheses concerning the mechanism of cytopenia: 1) The hypothesis of intrasplenic trapping [15]. After the formation of splenomegaly, blood volume in the spleen increases, and a great number of leukocytes, erythrocytes and platelets are trapped in the spleen. The ratio of trapped hemocytes compared with that in the normal spleen is 5.5- to 20-fold, resulting in hemocytopenia in the peripheral blood. 2) The hypothesis of cytophagy: There are a large number of mononuclear-macrophages in the spleen. Under pathological circumstances, mononuclear-macrophages demonstrate hyperfunctioning in cytophagy and destruction of hemocytes, especially erythrocytes [16]. Recently, a study using erythrocyte creatine (EC), the life-span sensitive marker of erythrocytes, revealed that the EC level was significantly increased in patients with splenomegaly due to post-necrotic cirrhosis compared with patients with hepatic cirrhosis with normal spleens ($P <$ 0.05). In addition, the same was observed compared with the normal control group but without a significant difference [17]. This suggested that splenomegaly accelerated the destruction of erythrocytes and the determination of the EC value could be used to evaluate the severity of cirrhotic splenomegaly [18]. 3) The spleen can produce excessive "splenic hormones" to suppress the hemopoietic function of the bone marrow, and accelerate the destruction of trap produced hemocytes to prevent them from entering into blood circulation [19]. 4) The hypothesis of autoimmunity: The spleen is a large lymph organ that produces antibodies. Antigens unprocessed by the liver enter the marginal zones of splenic lymph follicles (splenic nodule) and activate the pro-lymphocytes and plasma cells to generate antibodies. These antibodies can destroy hemocytes causing hemocytopenia in the peripheral blood.

2.5. Gastrointestinal Bleeding

Gastroesophageal fundus varices bleeding is a common complication for patients with cirrhotic portal hypertension. Gastrointestinal bleeding of any cause can directly lead to a decreased amount of hemocytes in the effective circulatory blood volume. Usually, these theories coexist, and rarely only one theory comes into play [20].

Chronic gastrointestinal bleeding can result in iron, folic acid and vitamin B12 deficiencies, and insufficient material for the synthesis of erythrocytes. Massive loss of erythrocytes can lead to anemia in patients. A Cr^{51} labeled-erythrocyte test demonstrated that only 20% of patients with cirrhosis complicated with anemia had increased erythrocytes in their spleens.

2.6. Malnutrition

Portal hypertensive gastropathy can cause malabsorption of hematopoietic growth factors and non-visible loss of nutrients necessary for hematopoiesis. Additionally, the lack of iron, folic acid and vitamin B_{12} results in insufficient materials for the synthesis of erythrocytes, leading to decreased hematopoiesis.

3. Clinical Significances

The significance of exploring the causes of hemocytopenia in the peripheral blood in the patients with cirrhotic portal hypertension lies in its guidance for treatment and evaluation for therapeutic effects [21]. If hemocytopenia is caused by splenomegaly or hypersplenism, whether monolineage or multi-lineage, the decreased hemocytes will rise significantly after a splenectomy ($P < 0.01$) [22]. The most sensitive hemocyte is the platelet, which will increase half an hour after the operation, and reach the highest level in 2 weeks; afterwards it will decrease gradually and remain at a normal level. Leukocytes and erythrocytes would increase following the platelets. Hemocytopenia in the peripheral blood caused by non-splenic factors does not lead to a definite increase in hemocytes after splenectomy.

References

[1]   Lv, Y.-F., Li, X.-Q. Huang, W.-W., *et al.* (2007) Peripheral Blood Cytopenia in Patients with Hypersplenism Due to Portal Hypertension. *Chinese Journal of General Surgery*, **22**, 702.

[2]   Bashour, F.N., Teran, J.C. and Mullen, K.D. (2000) Prevalence of Peripheral Blood Cytopenias (Hypersplenism) in Patients with Nonalcoholic Chronic Liver Disease[J]. *Gastroenterology*, **95**, 2936-2939.

[3]   Van, E., Niele, A.M. and Kroes, A.C. (1999) Human Parvovirus B19: Relevance in Internal Medicine[J]. *The New England Journal of Medicine*, **54**, 221-230. http://dx.doi.org/10.1016/S0300-2977(99)00011-X

[4]   Kevin, E., Brown, J.T., Barrett, A.J., *et al.* (1997) Hepatitis-Associated Aplastic Anemia[J]. *The New England Journal of Medicine*, 1059-1064.

[5]   Dilloo, D., Vohringer, R., Josting, A., *et al.* (1995) Bone Marrow Fibroblasts from Children with Aplastic Anemia Exhibit Reduced Interlukin-6 Production in Response to Cytokines and Viral Challenge[J]. *Pediatric Research*, **38**, 716-721. http://dx.doi.org/10.1203/00006450-199511000-00014

[6]   Young, N.S. and Maciejewski, J. (1997) The Pathophysiology of Acquired Aplastic Anemia[J]. *The New England Journal of Medicine*, **336**, 1365-1372. http://dx.doi.org/10.1056/NEJM199705083361906

[7]   Jacobs, J.S. and Miller, M.W. (2001) Proliferation and Death of Cultured Fetal Neocortical Neurons: Effects of Ethanol on the Dynamics of Cell Growth[J]. *Journal of Neurocytology*, **30**, 391-401. http://dx.doi.org/10.1023/A:1015013609424

[8]   Hao, L.P., Hu, X.F., Pang, H., *et al.* (2006, The Study on Apoptosis and Its Molecular Mechanism in Mouse Insulinama Cells Induced by Ethanol[J]. *Journal of Toxicology*, **20**, 138-140.

[9]   Neuman, M.G., Haber, J.A., Malkiewicz, I.M., *et al.* (2002) Ethanol Signals for Apoptosis in Cultured Skin Cells[J]. *Alcohol*, **26**, 179-190. http://dx.doi.org/10.1016/S0741-8329(02)00198-2

[10] Djordjević, J., Svorcan, P., Vrinić, D. and Dapcević, B. (2010) Splenomegaly and Thrombocytopenia in Patients with Liver Cirrhosis. *Vojnosanit Pregl*, **67**, 166-169.

[11] Sezai, S., Kamisaka, K., Ikegami, F., *et al.* (1998) Regulation of Hepatic Thrombopoietin Production by Portal Hemodynamics in Liver Cirrhosis[J]. *The American Journal of Gastroenterology*, **93**, 80-82. http://dx.doi.org/10.1111/j.1572-0241.1998.080_c.x

[12] Lu, Y.F., Yue, J., Gong, X.G., *et al.* (2009) Anaemia of Cirrhotic Portal Hypertension with Hypersplenism. *Journal of Surgery: Concepts & Practice*, **14**, 669-670.

[13] Wolber, E.M., Ganschow, R., Burdelski, M., *et al.* (1999) Hepatic Thrombopoietin mRNA Levels in Acute and Chronic Liver Failure of Childhood[J]. *Hepatology*, **29**, 1739-1742. http://dx.doi.org/10.1002/hep.510290627

[14] Forbes, S.J., Russo, F.P., Rey, V., *et al.* (2004) A Significant Proportion of Myofibroblasts Are of Bone Marrow Origin in Human Liver Fibrosis[J]. *Gastroenterology*, **126**, 955-963. http://dx.doi.org/10.1053/j.gastro.2004.02.025

[15] Shah, S.H., Hayes, P.C., Allan, P.L., *et al.* (1996) Measurement of Spleen Size and Its Relation to Hypersplenism and Portal Hemodynamics in Portal Hypertension Due to Hepatic Cirrhosis[J]. *The American Journal of Gastroenterology*,

91, 2580-2583.

[16] Jiao, Y.F., Okumiya, T., Saibara, T., Kudo, Y. and Sugiura, T. (2001) Erythrocyte Creatine as a Marker of Excessive Erythrocyte Destruction Due to Hypersplenism in Patients with Liver Cirrhosis. *Clinical Biochemistry*, **34**, 395-398. http://dx.doi.org/10.1016/S0009-9120(01)00242-9

[17] Friedman, L.S. (1999) The Risk of Surgery in Patients with Liver Disease. *Hepatology*, **29**, 1617-1623. http://dx.doi.org/10.1002/hep.510290639

[18] Zhou, Y.X. (2002) Modern Diagnostics & Therapeutics of Liver Cirrhosis. 1st Edition, People's Military Medical Press, Beijing, 247-249.

[19] Faeh, M., Hauser, S.P. and Nydegger, U.E. (2001) Transient Thrombopoietin Peak after Liver Transplantation for End-Stage Liver Disease. *British Journal of Haematology*, **112**, 493-498. http://dx.doi.org/10.1046/j.1365-2141.2001.02567.x

[20] Lv, Y.F., Li, X.Q., Han, X.Y., Gong, X.G. and Chang, S.W. (2013) Peripheral Blood Cell Variations in Cirrhotic Portal Hypertension Patients with Hypersplenism. *Asian Pacific Journal of Tropical Medicine*, **6**, 663-666. http://dx.doi.org/10.1016/S1995-7645(13)60115-7

[21] Lv, Y.F. (2009) Characteristics and Clilical Significance of Hypersplenism Secondary to Splenomegaly Caused by Cirrhotic Portal Hypertension. *World Chinese Journal of Digestology*, **17**, 2969-2971.

[22] Lv, Y.F., Li, X.Q., Gong, X.G., Xie, X.H., Han, X.Y. and Wang, B.C. (2013) Effect of Surgery Treatment on Hypersplenism Caused by Cirrhotic Portal Hypertension. *Minerva Chirurgica*, **68**, 409-413.

Role of Portal Hypertension in Prediction of Bacterial Infection in Decompensated Cirrhosis

Hasan Sedeek Mahmoud, Shamardan Ezz El-Din S. Bazeed

Department of Tropical Medicine and Gastroenterology, Qena Faculty of Medicine, South Valley University, Qena, Egypt
Email: hasan_sedeek@yahoo.com

Abstract

Background: Bacterial infection in cirrhotic patients is a fatal complication. The high incidence of bacterial infections in those patients may be related to several alterations in the defensive mechanisms against infections and increased intestinal permeability with bacterial translocation. Aim: To evaluate the role of portal hypertension (PH) in predicting the occurrence of bacterial infections in decompensated cirrhosis. Patients and Methods: In this retrospective cohort study, 99 patients—56 males and 43 females, with decompensated liver cirrhosis were included. Diagnosis of liver cirrhosis was based on clinical, laboratory and ultrasonographic examinations. Patients were classified according to the presence of bacterial infection into patients with infection—Group 1, and those without infection—Group 2. Laboratory, abdominal US and upper endoscopic data for all patients were collected. Logistic regression analysis was done to detect the independent factors for prediction of bacterial infection. Results: The mean age of patients was 50.5 ± 14.2 years. Bacterial infection was found in 41 patients (41.4%) and no infection in 58 patients (58.6%). Infected patients showed statistically significant higher values in the level of bilirubin, PT and Child-Pugh score (P value = 0.000) and lower values in the level of albumin, total serum protein and PC than those without infection (P value = 0.006, 0.000 and 0.000 respectively). Portal vein diameter (PVD) and splenic diameter (SD) showed statistically significant higher values in infected patients than in those without infection (P value = 0.028 and 0.000 respectively), also infection was more significantly prevalent in patients with varices than those without varices (P value = 0.000). The independent predictors for bacterial infection were: the age, total serum bilirubin, serum albumin, PT, PC, child score, PVD, SD and the presence of varices. Conclusion: Presence of varices (as a complication of PH) is an independent risk factor for the development of bacterial infection in decompensated cirrhotic patients and reduction of PH by any way could decrease this fatal complication.

Keywords

Portal Hypertension; Bacterial Infection; Liver Cirrhosis

1. Introduction

Bacterial infections are a known complication of cirrhosis, with a reported incidence that ranges between 15% and 47% [1] [2].

Hospitalized patients with cirrhosis are at increased risk of developing bacterial infections and the most common causes are spontaneous bacterial peritonitis [SBP] and urinary tract infections. The independent predictors for the development of bacterial infections in those patients are poor liver synthetic function and gastrointestinal hemorrhage [3].

Cirrhotic patients who develop an infection have a significantly higher mortality than uninfected patients [4] [5]. Current evidence also suggests that infection predisposes to recurrent variceal hemorrhage [6] and is associated with failure to control variceal hemorrhage [7] [8]. Antibiotic prophylaxis in the setting of variceal hemorrhage significantly decreases the incidence of bacterial infections and improves survival [9].

Studies assessing the aetiology and types of bacterial infections in cirrhotic patients showed that the most common infections were community-acquired, mainly urinary-tract infections, SBP and pneumonia, 70% - 80% of which were caused by gram negative bacilli (GNB), mainly *Escherichia coli*, suggesting that the gut was the main source of bacteria. The spectrum of bacteria causing infection in cirrhosis in more recent series showed a significantly higher rate of Gram-positive cocci infections, probably due to an increase in the number of therapeutic invasive procedures [10] and the use of chronic antibiotic prophylaxis [11] [12]. However, the most common infections, SBP and urinary-tract infection, are still caused mainly by GNB [10]. Recent investigations suggest that the prevalence of infections caused by multiresistant bacteria is increasing in cirrhosis [13].

Bacterial translocation (BT), which increases by portal hypertension, is defined as the migration of viable microorganisms from the intestinal lumen to mesenteric lymph nodes (MLN) and other extra-intestinal organs and sites. BT in cirrhotic patients increases in conditions associated with a high risk of infections by GNB and multiple organ failure such as hemorrhagic shock, intestinal obstruction, major burn injury and serious trauma [14]. BT has been postulated as the main mechanism in the pathogenesis of SBP [15].

1.1. Aim of the Work

To assess the possible role of PH in the development of bacterial infections in decompensated cirrhotic patients.

1.2. Patients and Methods

99 patients with decompensated liver cirrhosis admitted to Qena University Hospital from April to October 2013; with Child-Pugh classes B or C were included in this retrospective cohort study. Diagnosis of liver cirrhosis was based on clinical, laboratory and ultrasonographic examinations. Patients with other immune-compromised diseases or receiving immune-suppressive drugs were excluded. Patients with autoimmune hepatitis were diagnosed at the hospital on admission for their first time and they were included before starting corticosteroid therapy but others under corticosteroid therapy were excluded.

2. Methods

Clinical and laboratory data for all patients including: complete blood count, liver function tests (ALT, AST, serum bilirubin, albumin and total protein, prothrombin time (PT) and concentration (PC), INR and serum alkaline phosphatase), serum electrolytes, glucose and serum creatinine were recorded. Abdominal US data were recorded including the following: the size of the liver and the spleen, the size of the portal and splenic vein and the presence and the degree of ascites. The presence or absences of bacterial infections including their sites were detected. Patients were classified into: Group 1; patients without bacterial infections and Group 2; patients with bacterial infections. The types of infections were defined according to the following standard criteria: Pneumonia was diagnosed in the presence of infiltrates on chest x-ray with concurrent fever, cough, and neutrophilic leukocytosis; spontaneous bacterial peritonitis was diagnosed in the presence of a neutrophil leukocyte count in the ascitic fluid >250 cells/mm^3 without any evidence of surgically treatable sources of infections; urinary tract infection was diagnosed when fever and urinary symptoms were associated with bacteriuria, leukocyturia, and positive urine culture; GI tract infection was diagnosed when vomiting, diarrhea, fever, and abdominal pain were associated with neutrophilic leukocytosis and positive stool culture; skin and soft tissue infections were diagnosed when fever and cellulitis were associated with neutrophilic leukocytosis. Data of upper endoscopy (using Olympus, GIF-XQ260 instrument) for all patients including presence or absence of varices (esophageal or gastric) and/or portal hypertensive gastropathy (PHG) were recorded. The size of esophageal varices was graded as described by Beppu *et al.* (1981) into Grade 1: enlarged but straight varices, Grade 2: enlarged tortuous varices and Grade 3: coiled shaped markedly enlarged varices [16]. If PHG was present, it was described as either mild or severe as described by Mc Cormack *et al.*, (1985) [17].

Statistical Analysis

Data entry and analysis were done using statistical package of social science (SPSS) version 16. The data are presented as means ± SD. Statistical methods included independent-t-tests; used for comparison between the two groups in case of continuous variables and the Chi-square test for comparison between categorical variables. Logistic regression analysis was done to detect the independent predictors for statistically significant variables. P value < 0.05 was considered statistically significant.

3. Results

3.1. Patients

99 patients with liver cirrhosis; 56 male (56.6%) and 43 female (43.4%), with their mean age was 50.5 ± 14.2 years were included in the current study. The aetiology of liver cirrhosis was due to chronic HCV in 78 (78.8%), HBV in 13 (13.2%), autoimmune hepatitis in 2 (2%) and cryptogenic cause in 6 (6%) patients.

3.1.1. Clinical Data for All Patients

History of hematemesis was encountered in 20 patients (20.2%). Ascites was found in 90 patients (91%); 49 of them were mild degree, 24 moderate and 17 patients presented by marked ascites. Hepatic encephalopathy as a manifestation of decompensated cirrhosis was found in 78 patients (78.8%); 58 of them were Grade I-II and 20 patients were Grade III-IV. Hepato-renal syndrome was detected in 4 patients (4%). As regard Child-Pugh score; 58 were in class B (58.6%) and 41 were in class C (41.4%). This is illustrated in **Table 1**.

3.1.2. Laboratory Data for All Patients

The mean values of the laboratory data for all patients are illustrated in **Table 2** and the mean value for child-Pugh score were 11.71 ± 2.05.

3.1.3. Sonographic Data of All Patients

Showed that the mean liver span was 10.5 ± 2.5 cm, PVD was 1.29 ± 0.25 cm and SD was 16.09 ± 2.45 cm; this is illustrated in **Table 3**.

3.1.4. Endoscopic Data for All Patients

No varices detected were in 40 (40.5%) patients and esophageal varices were found in 59 (59.5%) patients; 17 were Grade 1, 20 were Grade 2 and 22 were Grade 3. Gastric varices (G.V) were found in 6 (6.1%) patients and PHG was found in 78 (78.8%) patients; this is illustrated in **Table 3**.

3.2. Bacterial Infection

It was not found in 58 patients (58.6%); (Group 1) and was found in 41 patients (41.4%); (Group 2). The most common type of infection was SBP which was found in 16 (39% of infected patients), followed by respiratory tract infection in 12 patients (29.3%) then urinary tract infection in 8 (19.5%), soft tissue infection in 3 (7.3%) and GIT infection in 2 patients (4.9%).

Table 1. Clinical data for all patients.

Parameter	Value
Age	50.5 ± 14.2
Gender	
Male	56 (56.6%)
Female	43 (43.4%)
Hematemesis	20 (20.2%)
Ascites	
No	9 (9.1%)
Mild	49 (49.5%)
Moderate	24 (24.2%)
marked	17 (17.2%)
HE	
No	21 (21.2%)
Grade I-II	58 (58.6%)
Grade II-IV	20 (20.2%)
HRS	4 (4%)
Child class	
B	58 (58.6%)
C	41 (41.4%)

All data are expressed as number (%) or mean ± SD. HE: hepatic encephalopathy, HRS: hepato-renal syndrome.

Table 2. Laboratory data for all patients.

Parameter	Value
Bilirubin (mg/dl)	2.57 ± 1.34
ALT (U/L)	87.7 ± 153.5
AST (U/L)	54.2 ± 74
Albumin (g/dl)	2.55 ± 0.89
Protein (g/dl)	6.42 ± 1.04
PT (sec)	16.81 ± 2.99
PC (%)	56.15 ± 17.85
INR	1.66 ± 0.74
ALP	105.6 ± 72.9
Child score	11.71 ± 2.05
Hemoglobin (g/dl)	10.3 ± 2.47
WBCs ($10^3/\mu l$)	10.2 ± 3.05
PLT ($10^3/\mu l$)	122.5 ± 92.1
Creatinine (mg/dl)	0.90 ± 0.86

All data are expressed as mean ± SD or Number (%).

Table 3. Sonographic and endoscopic data for all patients.

Parameter	Value
Liver span	10.5 ± 2.5
PVD (cm)	1.29 ± 0.25
SD (cm)	16.09 ± 2.45
Endoscopy	
O.V (%)	59 (59.5%)
G.V (%)	6 (6.1%)
PHG (%)	78 (78.8%)
Esophageal V.	
G1	17 (29%)
G2	20 (34%)
G3	22 (37%)

All data are expressed as mean ± SD or number (%).

3.3. Comparison between Both Groups

As regard parameters of liver function; patients with infection showed statistically significant higher values in the level of bilirubin, PT and Child-Pugh score (P value = 0.000) and a lower values in the level of albumin, total serum protein and PC than those without infection (P value = 0.006, 0.000 and 0.000 respectively). As regard sonographic findings; PVD and SD showed statistically significant higher values in patients with infection than in those without infection (P value = 0.028 and 0.000 respectively), also infection was more significantly prevalent in patients with esophageal & gastric varices than those without varices (P value = 0.000); this is shown in **Figure 1**. Total leucocytic count was significantly higher in patients with infection than in those without infection (P value = 0.04), this is illustrated in **Table 4**.

Logistic regression analysis for the significant parameters between both groups, which was done to detect the independent factors predicting the presence of infection, showed that the independent factors were: the age, total serum bilirubin, serum albumin, PT, PC, Child score, PVD, SD and presence of varices; this is illustrated in **Table 5**.

4. Discussion

Bacterial infection is a common complication in cirrhotic patients with high incidence of mortality. Liver dysfunction and PH are the two major sequels of this disease, so the current study aimed to define the possible risk factors for developing bacterial infection for possible avoidance of this fatal complication.

The independent factors predicting the occurrence of bacterial infection in the current study were the age, parameters of liver dysfunction including Child-Pugh score; bilirubin, albumin, PT, PC, INR and parameters of portal hypertension including presence of varices, PVD and SD. This result coincide with that obtained with Garsia-Tsao et al., (1993), who suggested that PH alone may not be a major factor in the development of spontaneous infections in cirrhosis and that other mechanisms, such as a defective immune system, may be more important [18].

The high incidence of bacterial infections in cirrhotic patients may be explained by the presence of several alterations in the defensive mechanisms against infections, small intestinal bacterial overgrowth, depression of hepatic monocyte macrophage functions and reduction of serum and ascitic fluid complement levels [19].

Advanced liver disease may contribute to the observed increase in BT by different mechanisms. PH is associated with characteristic structural changes in the small intestinal wall [20] and functional abnormalities such as protein-losing enteropathy [21], reduced small bowel motility [22] and small intestinal bacterial overgrowth [23]. Clearance of translocated organisms from MLN may also be impaired, given the impaired chemotaxis, phago-

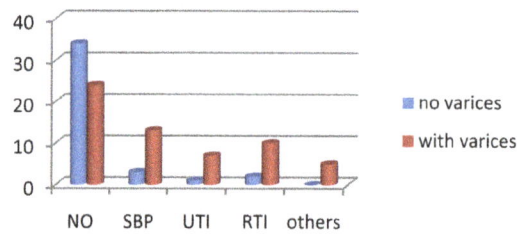

Figure 1. Types of bacterial infections in patients with and without varices.

Table 4. Laboratory, sonographic and endoscopic data for patients with and without bacterial infection.

Parameter	No infection 58 (58.6%)	Infection 41 (41.4%)	P value
Bilirubin (mg/dl)	1.68 ± 1.46	3.47 ± 1.22	0.000
Albumin (g/dl)	2.77 ± 0.96	2.33 ± 0.83	0.006
Protein (g/dl)	6.96 ± 0.88	5.88 ± 1.2	0.000
P.T (sec)	15.01 ± 1.7	18.6 ± 4.28	0.000
P.C (%)	63.4 ± 14.9	48.9 ± 20.8	0.000
Child score	10.40 ± 2.2	13.03 ± 1.9	0.000
WBCs ($10^3/\mu l$)	8.5 ± 3.4	11.9 ± 2.7	0.04
PLT ($10^3/\mu l$)	118.57 ± 78.8	127.88 ± 105.4	0.61
PVD (cm)	1.25 ± 0.27	1.34 ± 0.24	0.028
SD (cm)	14.35 ± 2.77	17.84 ± 2.14	0.000
Varices			
Yes (%)	24 (41.4%)	35 (85.4%)	
No (%)	34 (58.6%)	6 (14.6%)	0.000[*]

All data are expressed as mean ± SD and number (%). Independent t test were used, [*]Chi^2 test was used.

Table 5. Logistic regression analysis for the presence of infection.

Independent variables	P value	Odds ratio
Age	0.003	0.951
Bilirubin	0.001	1.39
Albumin	0.008	0.463
PT	0.000	1.5
PC	0.000	0.955
Child score	0.004	1.45
PVD	0.032	1.8
S.D	0.000	1.77
varices	0.039	1.67

P value < 0.05 = significant. CI = 95%.

cytosis and intracellular killing by polymorphonuclear leukocytes and monocytes associated with advanced liver disease [19].

Cirera *et al.*, [2001] also in their study postulated that the degree of PH was not associated with a higher prevalence of BT and suggested that, in addition to PH, the simultaneous occurrence of other factors favoring BT and the presence of several abnormalities in the defense mechanisms against the infection are probably required to allow enteric bacteria to translocate to MLN and to cause systemic infection in cirrhosis [24].

So, from this current study, we can postulate that reduction of PH by any way could reduce the translocation of bacteria through porto-systemic shunts by passing the liver and so could decrease the incidence of bacterial infection in those patients. In accordance with our results, Giannelli *et al.*, (2014) study; which is the most recent one, postulated that carvidolol as a β-blocker, in addition to reducing portal pressure, it can also reduce bacterial translocation and so reduce bacterial infections in those patients adding more beneficial effect to the usefulness of these agents [25].

5. Conclusion

Presence of varices (as a complication of PH) is an independent risk factor for the development of bacterial infection in decompensated cirrhotic patients and reduction of PH by any way could decrease this fatal complication. Further studies on larger sample size are needed.

References

[1] Yoshida, H., Hamada, T., Inuzuka, S., *et al.* (1993) Bacterial Infection in Cirrhosis, with and without Hepatocellular Carcinoma. *American Journal of Gastroenterology*, **88**, 2067-2071.

[2] Caly, W.R. and Strauss, E. (1993) A Prospective Study of Bacterial Infections in Patients with Cirrhosis. *Journal of Hepatology*, **18**, 353-358. http://dx.doi.org/10.1016/S0168-8278(05)80280-6

[3] Garcia-Tsao, G. (2004) Bacterial Infections in Cirrhosis. *Canadian Journal of Gastroenterology*, **18**, 405-406.

[4] Borzio, M., Salerno, F., Piantoni, L., *et al.* (2001) Bacterial Infection in Patients with Advanced Cirrhosis: A Multicentre Prospective Study. *Digestive and Liver Disease*, **33**, 41-48. http://dx.doi.org/10.1016/S1590-8658(01)80134-1

[5] Bleichner, G., Boulanger, R., Squara, P., *et al.* (1986) Frequency of Infections in Cirrhotic Patients Presenting with Acute Gastrointestinal Hemorrhage. *British Journal of Surgery*, **73**, 724-726. http://dx.doi.org/10.1002/bjs.1800730916

[6] Bernard, B., Cadranel, J.F., Valla, D., *et al.* (1995) Prognostic Significance of Bacterial Infection in Bleeding Cirrhotic Patients: A Prospective Study. *Gastroenterology*, **108**, 1828-1834. http://dx.doi.org/10.1016/0016-5085(95)90146-9

[7] Goulis, J., Armonis, A., Patch, D., *et al.* (1998) Bacterial Infection Is Independently Associated with Failure to Control Bleeding in Cirrhotic Patients with Gastrointestinal Hemorrhage. *Hepatology*, **27**, 1207-1212. http://dx.doi.org/10.1002/hep.510270504

[8] Vivas, S., Rodriguez, M., Palacio, M.A., *et al.* (2001) Presence of Bacterial Infection in Bleeding Cirrhotic Patients Is Independently Associated with Early Mortality and Failure to Control Bleeding. *Digestive Diseased Sciences*, **46**, 2752-2757. http://dx.doi.org/10.1023/A:1012739815892

[9] Soares-Weiser, K., Brezis, M., Tur-Kaspa, R. and Leibovici, L. (2002) Antibiotic Prophylaxis for Cirrhotic Patients with Gastrointestinal Bleeding. *Cochrane Database of Systematic Reviews*, **2**, CD002907.

[10] Fernandez, J., Navasa, M., Gomez, J., *et al.* (2002) Bacterial Infections in Cirrhosis: Epidemiological Changes with Invasive Procedures and Norfloxacin Prophylaxis. *Hepatology*, **35**, 140-148. http://dx.doi.org/10.1053/jhep.2002.30082

[11] Campillo, B., Richardet, J.P., Kheo, T. and Dupeyron, C. (2002) Nosocomial Spontaneous Bacterial Peritonitis and Bacteremia in Cirrhotic Patients: Impact of Isolate Type on Prognosis and Characteristics of Infection. *Clinical Infectious Diseases*, **35**, 1-10. http://dx.doi.org/10.1086/340617

[12] Campillo, B., Dupeyron, C., Richardet, J.P., *et al.* (1998) Epidemiology of Severe Hospital-Acquired Infections in Patients with Liver Cirrhosis: Effect of Long-Term Administration of Norfloxacin. *Clinical Infectious Diseases*, **26**, 1066-1070. http://dx.doi.org/10.1086/520273

[13] Merli, M., Lucidi, C., Giannelli, V., Giusto, M., Riggio, O., Falcone, M., *et al.* (2010) Cirrhotic Patients Are at Risk for HCA Bacterial Infections. *Clinical Gastroenterology and Hepatology*, **8**, 979-985. http://dx.doi.org/10.1016/j.cgh.2010.06.024

[14] Wiest, R. and Rath, H.C. (2003) Gastrointestinal Disorders of the Critically Ill. Bacterial Translocation in the Gut. *Best Practice and Research Clinical Gastroenterology*, **17**, 397-425. http://dx.doi.org/10.1016/S1521-6918(03)00024-6

[15] Garcia-Tsao, G. (1992) Spontaneous Bacterial Peritonitis. *Gastroenterology Clinics of North America*, **21**, 257-275.

[16] Beppu, K., Inokuchi, K., Koyanagi, N., *et al.* (1981) Prediction of Variceal Haemorrhage by Oesophageal Endoscopy. *Gastrointestinal Endoscopy*, **27**, 213-218. http://dx.doi.org/10.1016/S0016-5107(81)73224-3

[17] Mc Cormack, T.T., Sims, J., Eyre-Brook, I., *et al.* (1985) Gastric Lesions in Portal Hypertension: Inflammatory Gastritis or Congestive Gastropathy? *Gut*, **26**, 1226-1229. http://dx.doi.org/10.1136/gut.26.11.1226

[18] Garcia-Tsao, G., Albillos, A., Barden, G. and Brian West, A. (1993) Bacterial Translocation in Acute and Chronic Portal Hypertension. *Hepatology*, **17**, 1081-1085. http://dx.doi.org/10.1002/hep.1840170622

[19] Rimola, A. and Navasa, M. (1999) Infections in Liver Disease. In: Bircherm, J., Benhamou, J.P., McIntyre, N., Rizzetto, M. and Rodes, J., Eds., *Oxford Textbook of Clinical Hepatology*, 2nd Edition, 1861-1874.

[20] Viggiano, T.R. and Gostout, C.J. (1992) Portal Hypertensive Intestinal Vasculopathy: A Review of the Clinical, Endoscopic, and Histopathologic Features. *American Journal of Gastroenterology*, **87**, 944-954.

[21] Stanley, A.J., Gilmour, H.M., Ghosh, S., Ferguson, A. and McGilchrist, A.J. (1996) Transjugular Intrahepatic Portosystemic Shunt as a Treatment for Protein-Losing Enteropathy Caused by Portal Hypertension. *Gastroenterology*, **111**, 1679-1682. http://dx.doi.org/10.1016/S0016-5085(96)70033-1

[22] Chesta, J. and Deflippi, C. (1993) Abnormalities in Proximal Small Bowel Motility in Patients with Cirrhosis. *Hepatology*, **17**, 828-832.

[23] Shindo, K., Machida, M., Miyakawa, K. and Fukumura, M. (1993) A Syndrome of Cirrhosis, Achlorhydria, Small Intestinal Bacterial Overgrowth, and Fat Malabsorption. *American Journal of Gastroenterology*, **88**, 2084-2091.

[24] Cirera, L., Bauer, T., Navasa, M., Vila, J., Grande, L., Taura, P., Fuster, J., Garcia-Valdecasas, J., *et al.* (2001) Bacterial Translocation of Enteric Organisms in Patients with Cirrhosis. *Journal of Hepatology*, **34**, 32-37. http://dx.doi.org/10.1016/S0168-8278(00)00013-1

[25] Giannelli, V., Lattanzi, B., Thalheimer, U. and Merli, M. (2014) Beta-Blockers in Liver Cirrhosis. *Annals of Gastroenterology*, **27**, 20-26.

Variables Associated with Cirrhosis Diagnosis in Patients with Chronic Hepatitis C: A Case-Control Study

Gilmar Amorim de Sousa[1], Iris do Céu Clara Costa[2], Dyego Leandro Bezerra de Souza[2],
Fabia Barbosa de Andrade[2], Lívia Medeiros Soares Celani[3], Ranna Santos Pessoa[4],
Marlon César de Souza Filho[4], Daniel Fernandes Mello de Oliveira[4],
Luana Lopes de Medeiros[4], Lucila Samara Dantas de Oliveira[4], Maria Flávia Monteiro[4]

[1]The Integrated Department of Medicine, Federal University of Rio Grande do Norte, UFRN, Natal, Brazil
[2]The Post-Graduate Program in Collective Health of UFRN, Natal, Brazil
[3]Gastroenterology Program of the University Hospital Onofre Lopes/UFRN, Natal, Brazil
[4]The Medical School of UFRN, Natal, Brazil
Email: gilamorimdesousa@gmail.com, irisdoceu.ufrn@gmail.com, dysouz@yahoo.com.br,
fabiabarbosabr@yahoo.com.br, medeiroslivinha@gmail.com, rannaspessoa@gmail.com,
marlonsouzafilho@gmail.com, danielfernandesmo@gmail.com, luanalopesmedeiros@gmail.com,
lucilasamara@yahoo.com.br, mari_flaviaa@hotmail.com

Abstract

The diagnosis of liver cirrhosis in patients with chronic hepatitis C has not always been easy, since the gold standard method is the liver biopsy, which is an invasive procedure with interobserver accuracy problems and there have been reports of complications including records of deaths due to hemoperitoneum. Cirrhosis changes the prognosis of the subject with hepatitis C and requires a different clinical management. This study aimed to identify clinical and laboratory variables associated with the diagnosis of cirrhosis in the ultrasonography of patients infected with hepatitis C. In a case-control study, we evaluated 70 cirrhotic patients with chronic hepatitis C compared to a control group of 70 non-cirrhotic people with positive HCV. The results showed, through logistic regression analysis, that the variables blood donor and professional athlete, adjusted for alcohol consumption, showed OR 0.24 and 0.18, with p values of 0.044 and 0.035, respectively. We conclude that the diagnosis of cirrhosis in patients with chronic hepatitis C remains challenging, but the patients with the condition of blood donor or professional athlete prove to be less likely to cirrhosis in ultrasonography in the initial consultation.

Keywords

Cirrhosis, Biomakers, Fibrosis, Hepatitis C Virus, Clinical Diagnosis, Biopsy, Non-Invasive, Methods,

Scores, Ultrasonography

1. Introduction

Chronic hepatitis C (CHC) is the leading cause of advanced chronic liver disease in final stage of hepatocellular carcinoma (HCC), and of death related to liver disease in the Western world. It generally progresses slow and progressively, characterized by persistent inflammation and cirrhosis in approximately 10% to 20% of patients in the period of time from 20 to 30 years of infection. However, these evolutionary rates of progression to cirrhosis may vary widely depending on a number of factors from the etiologic agent, the host, and behavioral factors. Among these factors, it is mentioned that the age at which the patient has contracted the infection, alcohol consumption, obesity, insulin resistance, type 2 diabetes, the co-infections by hepatitis B-virus (HBV) and HIV, immunosuppressive therapy and genetic factors act in a multifactorial manner in the development and progression of fibrosis [1].

It is not uncommon that patients with chronic hepatitis C remain undiagnosed until the time that they develop complications from cirrhosis, since the natural history of the disease remains elusive [1] [2].

When cirrhosis is established, the evolution of the disease becomes unpredictable: it may remain indolent for many years or evolve aggressively, culminating with the development of HCC, hepatic decompensation and death [3] [4]. Once cirrhotic, the individual has the annual risk of 1% to 5% to develop HCC and of 3% to 6% develop decompensation. The evolutive monitoring of those presenting decompensation shows that the annual risk of death is estimated at 15% to 20%.

Various studies and meta-analyzes have shown that eradication of HCV with antiviral therapy reduces the risk of HCC in patients with chronic hepatitis C, regardless of the degree of fibrosis [5] [6]. However, morbidity and mortality rates have increased exponentially when cirrhosis develops. A large study of 838 German patients showed that the mortality rate in patients under 50 years old with chronic hepatitis C was 3.1, whereas in cirrhotic patients it was 26.2 for the same age [3].

D'amico *et al.* (2006) [4] in a systematic review of 118 studies with 23,797 patients showed an increased risk of death with the development of successive decompensations-varices, ascites, and variceal bleeding. Overall, survival rate was only 64%, with a median follow up of 31 months.

On the other hand, it is known that the progression rate of fibrosis to cirrhosis can be variable. A study of 2235 patients with non-treated chronic hepatitis C showed a median time of 30 years for the development of cirrhosis. However, the progression rate was shorter in alcohol users, in elderly, males and patients with high inflammatory activity rates diagnosed in biopsy [7]. Thus, it is of great importance that the diagnosis of cirrhosis is done accurately and as quickly as possible, since it implies a need for effective monitoring and appropriate management of complications of liver disease.

Liver biopsy is the gold standard for fibrosis staging and diagnosis of cirrhosis. It is an invasive procedure in which a sample of liver tissue is obtained through local puncture by a needle, under anesthesia. There are several fibrosis staging scoring systems, however, the two most commonly used are METAVIR and Ishak. The METAVIR evaluates the fibrosis in 5 rating scale points: F0 has no fibrosis and F4 is equal to cirrhosis. Above F2 is considered significant degree on fibrosis and F3—advanced fibrosis. The Ishak scoring system uses 7 points on the scale: F0 does not indicate fibrosis; F5—incomplete cirrhosis; F6—definite cirrhosis.

Many factors, however, may influence the accuracy of the stage of fibrosis on liver biopsy. We can mention, as an example, the size of the biopsy fragment obtained, considering that the accuracy would be correlated to the size of the biopsy specimen length. Another important aspect would be the interobserver variability, in addition to the difficulties inherent to the procedure, which as being invasive, though low risk, may not be available in all services, especially primary and secondary care [8] [9].

Liver biopsy complications have been reported in the literature in a large retrospective study and the mortality rate was estimated in about 9 per 100,000 procedures (around 0.01%). Deaths were attributed to hemoperitoneum and occurred exclusively in patients with cirrhosis or hepatocellular carcinoma. The complication rate in this study was 0.3%. Another retrospective study of 1000 biopsies showed a 5.9% complication rate with 5.3% of hospitalizations related to local pain or hypotension [10] [11].

Therefore, due to the limitations of the liver biopsy, a number of non-invasive techniques have been investigated for evaluation of fibrosis and cirrhosis diagnosis [12] [13]. A previous study has suggested that individual-based parameters such as age, albumin and stage of the disease have a direct influence on prognosis and consequently in the response to treatment.

Noninvasive markers of cirrhosis can be radiological or based on parameters collected in serum. Radiological techniques based on ultrasound, magnetic resonance imaging and elastography have been used to evaluate hepatic fibrosis. Biomarkers of cirrhosis have also been developed and classified into indirect and direct. Indirect biomarkers reflect liver function which may decline with development of cirrhosis. Direct biomarkers may reflect the renewal of extracellular matrix and include molecules involved in fibrogenesis process [14].

The high number of chronically infected patients, the prevalence of the disease and the absence of vaccination indicate that the treatment is the way of attempting to control the disease, whereas the majority of patients with persistent infection are asymptomatic. For this reason, this disease is known as the silent epidemic [1].

In order to identify clinical and laboratory variables associated with the diagnosis of cirrhosis in patients infected with the hepatitis C virus, we evaluated in a case-control study 70 patients with chronic hepatitis C diagnosed with cirrhosis through the ultrasonography (USG) and we performed paired comparison with the control group, 70 non-cirrhotic positive HCV people.

2. Methods

2.1. Ethical Aspects

The study was submitted to the Ethics Research Committee of the University Hospital Onofre Lopes (HUOL) of the Federal University of Rio Grande do Norte (UFRN) in the city of Natal, Rio Grande do Norte, Brazil, and was approved under the opinion number 448243/2013.

2.2. Study Design

It is a case-control study, identified through review of all medical records of patients treated at *Núcleo de Estudos do Fígado* (Liver Studies Center—NEF in Portuguese), which is a reference service in the care of patients with liver disease of the HUOL of the aforementioned University (UFRN) from May 1995 to December 2013. The HUOL is a hospital of medium and high complexity, with 240 active beds and 7247 admissions per year.

The NEF (reference service Liver disease HUOL) began its activities in May 1995, providing outpatient care and hospitalization to patients with liver disease. It receives patients from all hospitals and clinics of the public network of Natal, the capital of Rio Grande do Norte state, as well as all other cities in the state. It has an agreement with state blood banks and receives donors with positive serology for diagnosis. Patients are welcomed by nursing professionals and attended by specialist doctors, following in the anamnesis the script of a standard questionnaire, in which are included demographic, epidemiological, clinical and laboratory variables. Patients diagnosed with chronic liver disease are advised to return with varying intervals, depending on their clinical situation and with maximum interval of six months. For evaluation purposes in clinical research, patients who failed to return to that service within 12 months are considered as follow-up dropout.

2.3. Sample Selection

The sample was selected in retrospective case-control study of patients with chronic hepatitis C treated at NEF of the University Hospital Onofre Lopes of UFRN. A total of 10,304 medical records were examined in a retrospective study, of which 512 showed positive HCV RNA test. We selected 70 patients with confirmed diagnosis of hepatitis C and cirrhosis through ultrasound, which was the case group, compared with 70 patients with confirmed diagnosis of hepatitis C and who were not diagnosed with cirrhosis through ultrasound, which was the control group. Groups were matched by age and sex variables. The dependent variable was the ultrasound diagnosis of cirrhosis. As independent variables, we collected: sociodemographic variables—age, gender, marital status, profession, color and origin; variables associated with HCV infection—sexual promiscuity, intravenous drug use, whether the person is a health care professional or has profession related to risk, such as: manicure, policeman, fireman, butcher, hairdresser and barber, whether the person is or had been an athlete, or haemophiliac or performs hemodialysis, in addition to variables related to disease progression—genotype, higher viral load, alcohol consumption, diabetes mellitus, glutamic oxaloacetic transaminase (GOT), glutamic pyruvic transaminase (GPT), gamma-glutamyl transpeptidase (GGT) bilirubins, and prothrombin time activity (PTA), international normalized ratio (INR), albumin, ferritin and transferrin saturation. We also examined co-infections-hepatitis B and HIV infection.

2.4 Data Analysis

The results were statistically analyzed using the SPSS 17 software. To analyze the association of selected variables and the presence of cirrhosis, we calculated odds ratio (OR) and their respective confidence intervals (95% CI) as estimates of relative risks, considering the presence of cirrhosis as the dependent variable and the selected variables (those related to disease progression) as independent variables. The variables that were statistically significant in multivariate analysis or p less than or equal to 0.20 were included in the logistic regression model.

3. Results

Table 1 shows the sociodemographic and clinical variables. It is evidenced that age (stratified in 15 to 35 years old or older than 35 years old), gender, marital status, occupation (related to the risk of contracting hepatitis C), sexual promiscuity, being a health professional and discontinuation of treatment were not associated with statistical significance at diagnosis of cirrhosis through USG in univariate analysis. The duration of disease has not been studied.

With regard to statistical analysis, initially it was held the data tabulation using descriptive statistics. Then, using inferential statistics, we carried out bivariate analysis using the chi-square test (or Fisher's Exact Test). It was verified the magnitude of association through OR for each of the independent variables with respect to the dependent variable, the 95% significance level. The variables with (*p*-value) $p < 0.20$ were selected and analyzed with Logistic regression to build the multivariate model, through the Likelihood Ratio Test, absence of multicollinearity, as well as its ability to improve the model using the Hosmer-Lemeshow test. Finally, the variables associated in a statistically significant manner in the final model were Blood Donor and Athlete, adjusted for alcohol consumption (**Table 2**). The value of Hesmer-Lemeshow test was 0.960.

4. Discussion

As regards the social demographic variables, the athlete with hepatitis C would have less chance to present the diagnosis of cirrhosis (OR 0.179, with *p* equals to 0.018). This could be explained by the fact that the evolution of the disease would present a less aggressive behavior due to the physical activity practiced by athletes. It has been shown that patients with hepatitis C compared to blood donors would show a decrease in aerobic exercise capacity and consequent overweight, insulin resistance and hepatic steatosis [15].

As for comorbidities, none of the variables was statistically significant. It is known that alcoholism, diabetes, obesity and co-infections by virus B and HIV can accelerate the progression of the disease. However, these associations are challenged by different authors. Recent studies have shown that the development of liver fibrosis in hepatitis C is multifactorial and does not show a linear behavior [16]. White *et al.* [17] showed that HCV infection is associated with an increased risk of type 2 diabetes, compared to non-infected or infected patients with HBV. Mehta *et al.* [18] in 2003 showed that HCV is associated with type 2 diabetes in people aged over 40 years. HCV has been considered the major risk factor incident of type 2 diabetes in the post-transplant of liver or kidney, especially in obese and elderly people [19] [20].

Regarding the laboratory variables, bilirubin in the final consultation presented OR 4.773, with *p* = 0.004. Is must be highlighted that bilirubin is an important laboratory parameter and is part of all the scores used in Hepatology (Child-Pugh, Meld, Maddrey and Bonacini). In this study, it is noteworthy that the association with the diagnosis of cirrhosis to the ultrasound liver examination refers to the dosage of the final visit. The literature has given little attention the importance of bilirubin as an indicating parameter of cirrhosis in hepatitis C, often highlighting it in the context of cholestatic diseases.

In 2014, Shadid *et al.* [21] observed higher levels of bilirubin in patients with more severe fibrosis compared with those who were in early stages. Which pathophysiological hypothesis would explain this association due to cirrhosis by C virus? Would there be the development of a cholangiopathy as a result of the antiviral activity?

Platelets below 150,000 per cubic millimeter, in the initial and final consultation, are in turn important indi-

Table 1. Sociodemographic and clinical variables, of the case and control subjects, Natal, Rio Grande do Norte, Brazil, 2015.

	Cirrhosis				
	Yes	No	OR	CI	p
Age in Years					
> 35 a [*]	65	64	1.0		
15 to 35 a	5	6	1.219	0.364 - 4.196	0.753
Gender					
Male	53	54	1.0		
Female	17	16	0.924	0.423 - 2.017	0.842
Marital status					
Not married	27	27	1.0		
Married	43	42	0.977	0.494 - 1.932	0.946
Occupation					
Related to risk	4	6	1.0		
No related to risk	65	64	0.656	0.177 - 2.436	0.527
Sexual promiscuity					
Yes	11	6	1.0		
No	57	56	1.80	0.623 - 5.203	0.272
Healthcare professional					
Yes	3	6	1.0		
No	67	64	0.478	0.115 - 1.991	0.301
Athlete					
Yes	2	9	1.0		
No	67	54	0.179	0.037 - 0.864	0.018
Blood donor					
Yes	3	9	1.0		
No	67	61	0.303	0.079 - 1.173	0.070
Treatment dropout					
Yes	29	32	1.0		
No	41	36	0.796	0.406 - 1.560	0.506
Earlier transfusion					
Yes	22	23	1.0		
No	44	44	0.957	0.466 - 1.962	0.903

[*]Age at initial consultation by group: more than 35 years and 15 to 35 years.

cators of cirrhosis when undergoing ultrasound examination. Numerous factors can result in thrombocytopenia, such as infectious diseases, hematological disorders and advanced liver disease. It is known that thrombocytopenia happens as a result of portal hypertension by hypersplenism. It is also known that there may be an au-

Table 2. Multivariate analysis results of case-control study of patients with chronic hepatitis C diagnosed with cirrhosis through USG in Natal, Rio Grande do Norte, Brazil, 2015.

| | Cirrhosis | | | | | | | |
| | Yes | | No | | | | | |
	N	%	N	%	p	OR (CI: 95%)	p	Adjusted OR (CI: 95%)
Blood donor								
No	18	43.90	23	56.10	0.017*	1.00	0.044*	1.00
Yes	17	22.67	58	77.33		2.67 (1.18 - 6.06)		0.24 (0.60 - 0.96)
Atlhete								
No	51	31.10	113	68.10	0.591	1.00	0.035*	1.00
Yes	17	27.42	45	72.58		1.19 (0.62 - 2.28)		0.18 (0.33 - 0.88)
Alcohol consumption								
No	15	18.52	66	81.48	0.007	1.00	0.251	1.00
Yes	51	35.92	91	64.08		2.47 (1.28 - 4.67)		1.62 (0.70 - 3.73)

Hosmer-Lemeshow test = 0.960. *Blood donor p = 0.044 with statistic significance adjusted by alcohol consumption. **Professional Atlhete p = 0.035 with estatistic significance adjusted by alcohol consumption.

toimmune mechanism. Prevention in the production of thrombopoietin can develop after a long period of necrosis and hepatic fibrosis in patients with advanced chronic hepatitis C, contributing to the decrease of platelet production [22].

In 2006, Sheng-Nan Lu *et al.* [23] demonstrated that thrombocytopenia had 76.2% of sensitivity and 87.8% of specificity for the diagnosis of cirrhosis through USG in patients with chronic hepatitis C virus. Platelet accumulation in the liver with chronic hepatitis can be implied among stellate, and Kupffer cells decrease [24].

The INR is a parameter that characterizes liver function, so as its increase indicates deterioration of functional activity. It can be seen in this study that the INR of the final consultation is clearly an indicator of diagnosis of cirrhosis in undergoing USG examination. In this study, the INR presented odds ratio of 17.875. Only 2 patients of the sample showed INR < 1.3. The PTA of the final consultation with cutoff below 70%, was less precise to signal the presence of cirrhosis in USG.

These are important clinical parameters that help the healthcare professional to discern on the diagnosis of cirrhosis, when the biopsy is not available. It should be noted that the accuracy of the INR in this study was higher than the accuracy of PTA. It is possible that this difference is related to sampling quantitative or due to the heterogeneity of laboratory methods in both tests. The literature reports fully the changes in INR and PTA in patients with the diagnosis of cirrhosis [25]-[27].

In the case of the GOT of the final consultation greater than 60 IU (OR 3.267, p = 0.002) it is in accordance with the study of Sheet *et al.*, showing a GOT/GPT ratio greater than 1, where a histopathological worsening and clinical progression to cirrhosis occurs in patients with chronic C virus infection [28].

Among the variables that make up the images examinations block it is important to note that the computed tomography (CT) did not show p value with statistical significance. This result is compatible with that described in the literature and clearly demonstrates that CT is not routinely used for diagnosis of cirrhosis, since the accuracy is not greater than the USG, and it is more costly and exposes the patient to the risks inherent in the method [29].

Still in this set of imaging examinations, it was studied whether the presence of nodules in patients with chronic hepatitis C would have a positive association with the diagnosis of cirrhosis and no statistical significance was verified. Probably these results could be related to the small number of positive events found [29].

In this study, it was observed that of 67 cirrhotic patients studied, regarding the variable biopsy, 15 had undergone the procedure. The OR was estimated to be 0.402 with p > 0.05. This result may reflect the technical difficulties in performing these procedures in a population of patients with coagulation disorders, considering the risk of bleeding.

On the clinical progression, it was observed that 18 cirrhotic patients underwent antiviral treatment and 3 (16.6%) of these patients had virus negativity (SVR). This result is in agreement with the study HALT C-Trial Group-that showed a poorer clinical outcome in these patients and low response rates to treatment [30].

The limitations of this study are due to the fact that the selection of cases and controls was made from medical records analysis, though it was obtained retrospectively by systematized calls by completing a standard questionnaire. Another aspect is that it was not possible to perform liver biopsies in all cirrhotic patients, which can signal to operational difficulties to carry out this procedure in this patient population.

This study indicates that clinical and laboratory variables can be grouped into mathematical model for the construction of a clinical score that can support the non-invasive diagnosis of cirrhosis. The importance of this score in the diagnosis of cirrhosis may be proved a posteriori, with the completion of a diagnostic study. This score can be applied in real life, especially in patients with inadequate clinical conditions for performing liver biopsies.

4. Conclusion

Cirrhosis condition modifies the clinical course of patients with chronic hepatitis C, making it a disease with a greater chance of complications and painful treatment. This condition, particularly in its compensated forms, has few clinical signs and symptoms and the diagnosis often becomes difficult.

Among all the studied variables, we found that the chances of diagnosis of cirrhosis in patients with chronic hepatitis C would be lower (OR 0.24, $p = 0.044$) in blood donors through ultrasound, when compared to non-cirrhotic HCV positive controls in USG and in non-donors. Also, it was found that the chances of diagnosis of cirrhosis by USG in athletes were estimated to 0.18 with significant p of 0.035. Being a blood donor and athlete is protective factors that reduce the risk of cirrhosis, independent of alcohol consumption. Moreover, no other clinical or laboratory variable was associated with ultrasound diagnosis of cirrhosis in patients with chronic hepatitis C.

References

[1] Thein, H.H., Yi, Q., Dore, G.J. and Krahn, M.D. (2008) Estimation of Stage-Specific Fibrosis Progression Rates in Chronic Hepatitis C Virus Infection: A Meta-Analysis and Meta-Regression. *Hepatology*, **48**, 418-431. http://dx.doi.org/10.1002/hep.22375

[2] Tong, M.J., el-Farra, N.S., Reikes, A.R. and Co, R.L. (1995) Clinical Outcomes after Transfusion Associated Hepatitis C. *The New England Journal of Medicine*, **332**, 1463-1466. http://dx.doi.org/10.1056/NEJM199506013322202

[3] Niederau, C., Lange, S., Heintges, T., Erhardt, A., Buschkamp, M., Hürter, D., Nawrocki, M., Kruska, L., Hensel, F., Petry, W. and Häussinger, D. (1998) Prognosis of Chronic Hepatitis C: Results of a Large, Prospective Cohort Study. *Hepatology*, **28**, 1687-1695. http://dx.doi.org/10.1002/hep.510280632

[4] D'Amico, G., Garcia-Tsao, G. and Pagliaro, L. (2006) Natural History and Prognostic Indicators of Survival in Cirrhosis: A Systematic Review of 118 Studies. *Journal of Hepatology*, **44**, 217-231. http://dx.doi.org/10.1016/j.jhep.2005.10.013

[5] Morgan, R.L., Baack, B., Smith, B.D., Yartel, A., Pitasi, M. and Falck-Ytter, Y. (2013) Eradication of Hepatitis C Virus Infection and the Development of Hepatocellular Carcinoma: A Meta-Analysis of Observational Studies. *Annals of Internal Medicine*, **158**, 329-337. http://dx.doi.org/10.7326/0003-4819-158-5-201303050-00005

[6] Morgan, T.R., Ghany, M.G., Kim, H.Y., Snow, K.K., Shiffman, M.L., De Santo, J.L., *et al.* (2010) Outcome of Sustained Virological Responders with Histologically Advanced Chronic Hepatitis C. *Hepatology*, **52**, 833-844. http://dx.doi.org/10.1002/hep.23744

[7] Poynard, T., Bedossa, P. and Opolon, P. (1997) Natural History of Liver Fibrosis Progression in Patients with Chronic Hepatitis C. The OBSVIRC, METAVIR, CLINIVIR, and DOSVIRC Groups. *The Lancet*, **349**, 825-832.

[8] Poynard, T., Munteanu, M., Imbert-Bismut, F., Charlotte, F., Thabut, D., Le Calvez, S., Messous, D., Thibault, V., Benhamou, Y., Moussalli, J. and Ratziu, V. (2004) Prospective Analysis of Discordant Results between Biochemical Markers and Biopsy in Patients with Chronic Hepatitis C. *Clinical Chemistry*, **50**, 1344-1355. http://dx.doi.org/10.1373/clinchem.2004.032227

[9] Bedossa, P., Dargère, D. and Paradis, V. (2003) Sampling Variability of Liver Fibrosis in Chronic Hepatitis C. *Hepatology*, **38**, 1449-1457. http://dx.doi.org/10.1016/j.hep.2003.09.022

[10] Piccinino, F., Sagnelli, E., Pasquale, G. and Giusti, G. (1986) Complications Following Percutaneous Liver Biopsy. A

Multicentre Retrospective Study on 68,276 Biopsies. *Journal of Hepatology*, **2**, 165-173.

[11] Perrault, J., McGill, D.B., Ott, B.J. and Taylor, W.F. (1978) Liver Biopsy: Complications in 1000 Inpatients and Outpatients. *Gastroenterology*, **74**, 103-106.

[12] Imbert-Bismut, F., Ratziu, V., Pieroni, L., Charlotte, F., Benhamou, Y., Poynard, T. and Multivirc Group (2001) Biochemical Markers of Liver Fibrosis in Patients with Hepatitis C Virus Infection: A Prospective Study. *The Lancet*, **357**, 1069-1075.

[13] Fontana, R.J., Goodman, Z.D., Dienstag, J.L., Bonkovsky, H.L., Naishadham, D., Sterling, R.K., Su, G.L., Ghosh, M., Wright, E.C. and HALT-C Trial Group (2008) Relationship of Serum Fibrosis Markers with Liver Fibrosis Stage and Collagen Content in Patients with Advanced Chronic Hepatitis C. *Hepatology*, **47**, 789-798. http://dx.doi.org/10.1002/hep.22099

[14] Sharma, S., Khalili, K. and Nguyen, G.C. (2014) Non-Invasive Diagnosis of Advanced Fibrosis and Cirrhosis. *World Journal of Gastroenterology*, **20**, 16820-16830. http://dx.doi.org/10.3748/wjg.v20.i45.16820

[15] Yonossi, Z., Lynn, G., Fang, Y., Srishord, M., Winter, P., Kallman, J. and Moon, J. (2012) Disparities in Activity Level and Mutrition between Patients with Chronic Hepatitic C and Blood Donors. *The American of Physical Medicine and Rehabilitation*, **4**, 436-441.

[16] Dusheiko, G. and Westbrook, R.H. (2014) Natural History of Hepatitis C. *Journal of Hepatology*, **61**, 558-568.

[17] White, D.L., Ratziu, V. and El-Serag, H.B. (2008) Hepatitis C Infection and Risk of Diabetes: A Systematic Review and Meta-Analysis. *Journal of Hepatology*, **49**, 831-844. http://dx.doi.org/10.1016/j.jhep.2008.08.006

[18] Mehta, S.H., Brancati, F.L., Strathdee, A.S., Pankow, J.S., Netski, D., Coresh, J., Szklo, M. and Thomas, D.L. (2003) Hepatitis C Virus Infection and Incident Type 2 Diabetes. *Hepatology*, **38**, 50-56. http://dx.doi.org/10.1053/jhep.2003.50291

[19] Delgado-Borrego, A., Liu, Y.S., Jordan, S.H., Agrawat, S., Zhang, H., Christofi, M., *et al.* (2008) Prospective Study of Liver Transplant Recipients with HCV Infection: Evidence for a Causal Relationship between HCV and Insulin Resistance. *Liver Transplantation*, **14**, 183-201. http://dx.doi.org/10.1002/lt.21267

[20] Fabrizi, F., Messa, P., Martin, P. and Takkouche, B. (2008) Hepatitis C Virus Infection and Post-Transplant Diabetes Mellitus among Renal Transplant Patients: A Meta-Analysis. *International Journal of Artificial Organs*, **31**, 675-682.

[21] Shahid, M., Idrees, M., Nasir, B., Raja, A.J., Raza, S.M., Amin, I., Rasul, A. and Tayyab, G.U. (2014) Correlation of Biochemical Markers and HCV RNA Titers with Fibrosis Stages and Grades in Chronic HCV-3a Patients. *European Journal of Gastroenterology & Hepatology*, **26**, 788-794. http://dx.doi.org/10.1097/MEG.0000000000000109

[22] Shah, N.L., Northup, P.G. and Caldwell, S.H. (2015) Coagulation Abnormalities in Patients with Liver Disease. UPTODATE 18.0, February 2015.

[23] Lu, S.N., Wang, J.H., Liu, S.L., Hung, C.H., Chen, C.H., Tung, H.D., Chen, T.M., Huang, W.S., Lee, C.M., Chen, C.C. and Changchien, C.S. (2006) Thrombocytopenia as a Surrogate for Cirrhosis and a Marker for the Identification of Patients at High-Risk for Hepatocellular Carcinoma. *Cancer*, **107**, 2212-2222. http://dx.doi.org/10.1002/cncr.22242

[24] Kondo, R., Yano, H., Nakashima, O., Tanikawa, K., Nomura, Y. and Kage, M. (2013) Accumulation of Platelets in the Liver May Be an Important Contributory Factor to Thrombocytopenia and Liver Fibrosis in Chronic Hepatitis C. *Journal of Gastroenterology*, **48**, 526-534. http://dx.doi.org/10.1007/s00535-012-0656-2

[25] Lisman, T., Liebeek, F.W.G. and Groot, P.G. (2002) Haemostatic Abnormalities in Patients with Liver Disease. *Journal of Hepatology*, **37**, 280-287. http://dx.doi.org/10.1016/S0168-8278(02)00199-X

[26] Caldewell, S.H., Hoffman, M., Lisman, T., Marik, B.G., Northup, P.G., Reddy, K.R., *et al.* (2006) Coagulation Disorders and Hemostasis in Loiver Disease: Pathophysiology and Critical Assessment of Current Management. *Hepatology*, **44**, 1039-1046. http://dx.doi.org/10.1002/hep.21303

[27] Tripoli, A., Caldwell, S.H., Hoffman, M., Trotter, J.F. and Sanyal, A.J. (2007) Review Article: The Prothrombin Time Test as a Measure of Bleeding Risk and Prognosis in Liver Disease. *Alimentary Pharmacology & Therapeutics*, **26**, 1418.

[28] Shet, S.G., Flamm, S.L., Gordon, F.D. and Choopra, S. (1998) AST/ALT Ratio Predicts Cirrhosis in Patients with Chronic Hepatitis C Virus Infection. *American Journal of Gastroenterology*, **93**, 44-48. http://dx.doi.org/10.1111/j.1572-0241.1998.044_c.x

[29] Golbberg, E. and Chopra, S. (2015) Cirrhosis in Adults: Etiologies, Clinical Manifestations, and Diagnosis. UPTODATE, 18.0, February 2015.

[30] Ghany, M.G., Lok, A.S., Everhart, J.E., Everson, G.T., Lee, W.M., Curto, T.M., Wright, E.C., Stoddard, A.M., Steling, R.K., Di Bisceglie, A.M., Bonkovsky, H.L., Morishima, C., Morgan, T.R. and Dientag, J.L., HALT-C Trial Group (2010) Predicting Clinical and Histologic Outcomes Based on Standard Laboratory Tests in Advanced Chronic Hepatitis C. *Gastroenterology*, **138**, 136-146. http://dx.doi.org/10.1053/j.gastro.2009.09.007

Randomized, Placebo-Controlled Trial of Transdermal Rivastigmine for the Treatment of Encephalopathy in Liver Cirrhosis (TREC Trial)

Patrick P. Basu[1,2], Niraj James Shah[3], Mark M. Aloysius[2], Robert S. Brown[1]

[1]Columbia University College of Physicians & Surgeons, New York, USA
[2]King's County Hospital, New York, USA
[3]James J. Peters VA Medical Center, New York, USA
Email: mark.aloysius5@gmail.com

Abstract

Objectives: Cognitive dysfunction in patients with hepatic encephalopathy (HE) may be caused by alterations in cholinergic neurotransmission. The objective of the study was to evaluate the efficacy and safety of transdermal rivastigmine in improving cognitive function in patients with overt HE. Design: Randomized, controlled pilot study in which patients with grade 2 or 3 HE were treated with lactulose and randomized to receive either transdermal rivastigmine or placebo for 21 days. The modified encephalopathy scale (MES), object recognition test (ORT), trail test (TT), and serum ammonia were assessed at baseline weekly. Electroencephalography was performed at baseline and the final week of the study. Results: Patients were treated with lactulose (20 g/30 mL three times per day) and either transdermal rivastigmine (4.6 mg/d; n = 15) or placebo (n = 15). Transdermal rivastigmine significantly improved MES, ORT, and TT results compared with placebo ($P \leq 0.0001$ at all 3 weeks for all 3 assessments). Serum ammonia improved in both treatment groups, although there was significantly greater improvement with placebo than rivastigmine after 2 weeks of treatment ($P < 0.03$). There were no differences in electroencephalography results between treatment groups. Conclusions: Transdermal rivastigmine with concomitant lactulose significantly improved cognitive-function in patients with overt HE.

What is already known about this subject?
- Current approaches to the management of HE are primarily designed to reduce the levels of ammonia and other gut-derived toxins.
- Traditional strategies for HE treatment have included non-absorbable disaccharides (to decrease bowel transit time) or rifamixin (non-absorbable antibiotics to reduce ammoniogenic flora).

- No transdermal cholinomimetic agents have been used with oral lactulose to date, in HE.
  What are the new findings?
- **Transdermal rivastigmine is safe for use in patients with grade 2 & 3 HE.**
- **Transdermal rivastigmine in combination with oral lactulose in this study is far superior to lactulose alone in improving cognitive function.**
  How might it impact on clinical practice in the foreseeable future?
- **Transdermal rivastigmine in combination with oral lactulose can be used safely in clinical practice.**
- **Transdermal rivastigmine in combination with oral lactulose is efficacious in improving cognitive function in moderate HE (grade 2/3).**
  Further validation through large randomized clinical trials is required before this is adopted in universal clinical practice of treating HE.

Keywords

Acetylcholinesterase, Cholinergic Function, Hepatic Encephalopathy, Rivastigmine

1. Introduction

Hepatic encephalopathy (HE) is a neuropsychiatric complication of acute and chronic liver disease [1] that contributes to the mortality of patients with end-stage liver disease [2]. This potentially reversible condition [3] is characterized by a variety of symptoms, including changes in consciousness and cognition [4]. The precise pathogenic mechanisms leading to HE are unknown but are primarily believed to involve the accumulation of ammonia and other gut-derived toxins [5]. Such current approaches to the management of HE are primarily designed to reduce the levels of ammonia and other gut-derived toxins using therapies such as non-absorbable disaccharides (e.g., lactulose) and non-systemic antibiotics (e.g., rifaximin) [6].

Accumulation of ammonia and gut-derived toxins can lead to a variety of pathogenic changes, including alterations in neurotransmitter signaling systems [3] [6]. While the changes in neurotransmitter systems may be complex and heterogenous, [7] upregulation of the inhibitory GABAergic and serotonergic pathways and impairment of the excitatory glutamatergic and catecholaminergic pathways have been hypothesized to be involved in the pathogenesis of HE [6]. Alterations in the cholinergic neurotransmitter system have also been observed in both animal models of HE [8]-[12] and in postmortem brain tissue samples from patients with HE [7] [8] [13]. For example, the level of acetylcholine (ACh) and the neurotransmitter of the cholinergic system, were reduced, and the activity of acetylcholinesterase (AChE), the enzyme that hydrolyzes ACh, was increased in brain extracts from rats with experimentally induced HE [8]. Similarly, the activity of AChE was higher in postmortem brain samples from patients with HE compared with controls without cirrhosis [8]. Given the importance of cholinergic neurotransmission in cognitive function and consciousness [14] alterations such as those, which disrupt cholinergic signaling, may explain the deficits in cognitive function and consciousness observed in patients with HE. Thus, therapies that may restore cholinergic balance may be useful in the treatment of patients with HE.

Rivastigmine is a reversible AChE inhibitor used to treat dementia associated with Alzheimer's disease (AD) and Parkinson's disease [15]. Rivastigmine significantly improved learning in a rat model of HE [8], but, to date, there have been no reports of rivastigmine improving cognitive function in patients with HE. However, case reports of patients with HE treated with the AChE inhibitors neostigmine (with polyethylene glycol [PEG]) [16] [17] and physostigmine [18] have suggested that this class of drugs may be effective for the treatment of HE in humans. The objective of this randomized, placebo-controlled pilot study was to determine the efficacy and safety of transdermal rivastigmine for improving cognitive function in patients with grade 2 or 3 HE.

2. Methods

This was a randomized, placebo-controlled pilot study conducted between 2009 and 2010. The protocol was approved by the institutional review board at Finestein Institute. The study was performed in accordance with the Declaration of Helsinki, the International Conference on Harmonisation Good Clinical Practice Guidelines, and applicable local laws and regulations. Signed informed consent was obtained for each patient before study enrollment.

2.1. Patient Population

Patients were eligible if they had grade 2 or 3 HE. Hepatic encephalopathy grades were based on clinical and electroencephalographic criteria [19] [20]. Grade 2 HE was characterized by the presence of lethargy, inappropriate behavior, disorientation, asterixis, abnormal reflexes, and abnormalities on electroencephalogram (EEG) (*i.e.*, slowing, triphasic waves). Patients with grade 3 HE exhibited somnolence (but were rousable), loss of meaningful communication, asterixis, abnormal reflexes, and slowing, triphasic waves on EEG. Exclusion criteria included the recent or current use of antihistamines, metoclopramide, benzodiazepines, cannabinoids, or narcotics; acute gastrointestinal bleeding; sepsis; toxic metabolic syndrome; renal failure; and infection with human immunodeficiency virus. Urine was screened before the study and then weekly to detect the use of narcotics, benzodiazepines, and cannabinoids.

2.2. Treatments

Eligible patients were treated with oral lactulose (20 g/30 mL three times daily) and randomized to receive either the transdermal rivastigmine patch (Exelon® Patch; Novartis Pharmaceuticals Corporation, East Hanover, NJ; 4.6 mg/d) or a placebo patch for 21 days. Randomization was performed by electronic randomization. Dietary animal protein intake was restricted to 50 g/d. The use of oral or intravenous antibiotics was prohibited during the 21 days of the study.

2.3. Assessments

Hepatic encephalopathy was assessed using the modified encephalopathy score (MES), a measure of memory loss, confusion, sleep disturbance, and comprehension with a minimum score of 4 (mild HE) and a maximum score of 12 (severe HE; **Table 1**). All questions for the MES were administered in the patient's native language. Psychometric testing was performed using the trail test (TT) [21] and the object recognition test (ORT). The MES, TT, ORT, and serum ammonia levels were assessed at baseline and then weekly. Electroencephalography was performed at baseline and during the final week of the study. Safety was assessed

2.4. Statistical Analysis

T-test was used to compare transdermal rivastigmine to placebo by calculating p values for statistical significance (defined as $P \leq 0.05$) using SPSS version 20 for mac.

3. Results

3.1. Patients and Baseline Characteristics

A total of 72 patients were screened (**Figure 1**). Of 12 who did not participate 8 did not meet the inclusion criteria and 4 declined to participate after signing the consent. Thirty patients were randomized; 15 received the transdermal rivastigmine patch and 15 received a placebo patch.

The demographics of the treatment groups were similar with respect to age, sex, and body mass index (**Table 2**). There were more black patients and fewer white and Hispanic patients in the transdermal rivastigmine group

Table 1. Modified encephalopathy scoring system[a].

Variable	Mild (1 point)	Moderate (2 points)	Severe (3 points)
Memory loss	Past (30 days)	Recent past (7 days)	Recent (24 hours)
Confusion	Not confused	Mildly confused	Very confused
Sleep disturbance	Mild (changes in sleep pattern)	Moderate (fragmented sleep pattern)	Severe (less than 2 hours of sleep/night)
Comprehension	Well aware of surroundings	Impaired awareness of surroundings	Oblivious of surroundings

[a]Patients are rated from 1 to 3 for each variable, and scores are added to calculate the modified encephalopathy score. Total score can range from 4 (mild) to 12 (severe).

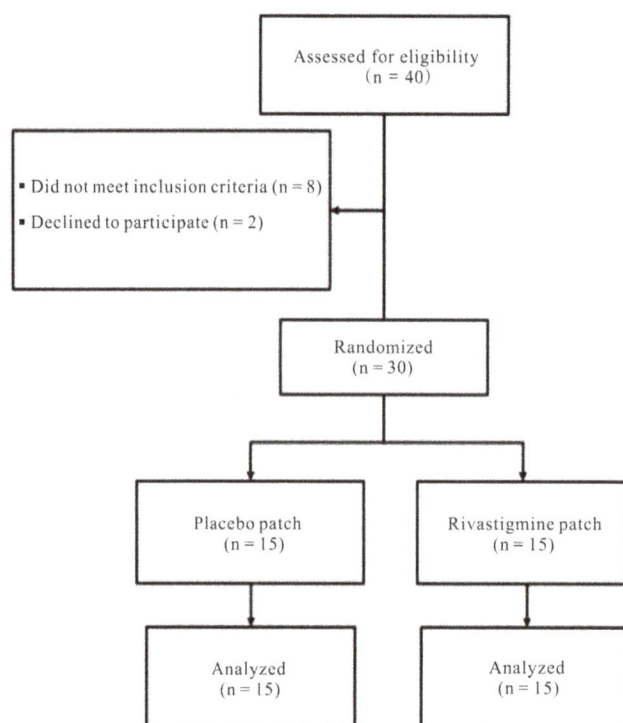

Figure. 1. Patient disposition.

than in the placebo group. Baseline disease characteristics were similar between the treatment groups with respect to model for end-stage liver disease score, EEG, MES, and performance on the TT and ORT (**Table 2**). More patients in the placebo group had alcoholic liver disease (40%) compared with the transdermal rivastigmine group (13%). Serum ammonia levels were slightly but significantly higher in the transdermal rivastigmine group compared with the placebo group ($P = 0.03$).

3.2. Modified Encephalopathy Score

Transdermal rivastigmine resulted in a significantly lower mean MES compared with placebo within 1 week of treatment (7 vs 10, respectively; $P < 0.0001$; **Figure 2**). Though the mean MES of both treatment groups leveled off during weeks 2 and 3, the significant effect of transdermal rivastigmine compared with placebo on the mean MES was sustained at weeks 2 and 3 of the study ($P < 0.0001$ at both weeks). These results suggest that transdermal rivastigmine was significantly more effective than placebo at improving the degree of cognitive dysfunction in HE in patients receiving concomitant lactulose.

3.3. Psychometric Tests

Patients who received transdermal rivastigmine had significantly better performance on both the ORT (**Figure 3(a)**) and TT (**Figure 3(b)**) during weeks 1 through 3 of the study compared with those who received placebo ($P = 0.0001$ at all 3 time points for both tests). In patients treated with transdermal rivastigmine, ORT results (45 seconds) were normal (<55 seconds) within 1 week of treatment with transdermal rivastigmine. The improvement in the ORT with transdermal rivastigmine was sustained as scores remained normal during weeks 2 and 3 of treatment (35 and 30 seconds, respectively). In contrast, in patients who received placebo, ORT results only slightly improved from 125 seconds at baseline to 98 seconds after 1 week and actually worsened during weeks 2 and 3 of the study (101 and 109 seconds, respectively).

Results of the TT were also substantially improved in patients who received transdermal rivastigmine, decreasing from a baseline score of 161 seconds to 90 seconds and 92 seconds after 1 and 2 weeks of treatment, respectively. By the third week of treatment, TT results (65 seconds) approached normal (<60 seconds). In contrast, TT results in the placebo group only slightly improved from 158 seconds at baseline to 154, 134, and 146

Table 2. Patient demographics and baseline disease characteristics.

Characteristic	Placebo (n = 15)	Transdermal rivastigmine (n = 15)
Mean age (range), y	56 (50 - 65)	56 (51 - 61)
Male, n (%)	10 (67)	9 (60)
Race, n (%)		
White	6 (40)	1 (7)
Black	3 (20)	11 (73)
Hispanic	6 (40)	3 (20)
Mean BMI (range)	28 (26 - 30)	28 (26 - 30)
Cirrhosis etiology, n (%)		
Alcohol	6 (40)	2 (13)
Hepatitis B	1 (7)	3 (20)
Hepatitis C	5 (33)	7 (47)
Nonalcoholic steatohepatitis	2 (13)	1 (7)
Other	1 (7)	2 (13)
HE grade, n (%)		
2	9	8
3	6	7
Mean MELD score (range)	20 (18 - 22)	20 (18 - 22)
Serum ammonia, mol/L[a]	239	245[b]
Mean MES	14	14
Mean ORT time, s[c]	122	125
Mean TT time, s[d]	158	161
EEG, n (%)		
Normal	9 (60)	10 (67)
Grade 1	3 (20)	2 (13)
Grade 2	2 (13)	3 (20)
Grade 3	1 (7)	0 (0)

BMI, body mass index; EEG, electroencephalogram; HE, hepatic encephalopathy; MELD, model for end-stage liver disease; MES, modified encephalopathy score; ORT, object recognition test; TT, trail test. [a]Normal range, 11 - 32 mol/L. [b]$P = 0.03$. [c]Normal, <55 seconds. [d]Normal, <60 seconds.

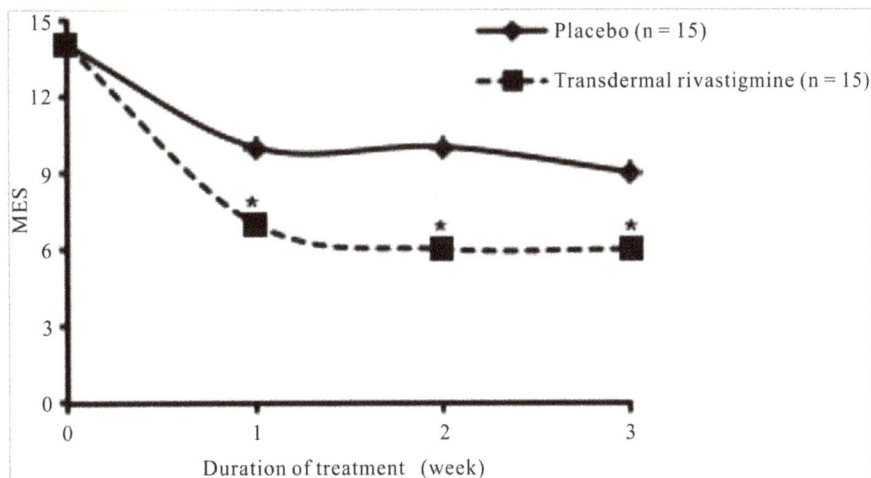

Figure 2. Modified encephalopathy scores over the 3 weeks of treatment with placebo or transdermal rivastigmine. [*]$P < 0.0001$. MES, modified encephalopathy score.

seconds after 1, 2, and 3 weeks of treatment, respectively.

3.4. Clinical Tests

For patients in both treatment groups, serum ammonia levels were progressively lower each of the 3 weeks of

(a)

(b)

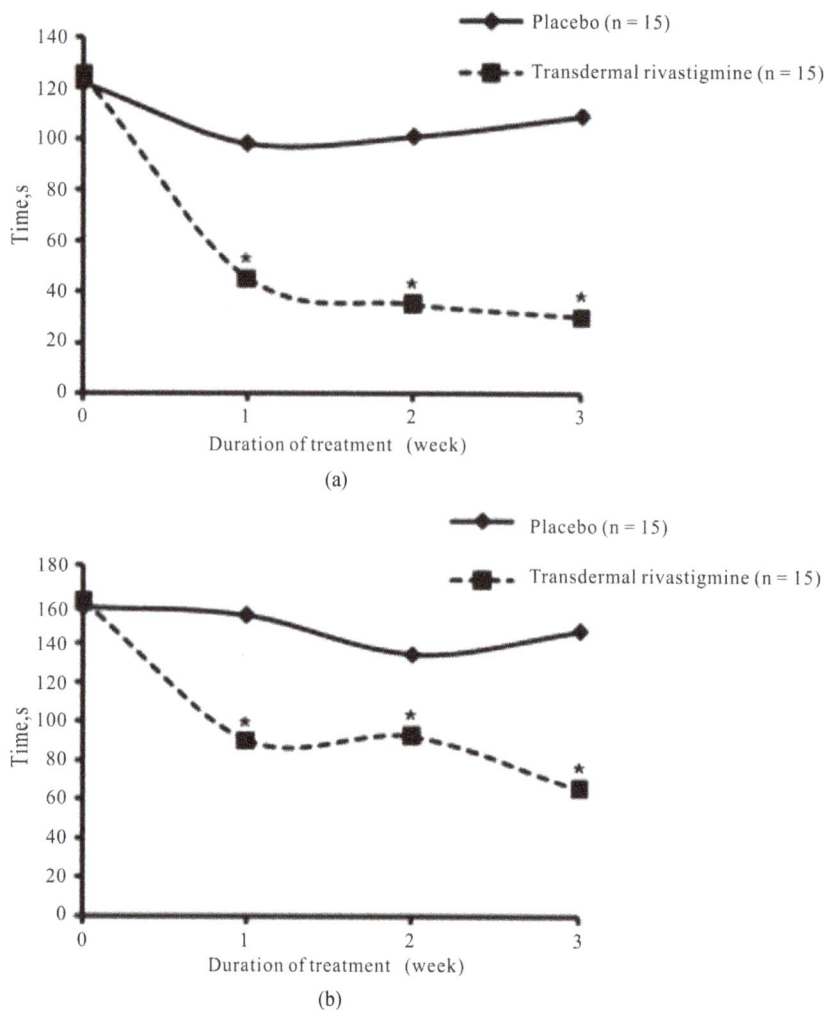

Figure 3. Psychometric tests. Scores for (a) object recognition test and (b) trail test over the 3 weeks of treatment with placebo or transdermal rivastigmine. Normal scores for object recognition test and trail testare less than 55 seconds and less than 60 seconds, respectively. $^{*}P = 0.0001$.

the study compared with baseline (**Table 3**). However, transdermal rivastigmine did not have a greater effect on serum ammonia levels than placebo. In fact, patients who received placebo had slightly but significantly greater reductions in serum ammonia levels compared with those who received transdermal rivastigmine after 2 weeks of treatment ($P = 0.03$). However, by the third week of treatment, thereduction in serum ammonia was not significantly different between the 2 groups. There were no differences in EEG results between the 2 treatment groups.

3.5. Safety

Overall, transdermal rivastigmine was well tolerated, and no reactions were reported at the patch-application site. In the transdermal rivastigmine group, 1 patient (7%) reported diarrhea and 1 (7%) reported dry mouth. In the placebo group, 2 patients (13%) reported diarrhea, 1 (7%) reported dry mouth, and 1 (7%) reported urinary retention (**Table 4**).

4. Discussion

The results of this randomized, placebo-controlled pilot study suggest that transdermal rivastigmine significantly improves cognitive function compared with placebo in patients with grade 2 or 3 HE who were receiving con-

Table 3. Serum ammonia levels.

| Treatment week | Mean change from baseline, mol/L | | |
	Placebo (n = 15)	Transdermal rivastigmine (n = 15)	P value
1	−92	−91	NS
2	−113	−107	0.03
3	−137	−131	0.05

NS = not significant.

Table 4. Side events.

Side Event	Placebo (n = 15)	Transdermal rivastigmine (n = 15)
Diarrhea	2	1
Dry mouth	1	1
Urinary Retention	1	0

improves cognitive function compared with placebo in patients with grade 2 or 3 HE who were receiving concomitant lactulose. These results support those of earlier case reports in which hepatic coma resolved in 3 patients with HE who were treated with other AChE inhibitors [16]-[18]. In 2 of the patients, treatment with neostigmine and PEG electrolyte solution to improve gastric motility and bowel cleansing, respectively, resulted in the resolution of hepatic coma after 2 days [16] [17]. The third patient was treated with physostigmine and immediately regained consciousness after administration of the AChE inhibitor [18].

The observation that inhibition of AChE enhances cognitive function is not unique to HE. Rivastigmine is also used to treat AD and Parkinson's disease dementia. The AChE inhibitors galantamine and donepezil are also used for the treatment of AD. Inhibition of AChE may improve cognitive function in patients with HE, AD, or Parkinson's disease dementia by increasing levels of ACh in the brain and prolonging ACh activity at neural synapses [15]. However, there is mounting evidence that AChE inhibitors may have an antiinflammatory effect [22] which has been observed in patients with AD [23]. Such a mechanism may be relevant to patients with HE because cerebral inflammation has been directly implicated in the pathogenesis of HE in a rat model of acute liver failure, [24] and systemic inflammation has been implicated in the pathogenesis of HE [25] and the impairment of neuropsychologic function [26] in patients with liver disease. In patients with HE, AChE inhibitors may also increase gastrointestinal motility and thereby enhance the effect of intestinal cathartics such as lactulose and PEG, which were used concomitantly in this study and the previously described case reports, respectively [16]-[18].

While rivastigmine is available in oral and transdermal formulations, the transdermal formulation may be more favorable in general and for patients with HE. First, transdermal rivastigmine generally offers improved tolerability compared with the oral formulation, particularly with respect to gastrointestinal adverse events such as nausea and vomiting [15] [27]. This may be because of the more gradual and smoother increase in absorption of transdermal rivastigmine compared with oral rivastigmine [15]. Second, the ease of dosing of the transdermal patch compared with oral formulations may be advantageous in patients with HE, particularly those with higher grades of HE who may be unable to easily ingest tablets or oral solutions. Lastly, oral rivastigmine is administered twice daily [28] while the patch is applied only once daily [29]. This dosing schedule may provide less interference with daily life. Overall, the advantages of transdermal delivery of rivastigmine may increase patient compliance and satisfaction for patients and caregivers. Indeed, caregivers of patients with AD preferred the transdermal rivastigmine patch compared with the oral capsule in a large (n = 1059 caregivers) 24-week study comparing the transdermal and oral formulations of rivastigmine in patients with AD [30].

One limitation to this study is that it was performed using a small population of patients (N = 30). Furthermore, patients in both treatment groups were receiving concomitant lactulose. The finding in this study that there was improvement of MES in the placebo group (albeit to a significantly lesser degree than the transdermal rivastigmine group) suggests that lactulose may have contributed to the improvement in MES observed in this study. However, patients who received placebo did not experience any substantial or sustained benefit in cognitive function as assessed using the TT and ORT. This suggests that transdermal rivastigmine had an independent effect on cognitive function in patients with grade 2 or 3 HE.

Interestingly, despite the improvement in cognitive function with transdermal rivastigmine, there was no improvement in serum ammonia levels with the AChE inhibitor compared with placebo. In fact, after 2 weeks of treatment, serum ammonia levels were reduced to a slightly but significantly greater degree in patients who received placebo compared with those who received transdermal rivastigmine. This suggests that reductions in serum ammonia levels are not clinically relevant in the context of improving the results of psychometric tests such as the TT and ORT. It is notable that serum ammonia levels can be influenced by several factors, including the conditions leading to HE (*i.e.*, precipitating factors), mild exertion by the patient before phlebotomy (including fist clenching), prolonged tourniquet application, and blood-sample drawing technique andstorage [1] [6] [31]-[33]. Thus, results may not be comparable among patients or even between samples taken at different times from an individual patient.

In this study, transdermal rivastigmine with lactulose was safe and well tolerated. Diarrhea, a common adverse event associated with lactulose treatment, [6] was reported by 1 patient who received transdermal rivastigmine and 2 patients who received placebo. Adverse events associated with AChE-inhibitor use were only reported by 1 patient in the transdermal rivastigmine group (dry mouth) and by 2 patients in the placebo group (dry mouth, urinary retention). Further, no skin irritation at the patch-attachment site was reported in this study. The favorable safety and tolerability profile reported here for transdermal rivastigmine in patients with HE is comparable to that reported in patients with AD, although a small percentage of patients with AD did develop some skin irritation [27].

5. Conclusion

In conclusion, although performed with a small number of patients, this randomized, placebo-controlled pilot study demonstrates that transdermal rivastigmine used offlabel with concomitant lactuloseis is safe and effective for improving cognitive function in patients with HE. Large, randomized clinical trials are needed to fully explore the benefits of transdermal rivastigmine in patients with HE. Because of the benefit of transdermal rivastigmine on cognitive function shown here, these trials should be conducted not only in patients with overt HE (grade ≥ 1) but also in patients with minimal HE (grade 0) in which the only symptom is cognitive dysfunction measurable only by psychometric or neurophysiologic testing [6].

References

[1]   Häussinger, D. and Schliess, F. (2008) Pathogenetic Mechanisms of Hepatic Encephalopathy. *Gut*, **57**, 1156-1165.

[2]   Stewart, C.A., Malinchoc, M., Kim, W.R. and Kamath, P.S. (2007) Hepatic Encephalopathy as a Predictor of Survival in Patients with End-Stage Liver Disease. *Liver Transplantation*, **13**, 1366-1371. http://dx.doi.org/10.1002/lt.21129

[3]   Blei, A.T. and Córdoba, J. (2001) The Practice Parameters Committee of the American College of Gastroenterology. Hepatic Encephalopathy. *The American Journal of Gastroenterology*, **96**, 1968-1976. http://dx.doi.org/10.1111/j.1572-0241.2001.03964.x

[4]   Ferenci, P., Lockwood, A., Mullen, K., Tarter, R., Weissenborn, K., Blei, A.T. and Members of the Working Party (2002) Hepatic Encephalopathy—Definition, Nomenclature, Diagnosis, and Quantification: Final Report of the Working Party at the 11th World Congresses of Gastroenterology, Vienna, 1998. *Hepatology*, **35**, 716-721. http://dx.doi.org/10.1053/jhep.2002.31250

[5]   Riordan, S.M. and Williams, R. (2010) Gut Flora and Hepatic Encephalopathy in Patients with Cirrhosis. *The New England Journal of Medicine*, **362**, 1140-1142. http://dx.doi.org/10.1056/NEJMe1000850

[6]   Mullen, K.D., Ferenci, P., Bass, N.M., Leevy, C.B. and Keeffe, E.B. (2007) An Algorithm for the Management of Hepatic Encephalopathy. *Seminars in Liver Disease*, **27**, 32-48. http://dx.doi.org/10.1055/s-2007-984576

[7]   Palomero-Gallagher, N., Bidmon, H.-J., Cremer, M., Schleicher, A., Kircheis, G., Reifenberger, G., *et al.* (2009) Neurotransmitter Receptor Imbalances in Motor Cortex and Basal Ganglia in Hepatic Encephalopathy. *Cellular Physiology and Biochemistry*, **24**, 291-306. http://dx.doi.org/10.1159/000233254

[8]   García-Ayllón, M.S., Cauli, O., Silveyra, M.-X., Rodrigo, R., Candela, A., Compañ, A., *et al.* (2008) Brain Cholinergic Impairment in Liver Failure. *Brain*, **131**, 2946-2956. http://dx.doi.org/10.1093/brain/awn209

[9]   McCandless, D.W., Looney, G.A., Modak, A.T. and Stavinoha, W.B. (1985) Cerebral Acetylcholine and Energy Metabolism Changes in Acute Ammonia Intoxication in the Lower Primate *Tupaia glis*. *Journal of Laboratory and Clinical Medicine*, **106**, 183-186.

[10] Méndez, M., Méndez-López, M., López, L., Aller, M.A., Arias, J. and Arias, J.L. (2010) Acetylcholinesterase Activity in an Experimental Rat Model of Type C Hepatic Encephalopathy. *Acta Histochemica*, **113**, 358-362.

[11] Song, G., Dhodda, V.K., Blei, A.T., Dempsey, R.J. and Rao, V.L.R. (2002) GeneChip® Analysis Shows Altered mRNA Expression of Transcripts of Neurotransmitter and Signal Transduction Pathways in the Cerebral Cortex of Portacaval Shunted Rats. *Journal of Neuroscience Research*, **68**, 730-737. http://dx.doi.org/10.1002/jnr.10268

[12] Swapna, I., Sathya SaiKumar, K.V., Murthy, Ch.R.K., Gupta, A.D. and Senthilkumaran, B. (2007) Alterations in Kinetic and Thermotropic Properties of Cerebral Membrane-Bound Acetylcholineesterase during Thioacetamide-Induced Hepatic Encephalopathy: Correlation with Membrane Lipid Changes. *Brain Research*, **1153**, 188-195. http://dx.doi.org/10.1016/j.brainres.2007.02.095

[13] Lal, S., Quirion, R., Lafaille, F., Nair, N.P., Loo, P., Braunwalder, A., *et al.* (1987) Muscarinic, Benzodiazepine, GABA, Chloride Channel and Other Binding Sites in Frontal Cortex in Hepatic Coma in Man. *Progress In Neuro-Psychopharmacology & Biological Psychiatry*, **11**, 243-250. http://dx.doi.org/10.1016/0278-5846(87)90067-4

[14] Woolf, N.J. and Butcher, L.L. (2010) Cholinergic Systems Mediate Action from Movement to Higher Consciousness. *Behavioural Brain Research*, **221**, 488-498.

[15] Darreh-Shori, T. and Jelic, V. (2010) Safety and Tolerability of Transdermal and Oral Rivastigmine in Alzheimer's Disease and Parkinson's Disease Dementia. *Expert Opinion on Drug Safety*, **9**, 167-176. http://dx.doi.org/10.1517/14740330903439717

[16] Kiba, T., Numata, K. and Saito, S. (2003) Neostigmine and Polyethylene Glycol Electrolyte Solution for the Therapy of Acute Hepatic Encephalopathy with Liver Cirrhosis and Ascites. *Hepatogastroenterology*, **50**, 823-826.

[17] Park, C.H., Joo, Y.E., Kim, H.S., Choi, S.K., Rew, J.S. and Kim, S.J. (2005) Neostigmine for the Treatment of Acute Hepatic Encephalopathy with Acute Intestinal Pseudo-Obstruction in a Cirrhotic Patient. *Journal of Korean Medical Science*, **20**, 150-152. http://dx.doi.org/10.3346/jkms.2005.20.1.150

[18] Kabatnik, M., Heist, M., Beiderlinden, K. and Peters, J. (1999) Hepatic Encephalopathy—A Physostigmine-Reactive Central Anticholinergic Syndrome? *European Journal of Anaesthesiology*, **16**, 140-142.

[19] Stewart, C.A. and Cerhan, J. (2005) Hepatic Encephalopathy: A Dynamic or Static Condition. *Metabolic Brain Disease*, **20**, 193-204. http://dx.doi.org/10.1007/s11011-005-7207-x

[20] Riordan, S.M. and Williams, R. (1997) Treatment of Hepatic Encephalopathy. *The New England Journal of Medicine*, **337**, 473-479. http://dx.doi.org/10.1056/NEJM199708143370707

[21] Conn, H.O. (1977) Trailmaking and Number-Connection Tests in the Assessment of Mental State in Portal Systemic Encephalopathy. *The American Journal of Digestive Diseases*, **22**, 541-550. http://dx.doi.org/10.1007/BF01072510

[22] Tabet, N. (2006) Acetylcholinesterase Inhibitors for Alzheimer's Disease: Anti-Inflammatories in Acetylcholine Clothing! *Age Ageing*, **35**, 336-338. http://dx.doi.org/10.1093/ageing/afl027

[23] Reale, M., Iarlori, C., Gambi, F., Lucci, I., Salvatore, M. and Gambi, D. (2005) Acetylcholinesterase Inhibitors Effects on Oncostatin-M, Interleukin-1β and Interleukin-6 Release from Lymphocytes of Alzheimer's Disease Patients. *Experimental Gerontology*, **40**, 165-171. http://dx.doi.org/10.1016/j.exger.2004.12.003

[24] Jiang, W., Desjardins, P. and Butterworth, R.F. (2009) Cerebral Inflammation Contributes to Encephalopathy and Brain Edema in Acute Liver Failure: Protective Effect of Minocycline. *Journal of Neurochemistry*, **109**, 485-493. http://dx.doi.org/10.1111/j.1471-4159.2009.05981.x

[25] Shawcross, D.L., Wright, G., Damink, S.W.M.O. and Jalan, R. (2007) Role of Ammonia and Inflammation in Minimal Hepatic Encephalopathy. *Metabolic Brain Disease*, **22**, 125-138. http://dx.doi.org/10.1007/s11011-006-9042-1

[26] Shawcross, D.L., Davies, N.A., Williams, R. and Jalan, R. (2004) Systemic Inflammatory Response Exacerbates the Neuropsychological Effects of Induced Hyperammonemia in Cirrhosis. *Journal of Hepatology*, **40**, 247-254. http://dx.doi.org/10.1016/j.jhep.2003.10.016

[27] Winblad, B., Cummings, J., Andreasen, N., Grossberg, G., Onofrj, M., Sadowsky, C., *et al.* (2007) A Six-Month Double-Blind, Randomized, Placebo-Controlled Study of a Transdermal Patch in Alzheimer's Disease—Rivastigmine Patch *versus* Capsule. *International Journal of Geriatric Psychiatry*, **22**, 456-467. http://dx.doi.org/10.1002/gps.1788

[28] (2006) Exelon. Novartis Pharmaceuticals Corporation, East Hanover.

[29] (2009) Exelon Patch. Novartis Pharmaceuticals Corporation, East Hanover.

[30] Winblad, B., Kawata, A.K., Beusterien, K.M., Thomas, S.K., Wimo, A., Lane, R., *et al.* (2007) Caregiver Preference for Rivastigmine Patch Relative to Capsules for Treatment of probable Alzheimer's Disease. *International Journal of Geriatric Psychiatry*, **22**, 485-491. http://dx.doi.org/10.1002/gps.1806

[31] Elgouhari, H.M. and O'Shea, R. (2009) What Is the Utility of Measuring the Serum Ammonia Level in Patients with Altered Mental Status? *Cleveland Clinic Journal of Medicine*, **76**, 252-254. http://dx.doi.org/10.3949/ccjm.76a.08072

[32] Howanitz, J.H., Howanitz, P.J., Skrodzki, C.A. and Iwanski, J.A. (1984) Influences of Specimen Processing and Sto-

rage Conditions on Results for Plasma Ammonia. *Clinical Chemistry*, **30**, 906-908.

[33] Barsotti, R.J. (2001) Measurement of Ammonia in Blood. *The Journal of Pediatrics*, **138**, S11-S20.
      http://dx.doi.org/10.1067/mpd.2001.111832

Role of Diabetes Mellitus in Liver Cirrhosis in the Anhui Region of China

Lifen Hu[1,2], Xihai Xu[1], Zhongsong Zhou[1], Guoshen Chen[1], Huafa Ying[1], Ying Ye[1*], Jiabin Li[1,3*]

[1]Department of Infectious Diseases, the First Affiliated Hospital of Anhui Medical University, Hefei, China
[2]Department of Center Laboratory, the First Hospital of Anhui Medical University, Hefei, China
[3]Department of Infectious Diseases, the Affiliated Chaohu Hospital of Anhui Medical University, Chaohu, China
Email: *13856980361@139.com, *lijiabin948@vip.sohu.com

Abstract

Aims: The aim of this study was to investigate the frequency of diabetes mellitus (DM) among cirrhotic inpatients with different etiologies and the impact of DM on the prognosis of these patients. Methods: A retrospective study of the association between DM and cirrhosis was performed on 672 cirrhotic inpatients at the First Affiliated Hospital of Anhui Medical University from January 1, 2012 to March 1, 2013. Data were assessed using SPSS 20.0. Results: The DM prevalences involving different etiologies were 45.08%, 40.43%, 42.85%, 41.67%, and 25.56% among 672 patients with alcoholic cirrhosis, cryptogenic cirrhosis, schistosomiasis, chronic hepatitis C, and chronic hepatitis B, respectively. Multivariate analysis indicated that relative to non-diabetic patients, patients with DM were older (OR 2.83, 95% CI 2.78 - 2.87, $p < 0.001$), had higher white blood cell levels (OR 3.01, 95% CI 2.83 - 3.22, $p = 0.001$) and were more frequently Child-Pugh class C (OR 1.28, 95% CI 1.14 - 1.55, $p < 0.001$); these differences were also statistically significant in univariate analysis. Univariate analysis suggested that the presence of DM was associated with a higher international normalized ratio ($p = 0.013$), a higher incidence of hepatic encephalopathy ($p = 0.014$), and a more frequent incidence of spontaneous bacterial peritonitis ($p < 0.001$). Conclusion: DM is extremely common among cirrhotic patients and is particularly prevalent among alcoholic cirrhosis patients in the examined region of China. The presence of DM in cirrhotic patients was strongly associated with an elevated Child-Pugh score and an accelerated progression of cirrhosis.

Keywords

Cirrhosis, Diabetes Mellitus, Child-Pugh Score, White Blood Cell, Encephalopathy, Spontaneous Bacterial Peritonitis

Subject Areas: Internal Medicine

---

*Corresponding authors.

1. Introduction

Liver cirrhosis which ultimately results in liver failure and portal hypertension is a common chronic liver disease and a major cause of death throughout the world, particularly in China. Because the liver plays an important role in carbohydrate metabolism and controls blood glucose levels through glycogenogenesis and glycogenolysis [1] [2], patients with cirrhosis may have alterations in glucose homeostasis; these patients often exhibit impaired glucose tolerance (IGT) or eventually develop diabetes mellitus (DM) [3]. In addition, increasing numbers of cirrhotic patients are complicated with DM; at present, DM is a common comorbidity among patients with liver cirrhosis. In cirrhotic patients, DM is reportedly associated with an increased risk for the continued development of cirrhosis [4]. Both liver cirrhosis and DM represent difficult challenges for clinicians and patients with respect to therapeutic and prognostic implications, particularly in China, which has large numbers of patients with both cirrhosis and DM. Several studies have reported a prevalence of DM among cirrhosis patients of 20% - 40% [3] [5]-[7], and this prevalence is reportedly higher among patients with hepatitis C virus (HCV)-related cirrhosis than among patients with cirrhosis caused by other factors [8]-[11].

However, little information is available regarding the impact of cirrhosis with different etiologies on the occurrence of DM among Chinese patients and the relationship between DM and complications of decompensated cirrhosis. The aims of this study were to explore the relationship between different etiologies of cirrhosis and the incidence of IGT or DM and to evaluate whether DM can accelerate the progression of liver disease among cirrhotic patients.

2. Materials and Methods

2.1. Patients and Methods

From January 1, 2012 to March 1, 2013, 672 patients with liver cirrhosis with a single etiology who visited the Infection Department and the Digestive System Department of the First Affiliated Hospital of Anhui Medical University were involved in this study. The study was conducted in accordance with the guidelines of the Declaration of Helsinki and Good Clinical Practice and was approved locally by the Ethics Committee of the First Affiliated Hospital of Anhui Medical University. Liver function tests were performed to get the values of total bilirubin, alanine aminotransferases (ALT), aspartate aminotransferase (AST), serum albumin, prealbumin (PA) and the international normalized ratio (INR) in patients, other Laboratory tests including viral markers (hepatitis B surface antigen and anti-HCV) by Polymerase chain reaction PCR, white blood cells (WBC) by routine blood test, and fasting blood glucose by glucose oxidase method were performed; in addition, an oral glucose tolerance test (OGTT) was performed as needed. Autoimmune markers, liver ultrasonography images and liver CT scans were also examined. Detailed clinical characteristics were recorded, including each patient's sex; age; body mass index (BMI); alcohol consumption; etiology of cirrhosis; and history of ascites, hepatic encephalopathy (HE) which were diagnosed according to West Haven criteria [12], spontaneous bacterial peritonitis (SBP) which were diagnosed by EASL clinical practice guidelines [13] and DM. The severity of liver cirrhosis was classified using Child-Pugh scores; in particular, cases with Child-Pugh scores of 5 - 6, 7 - 9 and 10 - 15 were categorized as class A, class B, and class C, respectively.

Diagnoses of cirrhosis were established using a combination of clinical features, laboratory tests, radiological and histological examinations, and liver biopsy findings (if available). The presence of DM was established based on a history of DM, the use of an oral hypoglycemic medication or insulin, fasting blood glucose levels > 126 mg/dL for two consecutive days, a blood glucose level > 200 mg/dL 2 h after an OGTT, or a random blood glucose level > 200 mg/dL. Patients were diagnosed with IGT if they exhibited a fasting glucose concentration ≥ 110 mg/dL and < 126 mg/dL or a glucose concentration ≥ 140 mg/dL and < 200 mg/dL at 2 h after an OGTT.

2.2. Statistical Analysis

Data were analyzed using SPSS, version 20.0 (SPSS Inc., USA). Quantitative variables were expressed as means ± standard deviation (SD) or as medians (range). Univariate analysis was performed by utilizing the independent Student's t-test or the Mann-Whitney test for between-group comparisons of continuous variables and chi-squared tests for between-group comparisons of qualitative data. Multiple logistic regression analysis was also used for between-group comparisons of differences in clinical and biochemical variables. A two-sided *p*-value < 0.05 was considered to be statistically significant.

3. Results

3.1. Prevalence of IGT and DM in Cirrhotic Patients

Among the 672 cirrhotic patients, 88 patients (13.1%) were diagnosed with IGT, and 210 patients (31.25%) were diagnosed with DM; thus, in total, 298 patients (44.35%) were diagnosed with IGT or DM. The prevelance of DM and IGT in cirrhotic patients with different etiologies were indicated in **Table 1**.

3.2. Clinical Characteristics of Patients with and without IGT or DM

Comparison of clinical and biochemical characteristics between cirrhotic patients with IGT or DM and without among 672 patients by univariate analysis are presented in **Table 2** and **Table 3**. The results of *t*-tests indicated that the DM group was significantly older than the patients without IGT or DM (55.95 ± 12.36 vs. 51.09 ± 12.36 years, $p < 0.001$) and the IGT group was also older than the patients without IGT or DM (52.88 ± 12.74 vs. 51.09 ± 12.36 years, $p = 0.226$). In comparisons of patients with DM and patients without DM or IGT, patients with DM not only had significantly higher WBC levels (5.76 ± 2.96 vs. 4.46 ± 3, $p < 0.001$) and INR levels (1.65 ± 0.62 vs. 1.50 ± 0.66, $p = 0.013$) as well as significantly higher incidences of HE (17.14% vs. 10.43%, $p = 0.014$) and SBP (25.23% vs. 15.24%, $p < 0.001$) but also were more frequently Child-Pugh class C (28.57% vs. 16.58%, $p < 0.001$). These comparisons revealed no associations between the presence of DM and sex, BMI, total bilirubin, ALT, AST, albumin, INR, or PA. Comparisons of patients with IGT and patients without DM or IGT revealed that patients with IGT had significantly higher WBC levels (5.68 ± 1.98 vs. 4.46 ± 3, $p < 0.001$) and were more frequently Child-Pugh class C (32.95% vs. 16.58%, $p < 0.001$).

As indicated in **Table 2** and **Table 3**, multiple logistic regression analysis that compared patients with DM and patients without IGT or DM indicated that older age (OR 2.83, 95% CI 2.78 - 2.87, $p < 0.001$), elevated WBC levels (OR 3.01, 95% CI 2.83 - 3.22, $p = 0.001$) and a greater frequency of Child-Pugh class C (OR 1.28, 95% CI 1.14 - 1.55, $p < 0.001$) remained significantly associated with DM; in contrast, the presence of DM was no longer significantly associated with INR level or the incidence of HE or SBP. In multiple logistic regression analysis, WBC levels (OR 3, 95% CI 2.79 - 3.26, $p = 0.008$) and the frequency of Child-Pugh class C (OR 1.31, 95% CI 1.13 - 1.78, $p = 0.008$) remained significantly associated with IGT.

4. Discussion

It has long been known that cirrhosis, glucose abnormalities, and DM are associated; several contributing factors to these associations, such as insulin resistance, hyperinsulinemia, and reduced glucose uptake by cirrhotic livers, have been identified [14] [15]. In addition, Perry recently reported that white adipose tissue-derived hepatic acetyl CoA is linked to inflammation-induced hepatic insulin resistance, which is associated with obesity and type 2 diabetes [16].

Various etiologies, such as alcoholic hepatitis, CHC, CHB and nonalcoholic fatty liver disease, are common causes of cirrhosis, and the different etiologies of liver cirrhosis may cause differences in the incidence rates of DM among patients with cirrhosis. Several studies have reported a high prevalence of DM among patients infected with HCV or HBV [3] [11] [17], and observations have indicated that diabetes occurs more frequently among HCV-infected patients than among HBV-infected patients. The results of this study also revealed that the

Table 1. Prevalences of diabetes mellitus and impaired glucose tolerance among cirrhotic patients with different etiologies.

Group	CHB	ALD*	CHC	Autoimmune Diseases	Schistosomiasis	Cryptogenic Cirrhosis†
N	403	122	24	62	14	47
DM, N (%)	103 (25.56)	55 (45.08)	10 (41.67)	19 (30.65)	6 (42.85)	19 (40.43)
IGT or DM, N (%)	155 (38.46)	75 (61.48)	10 (41.67)	26 (41.94)	6 (42.85)	27 (57.45)

DM: diabetes mellitus, IGT: impaired glucose tolerance, CHB: chronic hepatitis B virus, ALD: alcoholic liver disease, CHC: chronic hepatitis C virus. DM incidence comparison: *$p < 0.001$ in ALD vs. CHB, $p = 0.059$ in ALD vs. autoimmune-disease group, †$p = 0.03$ in CHB vs. the cryptogenic group. IGT or DM incidence comparison: *$p < 0.001$ in ALD vs. CHB, $p = 0.012$ in ALD vs. autoimmune-disease group, †$p = 0.012$ in CHB vs. the cryptogenic group.

Table 2. Comparison of clinical and biochemical variables among cirrhotic patients with DM and cirrhotic patients without IGT or DM.

Index		Without DM or IGT	DM	Univariate Analysis	Multivariate Analysis		
				p	OR	95%CI	p
N		374	210				
Age (years)		51.09 ± 12.36	55.95 ± 12.36	0.000	2.83	2.78 - 2.87	0.000
Sex: male, N (%)		241 (64.44)	142 (67.62)	0.401	2.1	1.63 - 3.05	0.149
BMI (kg/m^2)		22.92 ± 2.83	22.98 ± 3.02	0.881	2.71	2.55 - 2.89	0.906
Total bilirubin (umol/l)		31.9 (4.2 - 423)	28.4 (7.7 - 461)	0.755	2.71	2.70 - 2.72	0.110
ALT (units/l)		48 (15 - 1516)	46 (18 - 1744)	0.389	2.72	2.72 - 2.73	0.164
AST (units/l)		68 (28 - 1495)	57 (19 - 1586)	0.309	2.71	2.71 - 2.74	0.127
WBC (*10^9)		4.46 ± 3	5.76 ± 2.96	0.000	3.01	2.83 - 3.22	0.001
Albumin (g/l)		31.25 ± 7.51	30.35 ± 7.12	0.085	2.72	2.66 - 2.73	0.928
PA (mg/l)		85 (10 - 335)	94 (12 - 336)	0.134	2.72	2.71 - 2.73	0.589
INR		1.50 ± 0.66	1.65 ± 0.62	0.013	3.36	2.43 - 5.23	0.224
HE, N (%)		39 (10.43)	36 (17.14)	0.014	3.44	2.07 - 8.96	0.407
SBP, N (%)		57 (15.24)	59 (25.23)	0.000	3.71	2.25 - 8.37	0.270
Child-Pugh score	B, N (%)	98 (26.2)	63 (30.00)	0.325	1.68	1.36 - 2.42	0.016
	C, N (%)	62 (16.58)	60 (28.57)	0.000	1.28	1.14 - 1.55	0.000

Results are expressed as mean ± standard deviation (SD), frequencies (%) or medians (range). ALT: alanine aminotransferases, AST: aspartate aminotransferase, BMI: body mass index, DM: diabetes mellitus, IGT: impaired glucose tolerance, PA: prealbumin, INR: international normalized ratio, HE: hepatic encephalopathy, SBP: spontaneous bacterial peritonitis.

Table 3. Comparison of clinical and biochemical variables among cirrhotic patients with IGT and cirrhotic patients without IGT or DM.

Index		Without DM or IGT	IGT	Univariate Analysis	Multivariate Analysis		
				p	OR	95%CI	p
N		374	88				
Age (years)		51.09 ± 12.36	52.88 ± 12.74	0.226	2.76	2.71 - 2.82	0.116
Sex: male, N (%)		241 (64.44)	142 (67.62)	0.266	2.18	1.57 - 3.85	0.371
BMI (kg/m^2)		22.92 ± 2.83	22.38 ± 3.37	0.124	2.54	2.35 - 2.77	0.118
Total bilirubin (umol/l)		31.9 (4.2 - 423)	40 (5.1 - 434)	0.369	2.71	2.71 - 2.72	0.252
ALT (units/l)		48 (15 - 1516)	46 (19 - 1610)	0.6	2.72	2.72 - 2.73	0.372
AST (units/l)		68 (28 - 1495)	67 (23 - 1423)	0.664	2.72	2.71 - 2.72	0.461
WBC (*10^9)		4.46 ± 3	5.68 ± 1.98	0.004	3	2.79 - 3.26	0.008
Albumin (g/l)		31.25 ± 7.51	30.72 ± 8.26	0.349	2.72	2.66 - 2.79	0.905
PA (mg/l)		85 (10 - 335)	86.5 (13 - 367)	0.118	2.72	2.71 - 2.73	0.868
INR		1.50 ± 0.66	1.56 ± 0.52	0.386	2.33	1.68 - 3.95	0.498
HE, N (%)		39 (10.43)	12 (13.64)	0.389	2.82	1.63 - 9.10	0.924
SBP, N (%)		57 (15.24)	21 (23.86)	0.054	2.85	1.70 - 7.81	0.896
Child-Pugh score	B, N (%)	98 (26.2)	25 (28.41)	0.674	1.72	1.31 - 2.92	0.078
	C, N (%)	62 (16.58)	29 (32.95)	0.002	1.31	1.13 - 1.78	0.001

Results are expressed as mean ± standard deviation (SD), frequencies (%) or medians (range). ALT: alanine aminotransferases, AST: aspartate aminotransferase, BMI: body mass index, DM: diabetes mellitus, IGT: impaired glucose tolerance, PA: prealbumin, INR: international normalized ratio, HE: hepatic encephalopathy, SBP: spontaneous bacterial peritonitis.

incidence of DM was higher among patients with HCV-induced cirrhosis than among patients with HBV-induced cirrhosis. One concerning finding was the presence of a strong association between alcohol-related cirrhosis and DM. Among the examined etiologies of cirrhosis, alcohol-related cirrhosis was associated with the highest incidence of DM or IGT; this incidence was also higher than the corresponding incidences reported in other studies of alcohol-related cirrhosis. Excessive alcohol intake can lead to obesity, which is linked to inflammation-induced hepatic insulin resistance and to decreased insulin-mediated glucose uptake under acute conditions. Chronic alcoholism may damage pancreatic islet cells, which may lead to the development of DM or IGT.

For cirrhotic patients, the coexistence of DM with cirrhosis might increase the risk of adverse progression and cause severe complications. According to prior studies, increases in orocecal transit time and intestinal bacterial growth due to diabetes may increase blood ammonia levels, which could lead to a high prevalence of HE among cirrhotic patients with diabetes [18] [19]. However, in this study, although univariate analysis indicated that diabetes was significantly associated with the incidence of HE, the association between DM and the presence of HE did not reach statistical significance in multiple logistic regression analysis after adjusting for several other factors, such as age, BMI, WBC, INR, SBP, total bilirubin, ALT, AST, albumin, and PA. SBP is another important complication among cirrhotic patients that requires hospitalization and treatment with antimicrobial agents; this condition can lead to poor quality of life and shortened survival times [18] [20] [21]. A prior study reported that among patients with cirrhosis, the presence of DM in patients was associated with an increased risk of SBP [21]. In the present study, univariate analysis indicated that the prevalence of SBP was significantly higher among patients with DM than among patients without DM (25.23% vs. 15.43%, $p < 0.001$); however, this difference was not statistically significant in multivariate analysis after adjusting for several other factors.

In the present study, in both univariate and multivariate analysis, older age, higher WBC levels and elevated Child-Pugh scores were strongly and significantly associated with DM in cirrhotic patients. Higher WBC levels may imply an increased susceptibility to (bacterial) infections among cirrhotic patients with DM relative to cirrhotic patients without DM; this suggestion is consistent with previous observations that DM is associated with a greater incidence of not only SBP but also respiratory and urinary tract infections in patients with cirrhosis [22].

The Child-Pugh score is important for assessing the prognosis of cirrhosis. The percentage of patients who were Child-Pugh class C was significantly higher in both the DM group and the IGT group relative to non-diabetic cirrhosis patients; this finding may indicate that diabetes appears to result in the accelerating deterioration of liver disease and more severe liver function abnormalities in cases of decompensation cirrhosis. The data of this study revealed that in cirrhotic patients, diabetes does not significantly affect various biochemical parameters, including BMI, total bilirubin, ALT, AST, albumin, and PA levels. These results are similar to the findings of previous studies [23].

5. Conclusion

In conclusion, DM is extremely common among patients with liver cirrhosis. One noteworthy finding is that in the examined Chinese region, DM was particularly prevalent among patients with alcoholic cirrhosis. In addition, for cirrhotic patients, the presence of DM increased the risk of elevated Child-Pugh scores and accelerated the progression of cirrhosis. Therefore, clinicians should devote a great deal of attention to the incidence of diabetes in liver cirrhosis, particularly in cases involving alcoholic cirrhosis.

Acknowledgments

This study was supported by the Natural Science Foundation of China (No. 81172737 and No. 81101313).

References

[1]   Barthel, A. and Schmoll, D. (2003) Novel Concepts in Insulin Regulation of Hepatic Gluconeogenesis. *American Journal of Physiology - Endocrinology and Metabolism*, **285**, 685-692. http://dx.doi.org/10.1152/ajpendo.00253.2003

[2]   Picardi, A., D'Avola, D., Gentilucci, U.V., *et al.* (2006) Diabetes in Chronic Liver Disease: From Old Concepts to New Evidence. *Diabetes/Metabolism Research and Reviews*, **22**, 274-283. http://dx.doi.org/10.1002/dmrr.636

[3]   Hickman, I.J. and Macdonald, G.A. (2007) Impact of Diabetes on the Severity of Liver Disease. *The American Journal of Medicine*, **120**, 829-834. http://dx.doi.org/10.1016/j.amjmed.2007.03.025

[4]   Quintana, J.O., Garcia-Compean, D., Gonzdlez, J.A., Pérez, J.Z., González, F.J., *et al.* (2011) The Impact of Diabetes Mellitus in Mortality of Patients with Compensated Liver Cirrhosis a Prospective Study. *Annals of Hepatology*, **10**, 56-62.

[5]   Zein, N.N., Abdulkarim, A.S., Wiesner, R.H., Egan, K.S. and Persing, D.H. (2000) Prevalence of Diabetes Mellitus in Patients with End-Stage Liver Cirrhosis Due to Hepatitis C, Alcohol, or Cholestatic Disease. *Journal of Hepatology*, **32**, 209-217. http://dx.doi.org/10.1016/S0168-8278(00)80065-3

[6]   Holstein, A., Hinze, S., Thiessen, E., Plaschke, A. and Egberts, E.H. (2002) Clinical Implications of Hepatogenous Diabetes in Liver Cirrhosis. *Journal of Gastroenterology and Hepatology*, **17**, 677-681. http://dx.doi.org/10.1046/j.1440-1746.2002.02755.x

[7]   Wlazlo, N., Beijers, H.J., Schoon, E.J., Sauerwein, H.P. and Stehouwer, C.D. and Bravenboer, B. (2010) High Prevalence of Diabetes Mellitus in Patients with Liver Cirrhosis. *Diabetic Medicine*, **27**, 1308-1311. http://dx.doi.org/10.1111/j.1464-5491.2010.03093.x

[8]   Huang, J.F., Yu, M.L., Dai, C.Y. and Chuang, W.L. (2013) Glucose Abnormalities in Hepatitis C Virus Infection. *Kaohsiung Journal of Medical Sciences*, **29**, 61-68. http://dx.doi.org/10.1016/j.kjms.2012.11.001

[9]   Sporea, I., Sirli, R., Hogea, C., Sink, A.A. and Serban, V. (2009) Diabetes Mellitus and Chronic HCV Infection. *Romanian Journal of Internal Medicine*, **47**, 141-147.

[10] Lonardo, A., Adinolfi, L.E., Petta, S., Craxì, A. and Loria, P. (2009) Hepatitis C and Diabetes: The Inevitable Coincidence? *Expert Review of Anti-Infective Therapy*, **7**, 293-308. http://dx.doi.org/10.1586/eri.09.3

[11] Elgouhari, H.M., Zein, C.O., Hanouneh, I., Feldstein, A.E. and Zein, N.N. (2009) Diabetes Mellitus Is Associated with Impaired Response to Antiviral Therapy in Chronic Hepatitis C Infection. *Digestive Diseases and Sciences*, **54**, 2699-2705. http://dx.doi.org/10.1007/s10620-008-0683-2

[12] Ferenci, P., Lockwood, A., Mullen, K., Tarter, R., Weissenborn, K. and Blei, A.T. (2002) Hepatic Encephalopathy: Definition, Nomenclature, Diagnosis, and Quantification: Final Report of the Working Party at the 11th World Congresses of Gastroenterology, Vienna, 1998. *Hepatology*, **35**, 716-721. http://dx.doi.org/10.1053/jhep.2002.31250

[13] Garcia-Compean, D., Jaquez-Quintana, J.O., Gonzalez-Gonzalez, J.A. and Maldonado-Garza, H. (2009) Liver Cirrhosis and Diabetes: Risk Factors, Pathophysiology, Clinical Implications and Management. *World Journal of Gastroenterology*, **15**, 280-288. http://dx.doi.org/10.3748/wjg.15.280

[14] European Association for the Study of the L (2010) EASL Clinical Practice Guidelines on the Management of Ascites, Spontaneous Bacterial Peritonitis, and Hepatorenal Syndrome in Cirrhosis. *Journal of Hepatology*, **53**, 397-417. http://dx.doi.org/10.1016/j.jhep.2010.05.004

[15] Nielsen, M.F., Caumo, A., Aagaard, N.K., Chandramouli, V., Schumann, W.C., Landau, B.R., *et al.* (2005) Contribution of Defects in Glucose Uptake to Carbohydrate Intolerance in Liver Cirrhosis: Assessment during Physiological Glucose and Insulin Concentrations. *American Journal of Physiology—Gastrointestinal and Liver Physiology*, **288**, 1135-1143. http://dx.doi.org/10.1152/ajpgi.00278.2004

[16] Perry, R.J., Camporez, J.P.G., Kursawe, R., *et al.* (2015) Hepatic Acetyl CoA Links Adipose Tissue Inflammation to Hepatic Insulin Resistance and Type 2 Diabetes. *Cell*, **160**, 1-14. http://dx.doi.org/10.1016/j.cell.2015.01.012

[17] Papatheodoridis, G.V., Chrysanthos, N., Savvas, S., *et al.* (2006) Diabetes Mellitus in Chronic Hepatitis B and C: Prevalence and Potential Association with the Extent of Liver Fibrosis. *Journal of Viral Hepatitis*, **13**, 303-310. http://dx.doi.org/10.1111/j.1365-2893.2005.00677.x

[18] Ampuero, J., Ranchal, I., Del Mar Díaz-Herrero, M., Del Campo, J.A., Bautista, J.D. and Romero-Gómez, M. (2013) Role of Diabetes Mellitus on Hepatic Encephalopathy. *Metabolic Brain Disease*, **28**, 277-279. http://dx.doi.org/10.1007/s11011-012-9354-2

[19] Butt, Z., Jadoon, N.A., Salaria, O.N., Mushtaq, K., Riaz, I.B., Shahzad, A., *et al.* (2013) Diabetes Mellitus and Decompensated Cirrhosis: Risk of Hepatic Encephalopathy in Different Age Groups. *Journal of Diabetes*, **5**, 449-455. http://dx.doi.org/10.1111/1753-0407.12067

[20] Poh, Z. and Chang, P.E. (2012) A Current Review of the Diagnostic and Treatment Strategies of Hepatic Encephalopathy. *International Journal of Hepatology*, **2012**, Article ID: 480309. http://dx.doi.org/10.1155/2012/480309

[21] Wlazlo, N., van Greevenbroek, M.M., Curvers, J., Schoon, E.J., Friederich, P., Twisk, J.W., *et al.* (2013) Diabetes Mellitus at the Time of Diagnosis of Cirrhosis Is Associated with Higher Incidence of Spontaneous Bacterial Peritonitis, but Not with Increased Mortality. *Clinical Science*, **125**, 341-348. http://dx.doi.org/10.1042/CS20120596

[22] Diaz, J., Monge, E., Roman, R. and Ulloa, V. (2008) Diabetes as a Risk Factor for Infections in Cirrhosis. *American Journal of Gastroenterology*, **103**, 248. http://dx.doi.org/10.1111/j.1572-0241.2007.01562_9.x

[23] Sigal, S.H., Stanca, C.M., Kontorinis, N., Bodian, C. and Ryan, E. (2006) Diabetes Mellitus Is Associated with Hepatic Encephalopathy in Patients with HCV Cirrhosis. *American Journal of Gastroenterology*, **101**, 1490-1496. http://dx.doi.org/10.1111/j.1572-0241.2006.00649.x

Predictors of Intra-Hospital Mortality in Patients with Cirrhosis

Iliass Charif[1,2*], Kaoutar Saada[1,2], Ihssane Mellouki[1,2], Mounia El Yousfi[1,2], Dafrallah Benajah[1,2], Mohamed El Abkari[1,2], Adil Ibrahimi[1,2], Nourdin Aqodad[1,2]

[1]Department of Gastroenterology and Hepatology, Hassan II University Hospital, Fez, Morocco
[2]Faculty of Medicine and Pharmacy of Fez, Sidi Mohammed Ben Abdellah University of Fez, Fez, Morocco
Email: *charifiliass82@hotmail.com

Abstract

Intra-hospital mortality in cirrhotic patients is variable depending on the studies reported in literature. Several studies have demonstrated independent predictors of mortality. The aim of this work is indeed to identify these predictors. Patients and Methods: We conducted a retrospective study of 1080 cirrhotic patients hospitalized in our department of gastroenterology and hepatology between January 2001 and August 2010. A descriptive study of the study population was performed, and a univariate analysis looking for an association between intra-hospital mortality, and clinical, biological, etiological and socio-demographic characteristics of our patients. Results: The average age of our patients was 54 years, with an equal number of men and women. 41.1% of patients had cirrhosis secondary to hepatitis C and 18.5% had cirrhosis secondary to hepatitis B. 26.1% of our patients were CHILD C. Intra-hospital mortality was 8.7% (97 deaths) with a mean of 23.4 ± 35.8 months. Univariate analysis showed that the intra-hospital mortality was significantly associated with higher age ($p = 0.049$) as well as the reasons for admissions like hepatic encephalopathy, and hematemesis ($p < 0.0001$), melena, jaundice and ascites ($p = 0.001$). Among the biological parameters analyzed in univariate analysis, significant associations with mortality were objectified for high white blood cell count ($p = 0.035$), and high serum bilirubin and creatinine ($p < 0.0001$); low rate of prothrombin time (PT) ($p < 0.0001$), of albumin ($p = 0.0001$) and of serum sodium ($p < 0.0001$). Among the complications analyzed, significant associations with mortality were objectified for jaundice, ascites ($p = 0.001$), hemorrhagic decompensation, hepatic encephalopathy, and spontaneous bacterial peritonitis ($p < 0.001$). Univariate analysis of the etiology of cirrhosis objectified significant associations for cirrhosis secondary to hepatitis B ($p = 0.001$) and hepatitis C ($p = 0.022$). Multivariate analysis objectified four independent predictors of mortality: hepatic encephalopathy, infection (hyper leukocytosis $\geq 10,000/mm^3$), renal failure (serum creatinine ≥ 15 mg/l) and hyponatremia. Conclusion: In our series, we identified four independent predictors of intra-hospital mortality in cirrhotic patients: hepatic encephalopathy, infection, renal failure and hyponatremia.

---

*Corresponding author.

Keywords

Cirrhosis; Portal Hypertension; Intra-Hospital Mortality

1. Introduction

The search for risk factors that stratify cirrhotic patients into subgroups with different survival rates is of great prognostic value for the clinician. Many studies have focused on the search for predictors of mortality in cirrhotic patients, and their use to develop a reliable model of survival. In these studies, the study populations were cirrhotic patients [1]-[5], patients with alcoholic cirrhosis [6] [7], and cirrhotic patients after an episode of variceal bleeding [8]-[12]. The Child-Turcott score [1] and its subsequent modifications by Pugh [8] are old empirical methods used to assess the degree of liver failure in candidate patients for porto-systemic shunt. Although the statistical accuracy of the Child-Pugh score (CPS) was not assessed, it was long considered to be an adequate method to determine the degree of liver failure, and the probability of survival [13]-[15]. However, two of its elements are very subjective (ascites and encephalopathy) [5]. In some studies, the prognostic value of CPS has been described as incomplete. In addition, several other clinical and biological variables not included in the CPS were demonstrated to have prognostic significance [16]. In our study, the principal etiology of cirrhosis in our patients is hepatitis B and C. It seemed, therefore, interesting to investigate the factors involved in the short-term survival in these cirrhotic patients.

2. Patients and Methods

• Study population

We present a retrospective study of cirrhotic patients admitted in our department of hepatology and gastroenterology at the University Hospital of Fez, between January 2001 and January 2010. The diagnosis of cirrhosis was based on the combination of clinical, biological, endoscopic and ultra-sonographic criteria.

• Variables studied

All patients had received a biological assessment within 24 hours of admission. The variables studied were: age, sex, presence of ascites, encephalopathy, presence of hemorrhage, jaundice, hepatocellular carcinoma (HCC), platelet count, white blood cells, transaminases, prothrombin time, bilirubin, albumin, creatinine, serum sodium, spontaneous bacterial peritonitis, the etiology of cirrhosis and CPS.

• Statistical Methods

A descriptive study of our population was conducted, as well as univariate and multivariate analysis looking for an association between mortality and clinical, biological and socio-demographic characteristics of our patients. We used standard descriptive statistics to characterize the population studied: mean, standard deviation, median, range. Comparison of two independent variables with normal distribution was performed using the Student test (t) whereas comparisons of more than two means were based on analysis of variance (ANOVA). A multivariate analysis was performed using a technique of logistic regression stepwise including all variables and significance level was defined as a value less than 0.2 in univariate analysis. In all statistical tests, the risk of error α was set at 0.05. Data were analyzed using Epi Info™ 3.5.1 software.

3. Results

Between January 2001 and January 2010, 1080 patients were included in this study. The clinical characteristics of patients are shown in **Table 1**. The average age was 54 years, sex ratio M/F was 1.05 with 555 (51.3%) men and 525 (48.6%) women. Ninety-six percent (N = 746) of patients had ascites, 46.5% (N = 503) had a hemorrhagic decompensation and 12.3% (N = 12.3) had hepatic encephalopathy. Forty-two percent (N = 463) of patients had cirrhosissecondary to hepatitis Cand 19.9% (N = 215) secondary to hepatitis B. Twenty-six percent of patients were Child C, and 7% (N = 75) had a HCC.

The hospital mortality was 8.7% (97 deaths).

In univariate analysis, factors associated with intra-hospital mortality in cirrhotic patients (**Tables 2** and **3**) were older age (p = 0.049), and reasons for admission: hepatic encephalopathy and hematemesis (p < 0.0001),

Table 1. General data of patients (N = 1080).

Mean age	54 ans
Men	555 (51.3%)
women	525 (48.6%)
Ascites	746 (96%)
Variceal bleeding	503 (46.5%)
Hepatic encephalopathy	133 (12.3%)
Cirrhosis secondary to hepatitis C B	 463 (42.8%) 215 (19.9%)
CHILD A B C	 19.2% 54.6% 26.1%
HCC	75 (6.9%)

Table 2. Comparison of sex and age between patients who died and survivors.

	Survivors (n = 984)	Deads (n = 96)	p
Sex ratio (M/F)	0.9	1.2	0.178
Age	52.2 ± 17.17	55.8 ± 14.39	0.049

Table 3. Comparison of reasons for hospitalization among patients who died and survivors.

	Survivors (n = 984)	Deads (n = 96)	p
Asthenia	17.1%	18.8%	0.678
Fever	12.3%	16.7%	0.219
Hematemesis	38.0%	59.4%	<0.0001
Melena	27.2%	42.7%	0.001
Jaundice	20.6%	35.4%	0.001
Ascites	53.7%	71.9%	0.001
Edema of the lower limbs	17.3%	21.9%	0.260
Hepatic encephalopathy	6.4%	21.9%	<0.0001

melena, jaundice and ascites (p = 0.001). Among the biological parameters (**Table 4**), significant associations with mortality were objectified for high levels of white blood cell count (p = 0.035), of serum bilirubin (p < 0.0001), and of serum creatinine (p < 0.0001); low rates of PT (p < 0.0001), of albumin (p = 0.000) and of serum sodium (p < 0.0001). Complications of cirrhosis were analyzed also in univariate analysis (**Table 5**), and significant associations with mortality were objectified for jaundice, ascites (p = 0.001), hemorrhagic decompensation, hepatic encephalopathy and spontaneous bacterial peritonitis (p < 0.0001). Concerning the etiology of cirrhosis (**Table 6**) significant associations with mortality were found forcirrhosis secondary to hepatitis B (p = 0.001) and hepatitis C (p = 0.022). The multivariate analysis (**Table 7**) has objectified four independent predictors of mortality, which are hepatic encephalopathy (OR = 14.9), infection (leukocytosis ≥ 10,000/mm^3) (OR = 6.9), renal failure (Creatinine ≥ 15 mg/l) (OR = 10.8) and hyponatremia (OR = 15.3).

4. Discussion

In our series, the intra-hospital mortality was 8.7%, a figure much lower than the literature data. However, the

Table 4. Comparison of biological parameters between patients who died and survivors.

	Survivors		Deads		p
	n	Average ± SD	N	Average ± SD	
Hb (g/l)	865	9.6 ± 3.05	87	9.1 ± 3.16	0.126
GB (10^3/mm^3)	918	6.2 ± 11.61	90	8.8 ± 5.43	0.35
Platelets (10^3/mm^3)	867	129.3 ± 99.92	85	133.8 ± 83.31	0.689
PT (%)	912	69 ± 20.93	81	56.1 ± 23.47	<0.0001
Albumine (g/l)	168	31.8 ± 7.55	17	24.5 ± 5.5	0.000
Bilirubin (mg/l)	541	26.6 ± 43.73	56	58.9 ± 70,79	<0.0001
Creatinine (mg/l)	740	9.3 ± 7.39	72	14.6 ± 11.91	<0.0001
Natremia (meq/l)	525	136.3 ± 5.74	58	131.5 ± 9.02	<0.0001

Table 5. Comparison of the occurrence of complications of cirrhosis between patients who died and survivors.

	Survivors (n = 984)	Deads (n = 96)	p
Variceal bleeding	44.8%	66.6%	<0.0001
Ascites	67.7%	84.3%	0.001
Hepatic encephalopathy	10.1%	36.4%	<0.0001
SBP	3.3%	12.5%	<0.0001
Jaundice	20.6%	35.4%	0.001
HCC	6.9%	9.3%	0.370

SBP: spontaneous bacterial peritonitis, HCC: Hepatocellular carcinoma.

Table 6. Comparison of etiology of cirrhosis between patients who died and survivors.

	Survivors (n = 984)	Deads (n = 96)	p
Cirrhosis secondary to hepatitis C	41.4%	58.7%	0.022
Cirrhosis secondary to hepatitis B	17.8%	42.2%	0.001
Alcoholic cirrhosis	2.1%	3.1%	0.530

Table 7. Multivariate analysis of in-hospital mortality in cirrhotic.

	OR (IC 95%)	p
Leucocytes		
<10,000/mm^3	1	
>10,000/mm^3	6.9 (1.39 - 34.07)	0.018
Creatinine		
<15 mg/l	1	
>15 mg/l	10.8 (1.9- 59.49)	0.006
Natremia		
>130 meq/l	1	
>130 meq/l	15.3 (3.13 - 74.93)	0.001
Hepatic encephalopathy		
Absent	1	
Present	14.9 (2.20 - 101.16)	0.006

medical structures of admission of cirrhotic patients with complications are variable, making it difficult to compare mortality figures observed. It may be a hepatology and gastroenterology service or intensive care units [17].

Several studies had studied prognostic factors in patients with cirrhosis, and elaborated a survival models, easy to use in routine practice.

According to the findings of the international consensus conference of Baveno IV [18], there is no adequate prognostic model of portal hypertension in patients with cirrhosis, and individual characteristics are insufficient to establish a prognosis. However, four clinical stages of portal hypertension, of increasing severity were identified.

In addition, the CPS, active bleeding at endoscopy, the portosystemic pressure gradient, infection, renal failure, the severity of the initial bleeding episode, vein thrombosis and HCC were identified as indicators of poor prognosis [19] [20].

The older age was significantly associated with mortality in our series, this was also reported in the series of Luca. A *et al.* [21], in contrast to several studies that have not objectified significant association between age and mortality [22] [23].

The circumstances of admission of cirrhotic patients with complications in intensive care units such as gastrointestinal bleeding, sepsis, impaired consciousness associated with encephalopathy, acute respiratory distress syndrome or acute renal failure have been associated with high mortality in these patients [24]. In our study, variceal bleeding, jaundice, ascites and hepatic encephalopathy were significantly associated with mortality, which was similar to the literature data [24].

Gastrointestinal bleeding due to rupture of esophageal varices is the second leading cause of cirrhosis mortality [25]. Twenty to 40% of deaths occur in the following year of the bleeding episode and the mortality rate is 15% to 30% at 6 weeks [26] and 50% in patients with Child C [27]. The severity of variceal bleeding cannot be dissociated from the severity of cirrhosis that is appreciated by the degree of liver failure. The bleeding episode may worsen liver function and the underlying liver disease. The death is not related to hemorrhage itself (initial blood loss), because there is rarely an acute and massive hypovolemia, but due to complications such as infections, hepatic encephalopathy, severity of liver failure and kidney failure. Survival after bleeding episode has improved in 40 years, going from 45% to 60%. This result was due to a rapid assessment of the gravity (identified prognostic factors), non-specific intensive care, and the early introduction of a specific medical treatment of variceal bleeding (vasoactive drogues) [28] [29]. In our series, hemorrhagic decompensation was significantly associated with mortality in univariate analysis.

Several parameters were associated with intra-hospital mortality in cirrhotic patients, the most important being the degree of liver failure, since the bilirubin and prothrombin time were associated with a greater risk of mortality in several studies [30]. In our study, a prothrombin time < 45%, serum bilirubin > 50 mg/l, and serum albumin < 28 g/l were significantly associated with mortality , confirming the data of literature.

The prospective study of Singh and al shows that liver failure assessed by CPS and the model for end stage liver disease (MELD) are the main prognostic factor for short-term survival and the occurrence of complications [31].

Other factors of severity seem to be independent prognostic factors; these factors are active bleeding, renal function, a massive transfusion of more than 5 red blood cells, an initial state of shock and tracheal intubation [32].

In our study, we found that kidney failure was a poor prognostic factor in patients with cirrhosis, and serum creatinine at admission was higher in cirrhotic patients who died than in survivors, findings that are consistent with previous studies [33].

According to some studies, hyponatremia is associated with higher mortality [22]. In our series, serum sodium ≤ 130 meq/l was a poor prognostic factor in these patients. We also identified leukocytosis > 10,000/mm^3 as a poor prognostic factor and this leukocytosis is usually incorporated in severe sepsis. In a meta-analysis [34], the occurrence of infection was associated with recurrent bleeding and a higher mortality.

The etiology of cirrhosis does not appear to be independent prognostic factors [33]. In our study, the viral origin of cirrhosis emerged as a predictor of mortality, which confirms its more virulent and aggressive character than other liver diseases.

Ascites occurs in 30% of patients with cirrhosis. The occurrence of ascites is a major event in the natural history of cirrhosis, and survival rates at 1 and 5 years, are respectively 50% and 20%. Ten percent of patients develop refractory ascites which is a reflection of severe liver failure. The survival of these patients is 40% to 60%

at 1 year and 20% - 40% at 2 years [35]. In our series, ascites was also associated with a high mortality in cirrhotic patients. This has been reported in many other studies [24] [36].

Spontaneous bacterial peritonitis is a frequent and serious complication. It occurs in 8% - 25% of patients hospitalized for ascites. The prognosis is much improved over the past 20 years. The rate of healing is about 80% and hospital mortality is below of 30% [35]. This complication was found in 4.1% (N = 45) of patients, and always appears as a predictor of mortality in our study.

Hepatocellular carcinoma has became the most common fatal complication of cirrhosis. In France, the mortality rate of HCC has been multiplied by 4. This is explained primarily by the conjunction of two factors: in one hand, a decrease in mortality due to other complications of cirrhosis (like spontaneous bacterial peritonitis and bleeding due to portal hypertension) and in the other hand an increase in frequency of HCC related to cirrhosis secondary to hepatitis C [37]. In our series, this complication was not identified as a predictor of intrahospital mortality in cirrhotic patients p = 0.370).

Unlike other complications, there are very few studies that have focused on specific prognosis of hepatic encephalopathy, although it is a common and potentially serious complication of cirrhosis. It is estimated that the survival of cirrhotic patients after a first episode of hepatic encephalopathy, with or without other complication, is about 40% at one year and 15% to 20% in three years [37]. The short-term prognosis of cirrhotic patients admitted to intensive care unit has not been specifically studied. The difficulty of conducting this type of study is that the hepatic encephalopathy is usually associated with one or more severe complications of cirrhosis, and the prognosis is then more directly related to intercurrent complications.

In our study, hepatic encephalopathy was significantly associated with intra-hospital mortality in univariate and multivariate analysis.

Several studies have highlighted in multivariate analysis independent predictors of mortality in patients with cirrhosis: kidney failure (serum creatinine), serum bilirubin, prothrombin time, leukocytosis, cirrhosis decompensation [33] [24]. In our series, we identified in multivariate analysis four independent predictors of intra hospital mortality in patients with cirrhosis, which are hepatic encephalopathy, infection, renal failure and hyponatremia. These result joined the data of literature [24].

5. Conclusion

In conclusion, advanced patient age, severity of liver failure, viral origin of cirrhosis, variceal bleeding, ascites, and spontaneous bacterial peritonitis, were predictors of intra-hospital mortality in cirrhotic patients. Adding to this, hepatic encephalopathy, infection, renal failure and hyponatremia were meanwhile, independent predictors of this mortality.

Conflicts of Interest

No conflicts of interest.

References

[1] Child, C.G. and Turcotte, J.G. (1964) Surgery and Portal Hypertension. In: Child, C.G., Ed., *The Liver and Portal Hypertension*, Saunders, Philadelphia, 50-64.

[2] Gines, P., Quintero, E., Arroyo, V., Teres, J., Bruguera, M., Rimola, A., Caballeri, J., Rodes, J. and Rozman, C. (1987) Compensated Cirrhosis: Natural History and Prognostic Factors. *Hepatology*, **7**, 122-128. http://dx.doi.org/10.1002/hep.1840070124

[3] Adler, M., Verset, D., Bouhdid, H., Bourgeois, B., Gulbis, B., Le Moine, O., Vanderstadt, J., Gelin, M. and Thiry, P. (1997) Prognostic Evaluation of Patients with Parenchymal Cirrhosis: Proposal of a New Simple Score. *Journal of Hepatology*, **26**, 642-649. http://dx.doi.org/10.1016/S0168-8278(97)80431-X

[4] Merkel, C., Bolognesi, M., Bellow, S., Bianco, S., Honisen, B., Lampe, H., Angeli, P. and Gatta, A. (1992) Aminopyrine Breath Test in the Prognostic Evaluation of Patients with Cirrhosis. *Gut*, **33**, 836-642. http://dx.doi.org/10.1136/gut.33.6.836

[5] Kamath, P.S., Wiesner, R.H., Malinchoc, M., Kremers, W., Therneau, T.M., Kosberg, C.L., D'Amico, G., Dickson, E.R. and Kim, W.R. (2001) A Model to Predict Survival in Patients with End-Stage Liver Disease. *Hepatology*, **33**, 464-470. http://dx.doi.org/10.1053/jhep.2001.22172

[6] Orrego, H., Israel, Y., Blake, J.E. and Medline, A. (1983) Assessment of Prognostic Factors in Alcoholic Liver Disease:

Toward a Global Quantitative Expression of Severity. *Hepatology*, **3**, 896-905.
http://dx.doi.org/10.1002/hep.1840030602

[7] Pingon, J.P., Poynard, T., Naveau, S., Marteau, P., Zourabien-Vilio, O. and Chaput, J.C. (1986) Analyse multidimensionnelle selon le modele de Cox dela survie de patients atteints de cirrhose alcoolique. *Gastroentérologie Clinique et Biologique*, **10**, 461-467.

[8] Pugh, R.N.H., Murray-Lyon, I.M., Dawson, J.L., Pietroni, M.C. and Williams, R. (1973) Transection of the Esophagus in Bleeding Oesophageal Varices. *British Journal of Surgery*, **60**, 648-652. http://dx.doi.org/10.1002/bjs.1800600817

[9] Teres, J., Baroni, R., Bordas, J.W., Visa, J., Pera, C. and Rodes, J. (1987) Randomized Trial of Portocaval Shunt Stapling Transection and Endoscopic Sclerotherapy in Uncontrolled Variceal Bleeding. *Journal of Hepatology*, **4**, 159-167. http://dx.doi.org/10.1016/S0168-8278(87)80075-2

[10] Rikkers, L.F., Burnett, D.A., Volentine, G.D., Buchi, K.N. and Cormier, R.A. (1987) Shunt Surgery versus Endoscopic Sclerotherapy for Longterm Treatment of Variceal Bleeding. Early Results of a Randomized Trial. *Annals of Surgery*, **206**, 201-271. http://dx.doi.org/10.1097/00000658-198709000-00004

[11] Sauerbruch, T., Ansari, H., Wotzka, R., Soehendra, N. and Kopcke, W. (1988) Prognostic Factors in Cirrhosis of the Liver, Variceal Bleeding and Sclerotherapy: Comparison of Prognosis Systems Obtained by Discriminant Analysis with the Child-Classification. *Deutsche Medizinische Wochenschrift*, **113**, 11-14. http://dx.doi.org/10.1055/s-2008-1067583

[12] LeMoine, O., Adler, M., Bourgeois, N., Delhaye, M., Deviere, J., Gelin, M., Vandermeeren, A., Van Gossum, A. and Vereerstraeten, P. (1992) Factors Related to Early Mortality in Cirrhotic Patients Bleeding from Varices and Treated by Urgent Sclerotherapy. *Gut*, **33**, 1381-1385. http://dx.doi.org/10.1136/gut.33.10.1381

[13] Infante-Rivard, C., Esnaola, S. and Villeneuve, J.P. (1987) Clinical and Statistical Validity of Conventional Prognostic Factors in Predicting Short-Term Survival among Cirrhotics. *Hepatolology*, **7**, 660-664. http://dx.doi.org/10.1002/hep.1840070408

[14] Ferro, D., Saliola, M., Quintarelli, C., Alessandri, C., Basili, S. and Violi, F. (1992) 1-Year Survey of Patients with Advanced Liver Cirrhosis: Prognostic Value of Clinical and Laboratory Indexes Identified by the Cox Regression Model. *Scandinavian Journal of Gastroenterology*, **27**, 852-856. http://dx.doi.org/10.3109/00365529209000153

[15] Hartmann, A.H., Bircher, J. and Creutzfeldt, W. (1989) Superiority of the Child-Pugh Classification to Quantitative Liver Function Tests for Assessing Prognosis of Liver Cirrhosis. *Scandinavian Journal of Gastroenterology*, **24**, 269-276. http://dx.doi.org/10.3109/00365528909093045

[16] Zauner, C., Schneeweiss, B., Schneider, B., Madl, C., Klos, H., Kranz, A., Ratheiser, K., Kramer, L. and Lenz, K. (2000) Short-Term Prognosis in Critically Ill Patients with Liver Cirrhosis: An Evaluation of a New Scoring System. *European Journal of Gastroenterology & Hepatology*, **12**, 517-522. http://dx.doi.org/10.1097/00042737-200012050-00007

[17] Robert, R. and Veinstein, A. (2003) Pronostic du Malade Atteint de Cirrhose en Reanimation. *Gastroentérologie Clinique et Biologique*, **27**, 877-881.

[18] De Franchis, R. (2005) Evolving Consensus in Portal Hypertension. Report of the Baveno IV Consensus Workshop on Methodology of Diagnosis and Therapy in Portal Hypertension. *Journal of Hepatology*, **43**, 167-176. http://dx.doi.org/10.1016/j.jhep.2005.05.009

[19] D'Amico, G. and De Franchis, R. (2003) Upper Digestive Bleeding in Cirrhosis. Post-Therapeutic Outcome and Prognostic Indicators. *Hepatology*, **38**, 599-612. http://dx.doi.org/10.1053/jhep.2003.50385

[20] Moitinho, E., Escorsell, A., Bandi, J.C., *et al.* (1999) Prognostic Value of Early Measurements of Portal Pressure in Acute Variceal Bleeding. *Gastroenterology*, **117**, 626-631. http://dx.doi.org/10.1016/S0016-5085(99)70455-5

[21] Luca, A., Angermayr, B., *et al.* (2007) An Integrated MELD Model including Serum Sodium and Age Improves the Prediction of Early Mortality in Patients with Cirrhosis. *Liver Transplantation*, **13**, 1174-1180. http://dx.doi.org/10.1002/lt.21197

[22] Londoño, M.C., Cárdenas, A., Guevara, M., *et al.* (2007) MELD Score and Serum Sodium in the Prediction of Survival of Patients with Cirrhosis Awaiting Liver Transplantation. *Gut*, **56**, 1283-1290. http://dx.doi.org/10.1136/gut.2006.102764

[23] Botta, F., Giannini, E., Romagnoli, P., Fasoli, A., *et al.* (2003) MELD Scoring System Is Useful for Predicting Prognosis in Patients with Liver Cirrhosis and Is Correlated with Residual Liver Function: A European Study. *Gut*, **52**, 134-139. http://dx.doi.org/10.1136/gut.52.1.134

[24] Mouelhi, L., Ben Hammouda, I., Salem, M., Moussa, A., *et al.* (2010) Mortalité hospitalière des patients cirrhotiques admis en milieu de soins intensifs: facteurs pronostiques et apport des scores de gravité. *Journal Africain d' Hépato-Gastroentérologie*, **4**, 17-21.

[25] Calès, P. and Pascal, J.P. (1988) Histoire naturelle des varices oesophagiennes au cours de la cirrhose (de la naissance à

la rupture). *Gastroentérologie Clinique et Biologique*, **12**, 245-254.

[26]   Carbonell, N., Pauwels, A., Serfaty, L., Fourdan, O., Lévy, V.G. and Poupon, R. (2004) Improved Survival after Variceal Bleeding in Patients with Cirrhosis over the Past Two Decades. *Hepatology*, **40**, 652-659. http://dx.doi.org/10.1002/hep.20339

[27]   De Dombal, F.T., Clarke, J.R., Clamp, S.E., Malizia, G., Kotwal, M.R. and Morgan, A.G. (1986) Prognostic Factors in Upper G.I. Bleeding. *Endoscopy*, **18**, 6-10. http://dx.doi.org/10.1055/s-2007-1018418

[28]   McCormick, P.A. and O'Keefe, C. (2001) Improving Prognosis Following a First Variceal Haemorrhage over Four Decades. *Gut*, **49**, 682-685. http://dx.doi.org/10.1136/gut.49.5.682

[29]   Levacher, S., Letoumelin, P., Pateron, D., Blaise, M., Lapandry, C. and Pourriat, J.L. (1995) Early Administration of Terlipressin plus Glyceryl Trinitrate to Control Active Upper Gastrointestinal Bleeding in Cirrhotic Patients. *Lancet*, **346**, 865-868. http://dx.doi.org/10.1016/S0140-6736(95)92708-5

[30]   Singh, N., Gayowski, T., Wagener, M.M. and Marino, I.R. (1998) Outcome of Patients with Cirrhosis Requiring Intensive Care Unit Support: Prospective Assessment of Predictors of Mortality. *Journal of Gastroenterology*, **33**, 73-79. http://dx.doi.org/10.1007/s005350050047

[31]   Chalasani, N., Kahi, C., Francois, F., Pinto, A., Marathe, A., Bini, E.J., *et al.* (2003) Improved Patient Survival after Acute Variceal Bleeding: A Multicenter, Cohort Study. *American Journal of Gastroenterology*, **98**, 653-659. http://dx.doi.org/10.1111/j.1572-0241.2003.07294.x

[32]   Attia, K.A., Ackoundou-N'guessan, K.C., N'dri-yoman, A.T., *et al.* (2008) Child-Pugh-Turcott versus Meld Score for Predicting Survival in a Retrospective Cohort of Black African Cirrhotic Patients. *World Journal of Gastroenterology*, **14**, 286-291. http://dx.doi.org/10.3748/wjg.14.286

[33]   Bernard, B., Grangé, J.D., Nguyen Khac, E., Amiot, X., Opolon, P. and Poynard, T. (1999) Antibiotic Prophylaxis for the Prevention of Bacterial Infections in Cirrhotic Patients with Gastrointestinal Bleeding: A Metaanalysis. *Hepatology*, **29**, 1655-1661. http://dx.doi.org/10.1002/hep.510290608

[34]   (2004) Conférence de consensus: Complications de l'hypertension portale. *Gastroenterology*, **28**, B324-B334.

[35]   Karoui, S., Hamzaoui, S., Sahli, F., *et al.* (2002) Mortalité au cours des cirrhoses: Prévalence, causes et facteurs prédictifs. *Tunisie Médicale*, **80**, 21-25.

[36]   Trinchet, J.C. (2002) Histoire naturelle de l'infection par le virus de l'hépatite C. *Gastroentérologie Clinique et Biologique*, **26**, B144-B153.

[37]   Bustamante, J., Rimola, A., Ventura, P.J., Navasa, M., Cirera, I., Reggiardo, V., *et al.* (1999) Prognostic Significance of Hepatic Encephalopathy in Patients with Cirrhosis. *Journal of Hepatology*, **30**, 890-895. http://dx.doi.org/10.1016/S0168-8278(99)80144-5

Glucose Metabolism Disorders in Cirrhosis: Frequency and Risk Factors in Tunisian Population

Rym Ennaifer[1,2]*, Myriam Cheikh[1,2], Rania Hefaiedh[1,2], Hayfa Romdhane[1,2], Houda Ben Nejma[1,2], Najet Bel Hadj[1,2]

[1]Department of Hepato-Gastro-Enterology, Mongi Slim Universitary Hospital, Tunis, Tunisia
[2]Faculty of Medicine, University of Tunis El Manar, Tunis, Tunisia
Email: *rym.ennaifer@yahoo.fr

Abstract

Background and aims: Alterations in carbohydrate metabolism are frequently observed in cirrhosis; to determine the frequency of diabetes mellitus and impaired glucose tolerance in Tunisian cirrhotic patients and identify risk factors. Patients and methods: Cross-sectional study; fasting plasma glucose levels were measured in consecutive patients with cirrhosis. Oral glucose tolerance test was performed if fasting plasma glucose level was normal. Glucose metabolism disorders were then classified as: impaired glucose tolerance and diabetes mellitus. Cirrhotics with glucose metabolism disorder were compared to those without. Results: Seventy-seven patients with cirrhosis were included: 68.8% were diagnosed as having glucose metabolism disorder; diabetes in 42.8% and impaired glucose tolerance in 26%. The tests were able to identify 60.4% of glucose metabolism disorders. Univariate analysis disclosed a higher proportion of female gender (p = 0.04) and more frequent familial history of diabetes mellitus (p = 0.005) in the group with glucose metabolism disorder. There were no statistically differences regarding age, etiology and severity of the cirrhosis, and dry body mass index. Multivariate analysis showed that familial history of diabetes was the only independent risk factor (OR = 5.1, p = 0.005). Conclusion: In our study, the frequency of glucose metabolism disorders was 68.8%. Oral glucose tolerance test allowed disclosing nearly half of them, pointing a high incidence of latent glucose metabolism disorders. In this way, it should be routinely evaluated in all patients with cirrhosis. Familial history of diabetes was the only independent risk factor, suggesting that other factors in addition to liver disease may play a role.

---

*Corresponding author.

Keywords

Cirrhosis, Diabetes, Impaired Glucose Tolerance

1. Introduction

Glucose metabolism disorders (GMD) are frequent in liver cirrhosis [1]. The liver plays a key role in blood glucose control, thus, in the presence of chronic liver disease, the metabolic homeostasis of glucose is impaired and results in glucose intolerance and diabetes mellitus (DM) type 2 [2]. There is a wide variability in the prevalence of DM and impaired glucose tolerance (IGT) according to the literature [3]. About 50% - 80% of cirrhotic patients have IGT and 30% - 40% develop DM [4]. Moreover, DM in cirrhosis may be subclinical, since fasting serum glucose may be normal. In these cases, it is necessary to perform an oral glucose tolerance test (OGTT) to detect an impairment of glucose metabolism [2]. Previous studies have shown that the DM increases the risk of complications of cirrhosis and reduces survival [5]. Risk factors for GMD in cirrhosis have been thoroughly investigated.

This study is the first one to determine the prevalence of DM and IGT in Tunisian cirrhotic patients and identify factors that might be potentially associated with GMD.

2. Patients and Methods

2.1. Patients

From August 2011 to July 2012, we consecutively included patients with cirrhosis admitted to the department of Gastroenterology and Hepatology of Mongi Slim University Hospital in Tunis. Inclusion criteria were: presence of cirrhosis diagnosed according to histological analysis (Fibrosis grade 4 of METAVIR) or a combination of clinical criteria (splenomegaly, ascites, collateral venous circulation), laboratory tests (decreased prothrombin and factor V levels, hypoalbuminemia, hypocholesterolemia) and imaging (portal venous dilatation, ascites, splenomegaly on ultrasound and endoscopic portal hypertension findings); age above 18 years. Exclusion criteria were: patients with chronic pancreatitis, chronic ingestion of corticosteroids, diabetes type 1, others specific types of diabetes and gestational diabetes. Patients with autoimmune hepatitis receiving corticosteroids were also excluded from this study.

2.2. Methods

Age, sex, familial history of DM were recorded. We could not measure insulin resistance because it was not available in our center. Complications of cirrhosis (variceal bleeding, encephalopathy, ascites, infections, hepatocellular carcinoma) were assessed. Child-Pugh and Meld scores were calculated. Dry body mass index (dry BMI) was calculated using the standard formula with dry weight in order to compare among patients with and without ascites and/or oedemas [6]. Patients receiving therapy with insulin or oral hypoglycemic medications were considered as diabetics. Other patients were tested for fasting plasma glucose (FPG) twice on different days after 12 hours overnight fast, and an OGTT was performed to patients with normal FPG (less than 5.6 mmol/l). The OGTT was performed as follows, according to the WHO criteria: patients fasted for at least 12 h, baseline glucose level was measured; they were given 75 g of oral glucose load, and after 2 h a second sample was taken (2 h PG).

According to the latest American Diabetes Association criteria (ADA) [7]:
- DM was diagnosed if FPG ≥ 126 mg/dl (7 mmol/l) or 2 h PG ≥ 200 mg/dl (11.1 mmol/l).
- IGT was diagnosed if 2 h PG was between 140 (7.8 mmol/l) and 199 mg/dl (11 mmol/l).

The study protocol conformed to the ethical guidelines of the 1975 Declaration of Helsinki as approval by the institution's review board.

2.3. Statistical Analysis

Analysis was performed using SPSS version 19. Results were expressed as mean +/− standard deviation. An univariate analysis searching for the factors possibly associated with GMD was firstly performed by comparison between cirrhotic with and without GMD using Student t test for continuous variables and the Chi-square or Fisher exact test for categorical variables. Then, a multivariate analysis based on a stepwise logistic regression model was used to assess the independent effect of variables found significant at the univariate analysis. A p-value of less than 0.05 was considered statistically significant.

3. Results

3.1. Patient Population

Seventy-seven patients (42 females, 35 males) with cirrhosis were included in the study. They had a mean age of 57.1 ± 13.8 years (range: 18 - 84). The main etiologies of liver disease were hepatitis C in 28 patients, hepatitis B in 15 patients and autoimmune in 14 patients. The median duration of liver cirrhosis was 36 months (range: 0 - 240 months). Regarding the severity of the liver disease, 49 patients were classified as Child-Pugh B and C and the mean Meld score was 14.5 ± 6 (range: 5 - 33). History of ascitic decompensation (76.6%), hepatic encephalopathy (37.7%) and bacterial infections (spontaneous peritoneal infection, urinary infection and bronchopulmonary infection) in 51.9% were the most frequent complications related to cirrhosis. Demographical and clinical characteristics are given in **Table 1**, **Table 2** and **Table 3**.

3.2. Frequency of GMD

Fifty-three (68.8%) patients were diagnosed as having GMD: DM in 33 patients (42.8%) and ITG in 20 patients (26%). The OGTT was able to identify 25 GMD (47.1%): 20 ITG and 5 DM. Among patients with DM, 12 (36.3%) were diagnosed at the same time or after the diagnosis of cirrhosis so they may be considered as having hepatogenous diabetes. **Figure 1** summarizes these findings.

3.3. Comparison between Patients with and without GMD

- Univariate analysis disclosed a higher proportion of female gender (p = 0.04) and more frequent familial history of DM (p = 0.005) in the group with GMD.

  There were no statistically differences regarding age, etiology (viral/non viral; cryptogenic/non-cryptogenic) of the cirrhosis, Child-Pugh and Meld scores, mean duration of liver disease, cirrhosis complications and dry BMI.

- Multivariate analysis showed that **family history of diabetes** was the only independent risk factor for GMD in cirrhosis: **OR = 5.1, CI = [1.6 - 16], p = 0.005**.

  Table 4 and **Table 5** summarize these findings.

Table 1. Demographical and clinical characteristics of patients.

Characteristics	N = 77
Age (y)	57.1 ± 13.8 (18 − 84)
Sex (females)	42 (54.5%)
Child-Pugh score	A = 28 (36.3%), B = 27 (35.1%), C = 22 (28.6%)
Meld score	14.5 ± 6 (5 − 33)

Table 2. Cirrhosis etiology.

Cirrhosis etiology	N = 77
- HCV	28 (36.3%)
- HBV	15 (19.5%)
- Auto-immune	14 (18.2%)
- Cryptogenic	11 (14.3%)
- Alcohol	3 (3.9%)
- NASH	1 (1.3%)
- Others	5 (6.4%)

HCV: hepatitis C virus; HBV: hepatitis B virus; NASH: non-alcoholic steato-hepatitis.

Table 3. Cirrhosis complications.

Cirrhosis complications	
- Ascitic decompensation	59 (76.6%)
- Oesophageal varices	59 (76.6%)
- Bleeding oesophageal varices	27 (35.1%)
- Refractory ascites	14 (18%)
- Hepatic encephalopathy	29 (37.7%)
- Hepatocellular carcinoma	17 (22.1%)
- Bacterial infections	40 (51.9%)

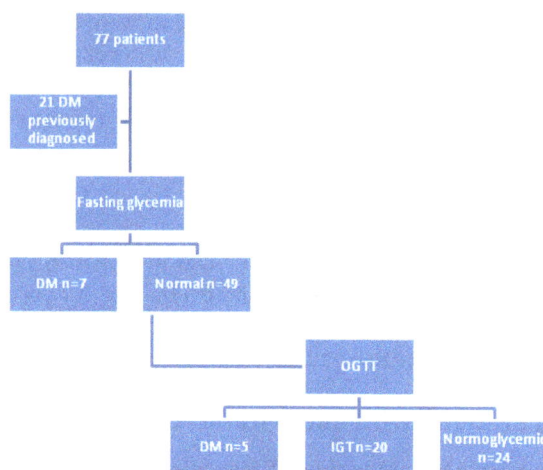

Figure 1. Results according to fasting glycemia and OGTT in the studied patients (DM: diabetes mellitus, OGTT: oral glucose tolerance test, IGT: impaired glucose tolerance).

Table 4. Clinical and epidemiological characteristics of cirrhotic patients with and without GMD.

	Cirrhosis with GMD (n = 53)	Cirrhosis without GMD (n = 24)	P	OR [CI 95%]
Age	56	58	0.58	
Female gender	33 (62%)	9 (37.5%)	0.04	1.8 [0.5 - 6]
Dry BMI (kg/m^2)	25	23	0.19	
Familial history of diabetes	30 (56.6%)	6 (25%)	0.005	3.84 [1.2 - 13]
Child Pugh score B or C	35 (66%)	(58.4%)	0.51	
Mean Meld score	14	15	0.45	
Etiology of cirrhosis				
- Viral/non viral	30/23	13/11	NS	
- Cryptogenic/non cryptogenic	8/45	4/20	NS	

BMI: body mass index, GMD: glucose metabolism disorders.

Table 5. Cirrhosis complications in patients with and without GMD.

Cirrhosis complications	Patients with GMD	Patients without GMD	P
- Ascitic decompensation	77.7%	75%	0.82
- Hepatic encephalopathy	37.7%	37.5%	0.98
- Bleeding oesophageal varices	36.5%	33.3%	0.78
- Hepatocellular carcinoma	21%	25%	0.67
- Bacterial infections	54.8%	45.8%	0.47

4. Discussion

IGT and DM are both frequently prevalent in cirrhosis. According to the literature, 50% - 80% of cirrhotic patients and even up to 96% have IGT while DM occurs in 30% - 40% [4] [8]. In most published studies, subclinical GMD are not routinely identified with OGTT, so the magnitude of the problem is often underestimated [8].

Our study is relevant because it prospectively assessed overt and subclinical GMD in a Tunisian population. We disclosed 68.8% of GMD: 42.8% of DM and 26% of IGT, which is about 4 times more than in the Tunisian general population (9.9% vs 42.8%), indicating that patients with cirrhosis are a high-risk population for GMD, although comparison is not adequate because the design of the 2 studies is different [9]. In our series, nearly half of GMD were subclinical and identified with OGTT, justifying its use as a routine in practice for this population. In our cohort, the greater prevalence of DM than described in previous studies may be related to the use of OGTT or because our patients had a more severe disease, although we did not found association between severity of liver disease using Child-Pugh and Meld scores. This finding does not agree with previous studies showing that the incidence of DM increases as liver function deteriorates [2] [6] [10].

It has been speculated that genetic and environmental factors and etiology of the liver disease, such hepatitis C and alcohol impair the insulin secretion [2]. Despite a high proportion of hepatitis C in our series, we did not found a statistically significant higher frequency of HVC in cirrhotic patients with GMD.

Several parameters are considered as possible risk factors associated with insulin resistance and DM in the general population, including age, female sex, family history of diabetes and overweight [10] [11]. However, in patients who present DM after the development of liver cirrhosis represent a different entity, called hepatogenous diabetes, classical risk factors are less frequently elicited, but this entity is not recognized by the ADA and the WHO [2]. In our study we could not always determine exactly if the onset of DM was before or after liver disease, firstly because in our population, cirrhosis is often diagnosed at the stage of complications, secondly because DM may be subclinical. In concordance with clinical characteristics of hepatogenous diabetes, we did not find a significant difference for age and BMI between diabetics and non diabetics cirrhotic patients, although weight was estimated subtracting ascites and oedema weight. On the other hand, familial history of diabetes was the only independent predictive factor for DM in cirrhosis, suggesting that they are possible adjunctive factor to liver disease in the arisen of DM. This is in agreement with the study of Zein *et al.* suggesting that cirrhosis alone do not always induce diabetes, and the cause of liver disease as well as environmental factors may play a role [12].

Retrospective studies have shown that DM (either type 2 or hepatogenous) increases the risks of complications of cirrhosis and reduces survival. Indeed, DM accelerates liver fibrosis and inflammation and enhance incidence of bacterial infection [5]. In our study, DM was not associated with a higher prevalence of cirrhosis complications. These results may be related to high rate of subclinical DM which may less contribute to liver injury.

Finally, it has been reported that hepatogenous diabetes is less associated with retinopathy, cardiovascular and renal complications and more frequently associated with hypoglycemic episodes as a result of impaired liver function. Liver disease abnormalities (low intravascular coagulability, low cholesterol, lower prevalence of hypertension) as well as shorter duration of DM may explain relatively lower rate of diabetic complication in chronic liver disease [13] [14].

The strength of our study is that it is the first one carried on a Tunisian population, allowing a preliminary determination of the prevalence and risk factors of GMD in cirrhosis and its design cross-sectional.

Nevertheless, our study had limitations:

1) In patients previously diagnosed as having DM, we could not clearly determine the chronology of the installation of diabetes in order to distinguish between DM type 2 and hepatogenous diabetes.

2) We did not determine the impact of the duration of cirrhosis on the prevalence of DM.

3) We did not include a control group without cirrhosis to compare the incidence of diabetes complications, but this was not the major aim of the study.

5. Conclusion

In our study, the frequency of GMD was 68.8%, with a majority of DM. OGTT disclosed nearly half of GMD, pointing a high incidence of latent GMD in cirrhosis. In this way, it should be routinely evaluated in all patients with cirrhosis. Familial history of diabetes was the only independent risk factor for GMD, suggesting that other factors in addition to liver disease may play a role in the development of GMD in cirrhosis.

References

[1] Buyse, S. and Valla, D. (2007) Carbohydrate Metabolism Dysregulation in Cirrhosis: Pathophysiology, Prognostic Impact and Therapeutic Implications. *Gastroentérologie Clinique et Biologique*, **31**, 266-273. http://dx.doi.org/10.1016/S0399-8320(07)89371-7

[2] Garcia-Compean, D., Jaquez-Quintana, J.O., Gonzales-Gonzales, J.A. and Maldonado-Garza, H. (2009) Liver Cirrhosis and Diabetes: Risk Factors, Pathophysiology, Clinical Implications and Management. *World Journal of Gastroenterology*, **15**, 280-288. http://dx.doi.org/10.3748/wjg.15.280

[3] Alavian, S.M., Hajarizadeh, B., Nematizadeh, F. and Larijani, B. (2004) Prevalence and Determinants of Diabetes Mellitus among Patients with Chronic Liver Disease. *BMC Endocrine Disorders*, **4**, 4. http://dx.doi.org/10.1186/1472-6823-4-4

[4] Wlazlo, N., Van Greevenbroek, M.M., Curvers, J., Schoon, E.J., Friederich, P., Twisk, J.W., *et al.* (2013) Diabetes Mellitus at the Time of Diagnosis of Cirrhosis Is Associated with Higher Incidence of Spontaneous Bacterial Peritonitis, but Not with Increased Mortality. *Clinical Science*, **125**, 341-348. http://dx.doi.org/10.1042/CS20120596

[5] Jaquez Quintana, J.O., Garcia-Compean, D., Gonzales Gonzales, J.A., Villarreal Perez, J.Z., Lavalle Gonzales, F.J., Munoz Espinosa, L.E., *et al.* (2011) The Impact of Diabetes Mellitus in Mortality of Patients with Compensated Liver Cirrhosis—A Prospective Study. *Annals of Hepatology*, **10**, 56-62.

[6] Bragança, A.C. and Alvares-da-Silva, M.R. (2010) Prevalence of Diabetes Mellitus and Impaired Glucose Tolerance in Patients with Decompensated Cirrhosis Being Evaluated for Liver Transplantation: The Utility of Oral Glucose Tolerance Test. *Archives of Gastroenterology*, **47**, 22-27.

[7] American Diabetes Association (2011) Diagnosis and Classification of Diabetes Mellitus. *Diabetes Care*, **34**, S62-S69. http://dx.doi.org/10.2337/dc11-S062

[8] Garcia-Compean, D., Jaquez-Quintana, J.O., Lavalle-Gonzales, F.J., Reyes-Cabello, E., Gonzales-Gonzales, J.A., Munoz-Espinosa, L.E., *et al.* (2012) The Prevalence and Clinical Characteristics of Glucose Metabolism Disorders in Patients with Liver Cirrhosis. A Prospective Study. *Annals of Hepatology*, **11**, 240-248.

[9] Bouguerra, R., Alberti, H., Salem, L.B., Rayana, C.B., Atti, J.E., Gaigi, S., *et al.* (2007) The Global Diabetes Pandemic: The Tunisian Experience. *European Journal of Clinical Nutrition*, **61**, 160-165. http://dx.doi.org/10.1038/sj.ejcn.1602478

[10] Gentile, S., Loguercio, C., Marmo, R., Carbone, L. and Del Vecchio Blanco, C. (1993) Incidence of Altered Glucose Tolerance in Liver Cirrhosis. *Diabetes Research and Clinical Practice*, **2**, 37-44. http://dx.doi.org/10.1016/0168-8227(93)90130-W

[11] Hickman, I.J. and Macdonald, G.A. (2007) Impact of Diabetes on the Severity of Liver Disease. *The American Journal of Medicine*, **120**, 829-834. http://dx.doi.org/10.1016/j.amjmed.2007.03.025

[12] Zein, N.N., Abdulkarim, A.S., Wiesner, R.H., Egan, K.S. and Persing, D.H. (2000) Prevalence of Diabetes Mellitus in Patients with End-Stage Liver Cirrhosis Due to Hepatitis C, Alcohol, or Cholestatic Disease. *Journal of Hepatology*, **32**, 209-217. http://dx.doi.org/10.1016/S0168-8278(00)80065-3

[13] Vidal, J., Ferrer, J.P., Esmatjes, E., Salmeron, J.M., Gonzales-Clemente, J.M., Gomis, R., *et al.* (1994) Diabetes Mellitus in Patients with Liver Cirrhosis. *Diabetes Research and Clinical Practice*, **25**, 19-25. http://dx.doi.org/10.1016/0168-8227(94)90157-0

[14] Moscatiello, S., Manini, R. and Marchesini, G. (2007) Diabetes and Liver Disease: An Ominous Association. *Nutrition, Metabolism and Cardiovascular Diseases*, **17**, 63-70. http://dx.doi.org/10.1016/j.numecd.2006.08.004

Carnitine Deficiency and Improvement of Muscle Cramp by Administration of Carnitine in Patients with Liver Cirrhosis

Naoki Hotta

Department of Internal Medicine, Division of Hepatology, Masuko Memorial Hospital, Aichi, Japan
Email: hotta4166@yahoo.co.jp

Abstract

Aim: We measured carnitine levels in patients with carnitine including dialysis patients, and examined whether administration of L-carnitine improved muscle symptoms. Methods: We measured carnitine levels in 27 patients with liver cirrhosis who were receiving treatment in our hospital, and administered L-carnitine (600 mg - 1800 mg) to patients having muscle cramps for approximately one month and examined the presence/absence of the symptom. We measured carnitine concentration before and after dialysis, before dialysis after the administration to eight dialysis patients, before and after the administration to 19 nondialytic patients. Results: The total carnitine levels before the dialysis of dialysis patients were an average of 42.2 μmol/L and fell to 17.7 μmol/L after more dialysis, but it was increased to 155 μmol/L after the administration of L-carnitine. In the nondialytic patients, the total carnitine levels were significantly increased from 71.7 μmol/L to 101.7 μmol/L after the administration of L-carnitine (P = 0.038). For symptomatic patients, significant improvement of muscle clamps was observed in the L-carnitine administrated group when compared with the non-administrated group (P = 0.0002). Conclusions: Total carnitine levels were low even before dialysis in the dialysis patients with liver cirrhosis in particular and they further decreased after the dialysis. Administration of L-carnitine increased the total carnitine levels and improved the symptom. Based on these results, we conclude that L-carnitine is useful for carnitine deficiency in patients with liver cirrhosis.

Keywords

Carnitine, Liver Cirrhosis, Dialysis Patient

1. Introduction

In recent years, we occasionally come across some reports that carnitine is useful for improving blood ammonia and cognitive function in hepatic cirrhosis patients with latent hepatic encephalopathy [1]-[3]. Recently, levo-carnitine chloride (L-Cartin® tablets 100 mg, 300 mg: Otsuka Pharmaceutical Co., Ltd., hereinafter referred to as LC) has also become available in Japan as a pharmaceutical agent for patients suspected of having carnitine deficiency. We measured serum carnitine concentrations in hepatic cirrhosis patients including dialysis patients and examined whether their cramps improved or not. Muscle spasms and hypotension during dialysis, atrophy of skeletal muscles and decrease in exercise capacity, anemia, and reduced cardiac function associated with decreased carnitine levels in dialysis patients have been reported [4]-[6]. Whereas a majority of the reports on the utility of carnitine have used supplements, we used LC for this investigation.

2. Methods

2.1. Patients

This was a randomized, double-blind, placebo-controlled study.

The serum carnitine concentrations in 27 patients with hepatic cirrhosis who are outpatients at our hospital were measured by using the enzyme cycling method (total carnitine reference value 45.0 - 91.0 μmol/L, free carnitine reference value 36 - 74 μmol/L, acylcarnitines reference value 6 - 23 μmol/L). LC (600 - 1800 mg) was administered to patients with cramps for 1 month, and the presence or absence of the symptom was examined. Serum carnitine concentrations were measured before and after dialysis and after LC administration in 8 dialysis patients, and before and after administration in 19 non-dialysis patients. In 17 patients with the symptom, the serum carnitine concentrations following administration were measured 1 month later. The study was reviewed and approved by the ethics committee established in the Masuko Memorial Hospital. The patients were given explanation on the study, for which written consents were obtained.

The 27 patients consisted of 8 dialysis and 19 non-dialysis patients, with no differences in the background factors (age, sex, cause, Child-Pugh grade, complication by liver cancer, implementation of dialysis, and with or without administration of branched-chain amino acids, hereinafter referred to as BCAAs, formulation) between the two groups. A significant difference was noted in serum total carnitine concentrations (**Table 1**). A significant difference was observed in serum total carnitine concentration between the dialysis patients prior to dialysis and the non-dialysis patients (42.2 ± 19.5 μmol/L vs. 70.5 ± 20.7 μmol/L, $P = 0.004$) (**Figure 1**). In the present study, cramps, one of the symptoms experienced by patients with hepatic cirrhosis, developed in 67% (18 of 27 patients) of the patients, consisting of 50% (4 of 8 patients) of the dialysis patients and 74% (14 of 19 patients) of the non-dialysis patients (**Table 2(a)**). The administration of LC caused the cramps to disappear in 92% (12 of

Table 1. The 27 patients consisted of 8 dialysis and 19 non-dialysis patients, with no differences in the background factors (age, sex, cause, Child-Pugh grade, complication by liver cancer, implementation of dialysis, and with or without administration of branched-chain amino acids, hereinafter referred to as BCAAs, formulation) between the two groups. A significant difference was noted in serum total carnitine concentrations.

Baseline participant characteristics			
	Hemodialysis n = 8	Non-hemodialysis n = 19	*P* value
Age mean ± S.D.	62 ± 11	66 ± 9	0.32*
Sex male/female	5/3	12/7	1.00**
Etiology (Alco/HBVHCV//nonBnonC/other)	0/0/7/0/1	6/2/10/1/2	0.10**
Child-pugh grade (A/B/C)	4/4/0	10/5/4	0.38**
Child-pugh score mean ± S.D.	6.5 ± 1.2	7.1 ± 2.7	0.43**
HCC complication (%)	7 (88)	12 (63)	0.36**
Total-carnitine mean ± S.D. μmol/L	42.2 ± 19.5	70.5 ± 20.7	0.004*
BCAA treatment (%)	4 (50)	11 (58)	1.00**
L-carnitine treatment (%)	5 (63)	11 (63)	1.00**
Cr mean ± S.D. mg/dL	8.9 ± 2.8	0.9 ± 0.4	<0.0001*

*Student's t test; **Fisher's exact test.

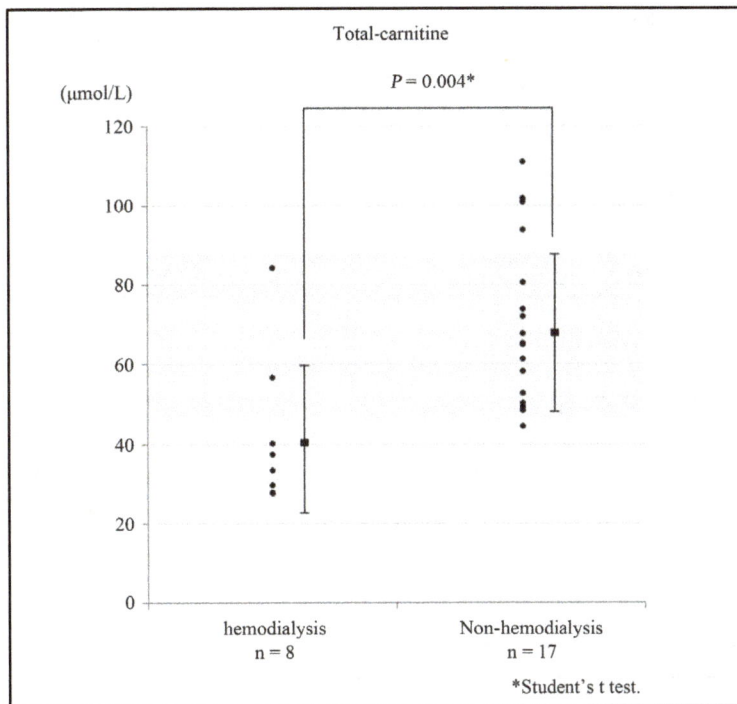

Figure 1. A significant difference was observed in serum total carnitine concentration between the dialysis patients prior to dialysis and the non-dialysis patients (42.2 ± 19.5 μmol/L vs. 70.5 ± 20.7 μmol/L, P = 0.004).

13 patients) of the patients, while the symptom did not disappear in the absence of LC administration, showing a significant difference between the two groups (P = 0.0002) (**Table 2(b)**).

2.2. Statistical Analysis

With respect to test methods, the paired t-test and Student's t-test were used for paired continuous variable data and unpaired continuous variable data, respectively, and the chi-square test or the Fisher's exact test was used for categorical variable data. All tests were two-sided, and difference levels of $p < 0.05$ were considered statistically significant.

3. Results

3.1. Dialysis Patients

The total carnitine concentration of 6 dialysis patients whose serum carnitine concentrations were measured was a subnormal 42.2 ± 19.5 μmol/L before dialysis and 17.7 ± 6.5 μmol/L after dialysis, showing significant decrease (P = 0.011). Free carnitine levels decreased significantly from 27.0 ± 13.2 μmol/L before dialysis to 11.2 ± 25.4 μmol/L after dialysis (P = 0.009); acylcarnitine levels decreased significantly from 15.2 ± 6.2 μmol/L to 5.8 ± 1.9 μmol/L (P = 0.021) (**Figure 2**). In 5 patients treated with LC, the levels of serum total carnitine, free carnitine, and acylcarnitine before and after administration increased significantly from 42.0 ± 24.3 μmol/L to 155.0 ± 114.8 μmol/L, 26.4 ± 15.3 μmol/L to 102.9 ± 80.1 μmol/L, and from 16.8 ± 10.1 μmol/L to 60.8 ± 34.3 μmol/L (**Figure 3**).

3.2. Non-Dialysis Patients

In the non-dialysis patients, no significant differences was observed in patient backgrounds (age, sex, cause, Child-Pugh grade, complication by liver cancer, total carnitine concentration, and with or without BCAAs administration) between the LC group of 12 patients and non-administration group of 7 patients. By Child-Pugh grades, serum total carnitine showed little variation (grade A, 67.4 ± 20.4 μmol/L; grade B, 51.6 ± 18.8 μmol/L;

Table 2. Cramps, one of the symptoms experienced by patients with hepatic cirrhosis, developed in 67% (18 of 27 patients) of the patients, consisting of 50% (4 of 8 patients) of the dialysis patients and 74% (14 of 19 patients) of the non-dialysis patients **(a)**. The administration of LC caused the cramps to disappear in 92% (12 of 13 patients) of the patients, while the symptom did not disappear in the absence of LC administration, showing a significant difference between the two groups (*P* = 0.0002) **(b)**.

(a)

		Incidence on muscle cramp		
	Total patients n = 27	Hemodialysis n = 8	Non-hempdialysis n = 19	*P* value
Muscle cramp (%)	18 (67)	4 (50)	14 (74)	0.3748

(b)

	Improvements on muscle cramp		
	L-carnitine n = 13	Non-L-carnitine n = 4	*P* value
Disappearance of muscle cramp (%)	12/13 (92)	0/4 (-)	0.0002*
Hemodialysis	3/3 (100)	0/1 (-)	0.2500
Non-hemodialysis	9/10 (90)	0/3 (-)	0.0140*

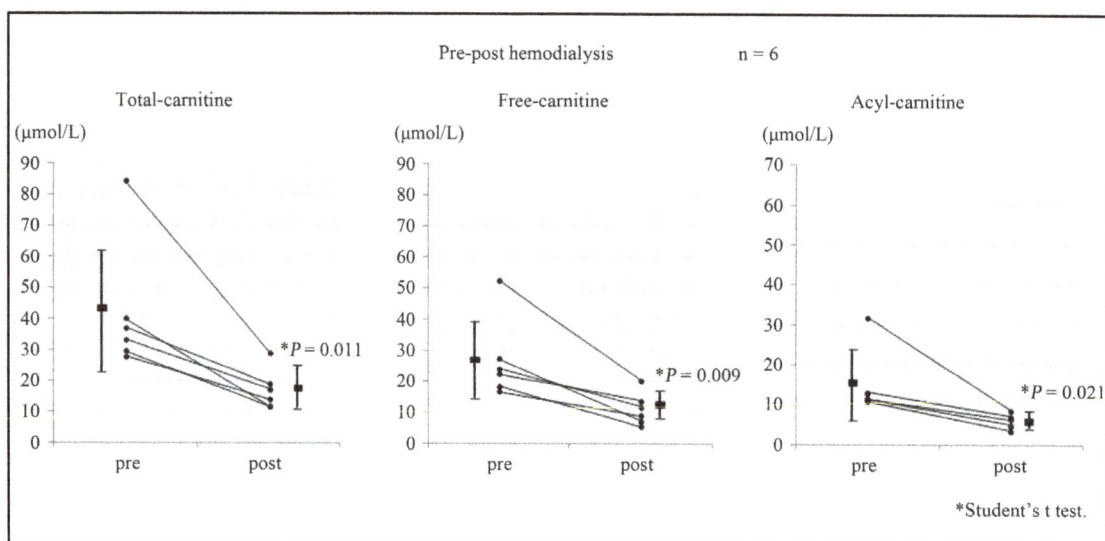

Figure 2. The total carnitine concentration of 6 dialysis patients whose serum carnitine concentrations were measured was a subnormal 42.2 ± 19.5 μmol/L before dialysis and 17.7 ± 6.5 μmol/L after dialysis, showing significant decrease (*P* = 0.011). Free carnitine levels decreased significantly from 27.0 ± 13.2 μmol/L before dialysis to 11.2 ± 25.4 μmol/L after dialysis (*P* = 0.009); acylcarnitine levels decreased significantly from 15.2 ± 6.2 μmol/L to 5.8 ± 1.9 μmol/L (*P* = 0.021).

grade C, 71.6 ± 28.1 μmol/L); free carnitine also exhibited little variation (grade A, 51.6 ± 16.8 μmol/L; grade B, 56.3 ± 10. 9 μmol/L; grade C, 55.0 ± 18.9 μmol/L); and acylcarnitine also showed little variation (grade A, 15.8 ± 8.9 μmol/L; grade B, 18.3 ± 9.9 μmol/L; grade C, 16.6 ± 10.5 μmol/L), neither of which showed a significant difference (**Figure 4**). In 12 patients treated with LC, the serum total carnitine levels before and after administrations increased significantly from 71.7 ± 22.8 μmol/L to 101.7 ± 45.5 μmol/L (*P* = 0.038); free carnitine increased significantly from 54.9 ± 15.4 μmol/L to 84.4 ± 36.2 μmol/L (*P* = 0.012); and acylcarnitine increased from 16.8 ± 9.1 μmol/L to 17.3 ± 9.8 μmol/L, showing an upward trend (**Figure 5**). Since no improvement was observed in 1 LC-treated case (cramps developed several times monthly), the dose of LC was increased, and consequently the symptom was improved 1 month later.

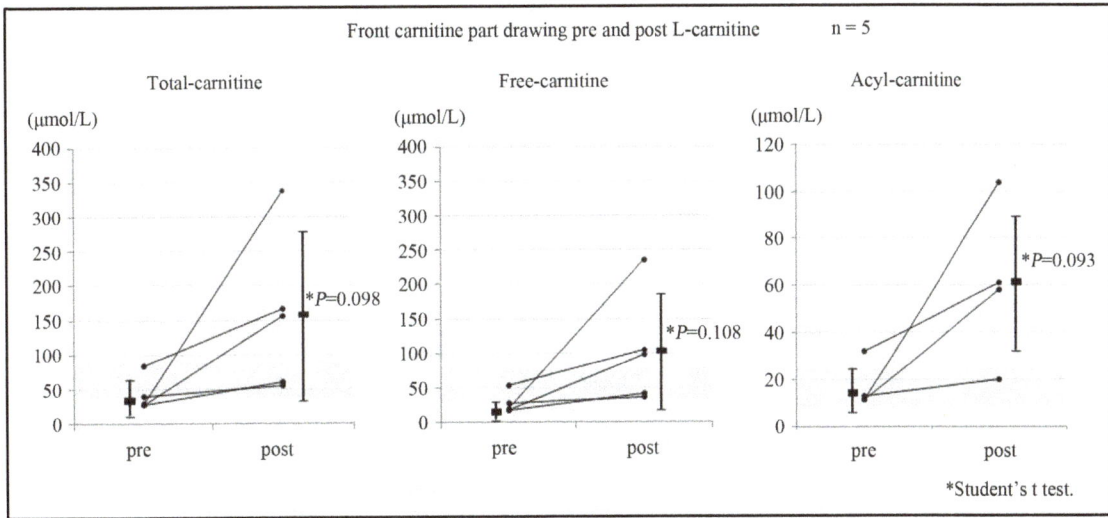

Figure 3. 5 patients treated with LC, the levels of serum total carnitine, free carnitine, and acylcarnitine before and after administration increased significantly from 42.0 ± 24.3 μmol/L to 155.0 ± 114.8 μmol/L, 26.4 ± 15.3 μmol/L to 102.9 ± 80.1 μmol/L, and from 16.8 ± 10.1 μmol/L to 60.8 ± 34.3 μmol/L.

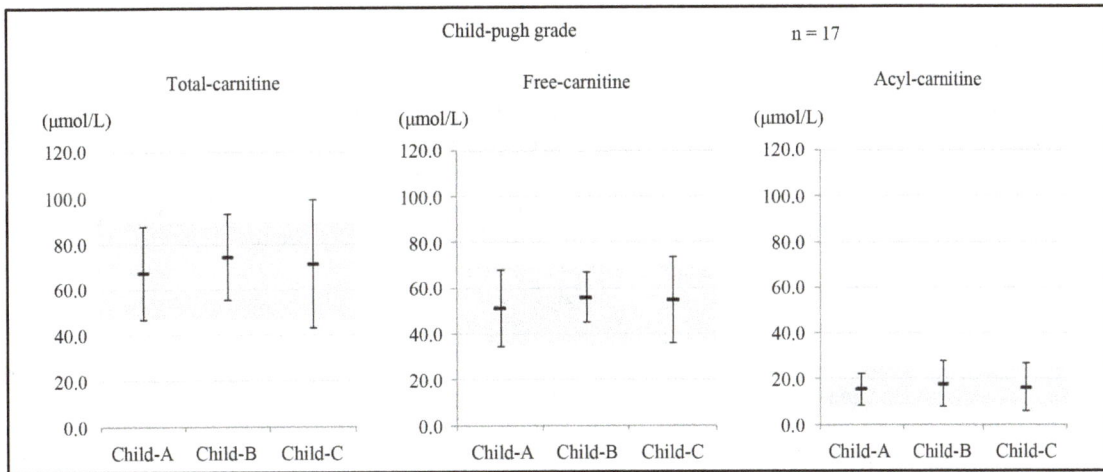

Figure 4. Child-Pugh grades, serum total carnitine showed little variation (grade A, 67.4 ± 20.4 μmol/L; grade B, 51.6 ± 18.8 μmol/L; grade C, 71.6 ± 28.1 μmol/L); free carnitine also exhibited little variation (grade A, 51.6 ± 16.8 μmol/L; grade B, 56.3 ± 10. 9 μmol/L; grade C, 55.0 ± 18.9 μmol/L); and acylcarnitine also showed little variation (grade A, 15.8 ± 8.9 μmol/L; grade B, 18.3 ± 9.9 μmol/L; grade C, 16.6 ± 10.5 μmol/L), neither of which showed a significant difference.

4. Discussion

Takayanagi reported that carnitine was an amino acid derivative with low molecular weight and played an important role in energy metabolism [7]. He presented the following three reasons for this: a) carnitine is essential for transportation of long-chain fatty acids to mitochondria; b) it adjusts the CoA/acyl-CoA ratio in mitochondria. Replacement of CoA with carnitine generates free CoA in the mitochondria; and c) carnitine removes cytotoxic acyl compounds from the cells as carnitine esters, which are excreted in the urine. Carnitine deficiency is classified into primary and secondary carnitine deficiency [8]. Primary carnitine deficiency is also known as congenital carnitine transporter deficiency or systemic carnitine deficiency. Secondary carnitine deficiency includes other inborn errors of metabolism and acquired medical conditions include: a) decrease in biosyntheses (hepatic cirrhosis, chronic kidney diseases, etc.); b) reduction in intake (long-term management of total parenteral nutrition, malnutrition, etc.); and c) reduction of body stores (pregnant and lactating women, very low birthweight infants, etc.) and those caused by medical interventions (dialysis- or drug-induced).

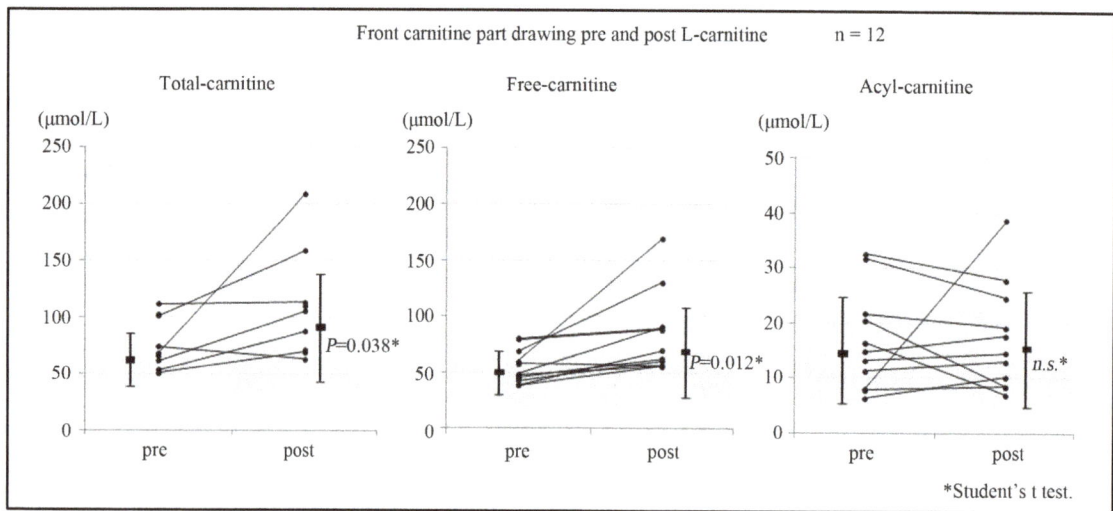

Figure 5. In 12 patients treated with LC, the serum total carnitine levels before and after administrations increased significantly from 71.7 ± 22.8 μmol/L to 101.7 ± 45.5 μmol/L ($P = 0.038$); free carnitine increased significantly from 54.9 ± 15.4 μmol/L to 84.4 ± 36.2 μmol/L ($P = 0.012$); and acylcarnitine increased from 16.8 ± 9.1 μmol/L to 17.3 ± 9.8 μmol/L, showing an upward trend.

The cases that we examined in this study were likely to have been those of secondary carnitine deficiency. The symptom of cramps in patients with hepatic cirrhosis is occasionally encountered in clinical practice, and switching therapeutic drugs from BCAA formulations or BCAA granules to oral nutrients for liver failure may improve the symptom [9]. Also, it has been reported that the causes for cramps are mechanism of peripheral neuropathy or myogenic, and decrease in blood taurine levels, which suppress abnormal excitation at the neuromuscular junction. Goto *et al.* discussed that BCAA formulations improved the symptom via decreased levels of free L-tryptophan caused by increased serum albumin levels, followed by facilitated ability to synthesize taurine through correction of amino acid imbalance [10]. We sometimes come across cases in which muscle symptoms occur as a result of carnitine deficiency in dialysis patients as well [11]. In 10 studies by Sakurauchi *et al.*, LC formulation was orally administered to 30 maintenance dialysis patients at a dose of 500 mg for 12 weeks, and the patients' somatic symptoms were assessed. As a result, it was reported that muscle symptoms (including fatigue, muscle spasm, and muscle pain) had improved in two-thirds of the patients [12]. In dialysis patients who develop cramps as a result of carnitine deficiency, it is said that muscle symptoms occur due to increased burden on each cell because of insufficient energy production in the muscle cells compared to healthy individuals by shifting of the energy sources from fatty acids to sugars and proteins, which in turn are caused by decreased blood and muscle carnitine levels. It is considered that LC administration allows the long-chain fatty acids to become sufficiently available again, and the acyl compounds accumulated in the cells to be washed out by carnitine, normalizing the cells, and these factors are involved in improvement of the muscle symptoms [13]. A meta-analysis by Lynch *et al.* that integrated the findings of 6 studies revealed that the use of LC significantly improved muscle spasms during dialysis in 2 cases [14]. Changes in the muscle fibers in dialysis patients and carnitine deficiency are also being examined; however, no characteristic changes have been noted [15] [16]. Administration of carnitine to these patients, however, increased the long diameters of type 1 muscle fibers. In these reactions, the type 1 muscle fibers contain many mitochondria, whose metabolism is aerobic. Therefore, what obtains benefits from carnitine administration is type 1 muscle fibers; uptake of fatty acids into the mitochondria increases, cell metabolism increases, the long diameters of the muscle fibers increase, and the proportion of atrophic muscle fibers to total muscle fibers decreases [17]. Although muscle biopsies were not carried out in any of the present cases, such mechanism is likely to be responsible for the improvement in the symptom. We need to consider the doses and duration of administration in future investigations. We also need to search for alternative test items in the future because serum carnitine fractionation is currently not covered by insurance.

References

[1]   Malguarnera, M., Vacante, M., Motta, M., *et al.* (2001) Acetly-L-Carnitine Improves Cognitive Functions in Severe

Hepatic Encephalopathy: A Randomized and Controlled Clinical Trial. *Metabolic Brain Disease*, **26**, 281-289, http://dx.doi.org/10.1007/s11011-011-9260-z

[2]   Malguarnera, M., Gargante, M.P., Cristaldi, E., *et al.* (2008) Acetyl-L-Carnitine Treatment in Minimal Hepatic Encephalopathy. *Digestive Diseases and Sciences*, **53**, 3018-3025. http://dx.doi.org/10.1007/s10620-008-0238-6

[3]   Malguarnera, M., Bella, R., Vacante, M., *et al.* (2011) Acetly-L-Carnitine Reduces Depression and Improves Quality of Life in Patients with Minimal Hepatic Encephalopathy. *Scandinavian Journal of Gastroenterology*, **46**, 750-759. http://dx.doi.org/10.3109/00365521.2011.565067

[4]   Moorthy, A.V., Rosenblum, M., Rajaram, R., *et al.* (1983) A Comparison of Plasma and Muscle Carnitine Levels in Patients on Peritoneal or Hemodialysis for Chronic Renal Falure. *American Journal of Nephrology*, **3**, 205-208. http://dx.doi.org/10.1159/000166711

[5]   Hiatt, W.R., Koziol, B.J., Shapiro, J.L., *et al.* (1992) Carnitine Metabolism during Exercise in Patients on Chronic Hemodialysis. *Kidney International*, **41**, 1613 -1619. http://dx.doi.org/10.1038/ki.1992.233

[6]   Ahmad, S., Robartson, H.T., Golper, T.A., *et al.* (1990) Multicenter Trial of L-Carnitine in Maintenance Hemodialysis Patients 2. Clinical and Biochemical Effects. *Kidney International*, **38**, 912-918. http://dx.doi.org/10.1038/ki.1990.290

[7]   Takayanagi, M. (2009) Abnormality. Carnitine Metabolism Child Internal Medicine, **41**, The Special Number 387-389.

[8]   Pons, R. and De Vivo, D.C. (1995) Primary and Secondary Carnitine Deficiency Syndrome. *Journal of Child Neurology*, **10**, S8-S24.

[9]   Kaneko, Y. and Tsuchiyama, H. (2012) Examination of the Usefulness of the Oral Nutrition Agent for Hepatic Insufficiency to Liver Cirrhosis. *Frontiers in Gastroenterology*, **17**, 94-101.

[10] Goto, N., Iida, K. and Hagisawa, Y. (2001) Usefulness of the Branched-Chain-Amino-Acid Tablet to the Leg Cramps Accompanying Liver Cirrhosis. *Liver*, **42**, 590-599.

[11] Casciani, C.U. and Caruso, U. (1982) Beneficial Effects of L-Carnirine in Post-Dialysis Sundrome. *Current Therapeutic Research*, **l32**, 116-127.

[12] Sakurauchi, Y., Matsumoto, Y., Shinzato, T., *et al.* (1998) Effects of L-Carnitine Supplementation on Muscular Symotpms in Hemodaialyzed Patients. *American Journal of Kidney Diseases*, **32**, 258-264. http://dx.doi.org/10.1053/ajkd.1998.v32.pm9708610

[13] Borum, P.R. and Taggart, E.M. (1986) Carnitine Nutriture of Dialysis Parients. *Journal of the American Dietetic Association*, **86**, 644-647.

[14] Lynch, K.E., Feldman, H.I., Berlin, J.A., *et al.* (2008) Effects of L-Carnitine on Dialysis-Related Hypotension and Muscle Cramps: A Meta-Analysis. *American Journal of Kidney Diseases*, **52**, 962-971. http://dx.doi.org/10.1053/j.ajkd.2008.05.031

[15] Siami, G., Clinton, M.E., Mrak, R., *et al.* (1991) Evaluation of the Effect of Intravenous L-Carnitine Therapy on Function, Structure and Fatty Acid Metabolism of Skeletal Muscle in Patients Receiving Chronic Hemodialysis. *Nephron*, **57**, 306-313. http://dx.doi.org/10.1159/000186280

[16] Ginvenali, P., Fenocchio, D., Montanari, G., *et al.* (1994) Selective Trophic Effect of L-Carnitine Type 1 and 2 Skeletal Muscle Fibers. *Kidney International*, **46**, 1616-1619. http://dx.doi.org/10.1038/ki.1994.460

[17] Takahashi, T., Hashimoto, Y. and Doi, T. (2000) No 2 20 Carnitine and Muscles. *Clinical Dialysis*, **16**, 201-206.

Predictors of Spontaneous Bacterial Peritonitis (SBP) in Liver Cirrhosis: Current Knowledge and Future Frontiers

Helen Ngo, Raymund Gantioque

Patricia A. Chin School of Nursing, California State University, Los Angeles, USA

Email: hngo22@calstatela.edu, rgantio@calstatela.edu

Abstract

Spontaneous bacterial peritonitis (SBP) in patients with cirrhotic liver disease is a serious complication that contributes to the high morbidity and mortality rate seen in this population. Currently, there is a lack of consensus amongst the research community on the clinical predictors of SBP as well as the risks and benefits of prophylactic antibiotic therapy in these patients. Pharmacological gastric acid suppression (namely with PPIs and H2RAs) are frequently prescribed for these patients, many times without a clear indication, and may contribute to gut bacterial overflow and SBP development. However, this remains controversial as there are conflicting findings in SBP prevalence between PPI/H2RA-users and non-users. In addition, studies show recent antibiotic use, whether for SBP prophylaxis or for another infectious process, appear to be associated with higher rates of SBP and drug-resistant organisms. Other researchers have also explored the link between zinc, platelet indices (MPV), and macrophage inflammatory protein-1 β (MIP-1β) levels in liver cirrhosis, all of which appear to be promising markers for classifying SBP risk and diagnosis. This literature review was limited by the number and quality of studies available as most are retrospective in nature. Thus, more ongoing, prospective studies and trials are needed to judge the true value of the findings in the studies reviewed in hopes that they can guide appropriate prevention, diagnosis, and management of SBP.

Keywords

Spontaneous Bacterial Peritonitis (SBP), Liver Cirrhosis, PPIs, H2RA, Antibiotic Prophylaxis, Antibiotic Resistance, Zinc, Inflammatory Biomarkers, Platelet Indices (MPV), Macrophage Inflammatory Protein-1 β (MIP-1β)

1. Introduction

Spontaneous bacterial peritonitis (SBP), an infection of the ascites fluid in the peritoneum that occurs in the absence of another infectious source, is a complication seen in patients with liver cirrhosis [1]. It carries potentially significant morbidity and mortality in this population due to their altered immunocompetency and overall disease burden [2]. SBP is generally theorized to be the result of gut bacteria translocation into the surrounding ascitic peritoneal fluid secondary to dysregulated local mucosal defense mechanisms and gastrointestinal hypomotility and is a sign of decompensated liver cirrhosis [3]. Clinical manifestations, though not always present, typically include fever, chills, and abdominal pain/discomfort, and may progress to mental status alterations and sepsis [4]. SBP may be accompanied by other signs of decompensation such as jaundice, ascites, portal hypertension (with or without resultant gastrointestinal bleeding), hepatic encephalopathy, and hepatorenal syndrome [1].

The prevalence of SBP in patients with liver cirrhosis ranges anywhere from twenty to fifty percent, depending on the study reviewed, with inpatient mortality rates as high as 32% [5]. True incidence and prevalence appear to be difficult to recognize as diagnostic ascitic fluid cultures can remain negative even in the presence of SBP [6]. In addition, some patients simply are asymptomatic through the course of the infection and would have otherwise remained missed cases if it were not for having a diagnostic or therapeutic paracentesis performed. And while those who stay asymptomatic are not burdened by the clinical manifestations that threaten their physiological state at the moment, studies suggest prior episodes of SBP may predispose them to more difficult to manage subsequent episodes [2] [5] [6].

The diagnosis of SBP requires a paracentesis to obtain an ascites fluid sample and is based on a positive ascites fluid culture and polymorphonuclear (PMN) leukocyte count greater than 250/mm3 [1] [7]. As with any invasive procedure, performing a paracentesis comes with certain risks, such as bleeding, infection, bowel perforation, and causing hemodynamically significant fluid shifts, and ultimately, the decision to proceed is that of clinical judgement and facility-based protocols, whether it is for diagnostic (i.e., suspicion of SBP in high risk individuals) or therapeutic (i.e., to ease work of breathing, relieve abdominal discomfort) purposes [8]. Some facilities may also routinely perform a diagnostic paracentesis for all admitted liver cirrhosis patients with ascites, regardless of SBP suspicion; however, this practice remains controversial due to the risks of the procedure [2].

Current guidelines for inpatient SBP treatment include the use of an intravenous third-generation cephalosporin (such as ceftriaxone) or a quinolone [9] [10]. Additionally, clinicians may also choose to prescribe oral ciprofloxacin or trimethoprim-sulfamethoxazole for SBP prophylaxis in high-risk patients in both inpatient and outpatient settings [2]. Norfloxacin, a quinolone previously popular for SBP prophylaxis but has since been discontinued in the United States in 2014, had the strongest evidence for its use but simultaneously appeared to be correlated with quinolone-resistant SBP [10]. As researchers delved more into this matter, recent findings suggest there is an increasing number of drug-resistant bacteria cases that implicate not only norfloxacin but also other agents including levofloxacin, ciprofloxacin, and cephalosporins [9] [10].

1.1. Scope of Problem

With the up-trending prevalence of antibiotic resistant SBP cases, treatment options will only continue to dwindle. Amongst one of the catalysts for this phenomenon is the poor and inappropriate diagnosis and management of SBP. Until around four years ago, there have been little to no distinction made by clinicians in approaching and treating community-acquired and nosocomial SBP, despite the involvement of different infectious flora between the two classifications [2]. In addition, nosocomial SBP infections are also more likely to implicate multi-drug resistant organisms (MRDOs), which only further complicate treatment strategies [9] [11] [12] [13].

The disease burden of SBP in liver cirrhosis patients greatly affect and increase morbidity and mortality amongst this group. Studies show SBP predisposes patients to recurrent episodes of SBP or infection of a different source (and vice versa), with subsequent infections more likely to be associated with more dire consequences due to involvement of drug-resistant organisms (DROs) [9] [11] [12] [13]. Those with a recent infection who are then discharged from the hospital have as high as a 41% risk of death or need for liver transplantation within six months [12]. There is also a subset of patients who become disqualified from liver transplantation while on the waitlist due to sepsis and multi-organ failure secondary to SBP. Better patient outcomes require both appropriate and timely antibiotic therapy prior to onset of hypotension and sepsis [14]. As for those who do proceed to liver transplantation, history of SBP occurrence pre-transplantation may be correlated with inferior graft function and even graft failure, with increased morbidity and mortality post-liver transplantation [15].

Lastly, the medical costs related to SBP annually place tremendous strain on the healthcare system. Based on the U.S. Nationwide Inpatient Sample (NIS) data, costs associated with ICU admission and care with presumed infection in this patient population alone approximates $3 billion annually [6].

1.2. Knowledge Gap

There is a current knowledge gap in managing liver cirrhosis patients at risk for and those who have SBP. The research community lacks a consensus regarding both prevention and treatment strategies. There are conflicting findings and opinions regarding the role and use of antibiotic therapy and pharmacological gastric acid suppression and their potential associations with SBP prevalence, disease process and progression. Some researchers are attributing the rise of antibiotic resistance organisms and poorer clinical outcomes to the absence of up-to-date standardized guidelines on SBP prevention and treatment. Lastly, the potential roles of specific trace elements and inflammatory biomarkers are growing areas of interest amongst researchers for its prospect in predicting SBP risk in hopes of avoiding unnecessary antibiotic therapy.

1.3. Aim of Literature Review

The aim of this article is to explore the current state of knowledge regarding independent predictors of SBP development in liver cirrhosis patients as well as the potential utilization of trace elements (particularly zinc) and inflammatory biomarkers to stratify SBP risk and vulnerability. This is all in efforts to better assist clinical judgment in prioritizing antibiotic prophylactic treatment and reduce the risk of SBP development.

2. Methods

The online databases resourced for articles reviewed in the paper included PubMed, The Cumulative Index to Nursing and Allied Health Literature (CINAHL), National Institutes of Health (NIH), and Google Scholar. Database searches were conducted in September and October 2017. The keywords used to search for articles reviewed in this paper included "spontaneous bacterial peritonitis", "SBP", "liver cirrhosis", "ascites", "end-stage liver disease", "ESLD", "predictors", "PPI", "H2RA", "antibiotic prophylaxis", "antibiotic resistance", "zinc", "mean platelet volume" and "macrophage inflammatory protein". The terms "spontaneous bacterial peritonitis" and "predictors" were initially searched in combination to identify predictors of interest. The terms in various combinations were then used to compile articles, which were subsequently reviewed for relevancy to topic. Such combinations included "spontaneous bacterial peritonitis", "ascites", "liver cirrhosis" or "end-stage liver disease" with each of the studied predictors ("PPI", "H2RA", "antibiotic prophylaxis", "antibiotic resistance", "zinc", "mean platelet volume" and "macrophage inflammatory protein"). Furthermore, the reference lists of relevant studies were reviewed in attempt to seek out additional pertinent studies not found in prior searches. Only articles available in the English language were included in this review, and the literature search was limited to articles published within the last five years (2012 to present).

3. Results

3.1. Pharmacological Gastric Acid Suppression

Proton-pump inhibitors (PPIs) and histamine-2-receptor antagonists (H2RAs), the two most common classes of pharmacological gastric acid suppression, are frequently prescribed for patients for gastrointestinal prophylaxis against ulcer development (particularly within the hospital setting), for treatment of gastric or duodenal ulcers, and to relieve symptoms for those with gastroesophageal reflux disease (GERD) [16]. In liver cirrhosis patients specifically, gastric acid suppression is undoubtedly a vital part of managing this disease process and preventing complications such as gastrointestinal bleeding [16]. However, lowering the acidity level of gastric contents may also negatively affect the native gut bacterial flora, allowing for overgrowth and subsequent transmigration to the surrounding peritoneal fluid in the presence of ascites [2] [16].

A number of studies show that PPI use is associated with higher prevalence of SBP in liver cirrhosis. In a retrospective cohort study of 7299 patients with decompensated cirrhosis from the U.S. Veterans' Health Administration database between the years 2001 and 2009, PPI use appeared to increase the rate of infection by 1.75 times compared to those who were not on PPIs [17]. Around 25.9% who used PPIs developed serious infections, with the majority (75%) of infections being acid-suppression associated infections, including SBP, C. difficile, and pneumonia. Of those who developed infections while taking PPIs, the leading sources and types of infection involved were SBP (30%), pneumonia (25%), skin infections (23%), spontaneous bacteremia and septicemia (16%), C. difficile (5%), and UTI (1%). Of note, the researchers found no clinically significant difference in infection rates between patients who were on H2RA therapy and those who were not on any form of pharmacological gastric acid suppression [17].

Another study by Goel et al. supported the findings of Bajaj et al. reported above [7] [17]. In this retrospective case-control study of 130 hospitalized patients, Goel et al. found the SBP-positive group had a higher incidence of PPI use within 7 days of diagnosis compared to a Child-Pugh score-matched SBP-negative control (71% vs. 41% respectively) (p < 0.001) [7]. Those who did not use PPIs in the last 90 days were almost 70% less likely to have SBP (p = 0.05). And those who have used PPIs within 90 days of hospitalization were 79% less likely to have SBP than those with PPI use within 7 days of hospitalization; there was no significant difference between no PPI use within 90 days and PPI use in the last 8-90 days but not within 7 days.

O'Leary et al. examined the risk factors of recurrent bacterial infections in a prospective study of 188 hospitalized liver cirrhosis patients across 12 United States centers enrolled in North American Consortium for the Study of End-Stage Liver Disease [12]. The authors performed a six-month follow-up after discharge from the hospital and found PPI use to be an independent predictor of subsequent infections. Around 45% were readmitted for infections within this period, and a higher proportion of these patients were older in age, used PPIs, or received prophylactic therapy for SBP.

While the previous studies suggest a correlation between PPI use and SBP/infection rates, these findings do not go unopposed. In fact, Terg et al. report no statistically significant association with PPI use and higher SBP prevalence [18]. In their prospective study of 519 decompensated liver cirrhosis patients across 23 hospitals in Argentina between March 2011 and April 2012, 24.7% of subjects developed SBP. The authors found similar rates of PPI use between those who developed SBP and those who remained SBP-free (44.3% vs. 42.8%). In addition, the duration of PPI use and rate of SBP development were not correlated. Amongst those who developed SBP during this period, there was little difference in the microbes seen between PPI and non-PPI users.

In contrast with the PPIs, H2RAs are less commonly prescribed and appear to show mixed results with regards to its part in SBP development. Bajaj et al. saw no significant difference in infection rates found in subjects who used H2RAs versus no gastric acid suppression at all [17]. On the contrary, Goel et al. did endorse the SBP-positive group had a slightly higher incidence of H2RA use within the past 90 days compared to those in the Child-Pugh score-matched SBP-negative control group (15% versus 2%) (p = 0.02) [7].

3.2. Antibiotic Therapy in SBP

Another prominent area of study in examining independent predictors of SBP development is antibiotic therapy use in this group. According to Tandon et al., recent antibiotic use is associated with higher rates of SBP [13]. Amongst the 115 unique bacterial infections seen in patients with cirrhosis who were admitted or developed a bacterial infection during hospitalization, 28 (24%) were SBP. Of the 70 patients with a positive ascitic fluid culture, 31 (44%) had prior exposure to one or more systemic antibiotics within 30 days of infection, 23 (33%) had no antibiotic exposure, and 16 (23%) had exposure to oral non-absorbed antibiotics alone.

Antibiotic resistance has been in the forefront of discussion in the recent decade across the field of medicine, and SBP in liver cirrhosis is certainly no exception. Multiple studies show a higher prevalence of drug-resistant organisms in SBP cases with recent antibiotic use, whether intended for SBP treatment, SBP prophylaxis, or treatment of non-SBP infections. DROs were found at higher rates in those with SBP as a subsequent infection rather than SBP as a primary/index infection (42 versus 7%, p = 0.02) [12]. In Ariza et al.'s study, 21.5% of the positive ascitic fluid cultures were found to have global resistance to third-generation cephalosporins, an antibiotic traditionally used for treatment, with resistance rates higher in nosocomial SBP cases [9]. Further analyses show that previous use of cephalosporins, history of diabetes, history of upper GI bleed, and low PMN in ascitic fluid were other pos-itive risk factors and predictors for DRO involvement [9]. Fernández et al. reported similar findings in their 2012 study, attributing recent beta-lactam use, long-term SBP prophylaxis with norfloxacin, and history of MDROs as risk factors for development of MDRO-related infections [11].

Tandon et al. also shed light on this topic in their 17-month study of 115 participants admitted to the liver unit of Yale New Haven Hospital [13]. They not only saw a higher prevalence of antibiotic resistance in those with recent systemic antibiotic use as the previous studies showed but also discovered 35% of these resistant infections were spontaneous infections (including SBP, spontaneous bacterial empyema, and spontaneous bacteremia). Of the 13 culture-positive SBP infections, 6 (46%) were resistant to both third-generation cephalosporins, the first-line empiric antibiotic used in SBP treatment, and ciprofloxacin, a quinolone commonly used for SBP prophylaxis. Further analysis of culture sensitivities showed there was no significant difference in the presence and rate of antibiotic resistant SBP between the specific systemic antibiotics the patients took or were taking, whether it was used for SBP prophylaxis (typically fluoroquinolones or trimethoprim-sulfamethoxazole) or for another infection.

Other researchers shifted their attention to rifaximin, an antibiotic commonly prescribed for acute hepatic encephalopathy in those with liver disease for its role in eliminating ammonia-producing bacteria in the intestinal tract [2] [19]. It is a poorly absorbed oral agent and thus has a relatively low risk of acquiring resistance [19]. Two recent studies support its use, stating rifaximin alone may be sufficiently effective in serving as a SBP prophylactic agent. In a retrospective study published by Hanouneh et al. including 404 patients, 89% of the liver cirrhotic ascites patients who received rifaximin remained free of SBP compared to 68% of those not on rifaximin; the rifaximin test group saw a 72% reduction in SBP development [19]. The study by Tandon et al. showed no significant correlation for antibiotic resistant infections with prior use of oral non-absorbed antibiotics (like rifaximin) when compared to traditional systemic antibiotics stated previously [13].

The implication of DROs in SBP-positive ascitic fluid is associated with poorer outcomes and survival, primarily due to limited treatment options. Those with ESBL-E and other MDROs involved have higher incidence of septic shock, rapid clinical deterioration and mortality [11]. Ariza et al. supported these findings, revealing that the presence of DROs in ascites fluid is linked to increased mortality rate, especially in the setting of hepatorenal syndrome [9].

3.3. Zinc

Zinc, a physiological trace element with known functions in the immune system, has recently received some attention for its potential role in SBP development. Zinc deficiency is a frequent finding in decompensated cirrhosis [20]. In a 2015 study performed by Mohammad et al., low zinc levels (defined as less than 60 µg/dL) were correlated with SBP development [21]. In this study, 35 of 54 (64.8%) SBP-positive subjects had a serum zinc level < 60 µg/dL, whereas only 45 of 122 (36.9%) SBP-negative subjects were found to have low serum zinc (p = 0.001). Sengupta et al. concluded in their 2015 study that serum zinc concentrations were inversely correlated with infection and ascites [20]. In addition, Sengupta et al. and Kar et al. examined and reported that zinc deficiency was linked to poorer clinical outcomes and shorter transplant-free survival [20] [22]. These latest findings indicate zinc may be an independent predictor of SBP in liver cirrhosis and have the potential to serve as a useful diagnostic and prognostic marker.

3.4. Platelet Indices and Inflammatory Biomarkers

Some researchers have turned to platelet activation and other inflammatory biomarkers as indicators for identifying the underlying inflammatory process involved in SBP. Higher mean platelet volume (MPV) levels are seen in those with ascitic fluid infections compared to liver cirrhosis patients without SBP and healthy controls alike, with no statistically significant difference between the latter two [1] [23]. The authors for both studies endorse MPV as an accurate diagnostic test and marker in predicting the presence of ascitic fluid infection in the liver cirrhosis population.

Other inflammatory biomarkers have been examined in its reliability in diagnosis and correlation with SBP. Specifically, macrophage inflammatory protein-1 beta (MIP-1β) have been examined in two studies by Khorshed et al. and Lesińska et al., both showing ascitic fluid MIP-1β to be significantly increased in the presence of SBP (p < 0.001 and p = 0.01 for the respective studies), with sensitivity and specificity for SBP diagnosis to be 80% and 76.1%; and 72.7 and 100% respectively [1] [24]. Perhaps of more value, a combined measurement of serum and ascitic fluid MIP-1β levels was found to have 100% sensitivity and specificity for SBP diagnosis [1]. Lesińska et al. also looked at procalcitonin levels but concluded it was not useful as a diagnostic marker [24].

4. Discussion

While ongoing further research is still necessary, the current state of knowledge regarding pharmacological gastric acid suppression therapy suggests there is stronger evidence relating PPI use with increased SBP incidence and overall poorer clinical outcomes in decompensated liver cirrhosis. Oftentimes, these agents are seen as fairly benign medications and can arguably be described as something clinicians prescribe out of habit or tradition rather than for a true indication [10] [25]. In fact, Ladato et al. argue there may be little efficacy of PPI use in the presence of hypertensive gastropathy, such as seen in liver cirrhosis patients [25]. By bringing more awareness to the potential adverse effects of pharmacologic gastric acid suppression particularly in this patient population, clinicians can better assess the risk-benefit ratio of its use prior to starting therapy, ensuring benefits outweigh the potential morbidity of SBP.

Based on the literature, recent antibiotic use is associated with higher rates of SBP infections and prevalence of DROs, presenting either through SBP or other infections. With the emergence of multi-drug resistant "superbugs", healthcare providers must be more conscientious about the risks related to inappropriate antibiotic prescription, particularly in caring for vulnerable groups such as in decompensated liver cirrhosis. Providers must also avoid indiscriminate antibiotic use and escalation in otherwise low-risk patients. The rise in antibiotic resistance has led to increased treatment failure with traditional methods and will continue to exhaust the availability of effective antibiotics, such as third-generation cephalosporins and ciprofloxacin in SBP treatment and prophylaxis respectively. It will then only be a matter of time before they and other agents are rendered useless, leaving these patients completely defenseless against infections.

In order to improve clinical management and establish updated guidelines for SBP diagnosis and treatment, further exploration of the role of zinc would increase the knowledge base of SBP pathophysiology. Zinc has known functions in immunocompetency and serves as an enzyme cofactor in a number of metabolic and cellular processes, plays a role in oxidative stress, and may also have antiinflammatory effects [21] [26]. With the rise of antibiotic resistance, it is becoming more difficult for clinicians to determine the risk-benefit ratio in prophylactically treating SBP. If further studies determine and support that zinc is indeed a significant independent predictor of SBP, perhaps zinc replacement can alter infection risk and/or overall outcomes. In essence, can zinc replacement delay or prevent decompensation, and if so, to what extent? With a greater understanding of such questions, providers can take a more proactive rather than reactive approach in hopes of significantly reducing morbidity and mortality in those with chronic liver disease and failure.

On a similar note, inflammatory biomarkers like MPV and MIP-1β appear to be promising indicators that can be utilized for SBP diagnosis. Use of inflammatory biomarkers to determine the need to initiate empiric antibiotic therapy in the absence of elevated PMN and/or while pending fluid culture reports is a topic worth exploring more. And if future research supports the utility of these biomarkers to assess for presence of ascitic fluid infections, a practice guideline change reflective of such may allow for earlier and proper diagnosis and treatment of SBP, with hopes of improving patient outcomes, reduce patient risk of need for more invasive diagnostic measures (i.e., paracentesis), and lower overall healthcare spending.

5. Conclusion

While this paper is limited by the relatively small number of studies available on this topic as well as the quality and type of studies (most are retrospective in nature), it sheds light on how much is still unknown regarding this disease process and how there is a need to change certain aspects of SBP clinical management in liver cirrhosis based on the high rates of morbidity and complications. In performing a literature search, it is clear more prospective, randomized controlled trials are required to better assess the risk versus protective factors for SBP development in liver cirrhosis. This is vital in judging the true value of the findings presented in the studies reviewed in this paper and will be a major milestone for the research community before updated evidence-based practice guidelines can be put forth.

Acknowledgements

The authors have no conflicts of interest that are relevant to the content of this article.

References

[1] Khorshed, S.E., Ibraheem, H.A. and Awad, S.M. (2015) Macrophage Inflammatory Protein-1 Beta (MIP-1β) and Platelet Indices as Predictors of Spontaneous Bacterial Peritonitis. Open Journal of Gastroenterology, 5, 94-102.

https://doi.org/10.4236/ojgas.2015.57016

[2] Wiest, R., Krag, A. and Gerbes, A. (2012) Spontaneous Bacterial Peritonitis: Recent Guidelines and Beyond. Gut, 61, 297-310.

https://doi.org/10.1136/gutjnl-2011-300779

[3] Ismail, M. and Rahman, M.A. (2015) Prevalence and Short Term Outcome of Spontaneous Bacterial Peritonitis of Known Chronic Liver Disease Patients. Medicine Today, 27, 15-19. https://doi.org/10.3329/medtoday.v27i1.25992

[4] Paul, K., Kaur, J. and Kazal, H.L. (2015) To Study the Incidence, Predictive Factors and Clinical Outcome of Spontaneous Bacterial Peritonitis in Patients of Cirrhosis with Ascites. Journal of Clinical and Diagnostic Research: JCDR, 9, OC09-OC12. https://doi.org/10.7860/JCDR/2015/14855.6191

[5] de Mattos, A.A., Costabeber, A.M., Lionço, L.C. and Tovo, C.V. (2014) Multi-Resistant Bacteria in Spontaneous Bacterial Peritonitis: A New Step in Management? World Journal of Gastroenterology: WJG, 20, 14079-14086.

https://doi.org/10.3748/wjg.v20.i39.14079

[6] Bajaj, J.S., O'Leary, J.G., Wong, F., Reddy, K.R. and Kamath, P.S. (2012) Bacterial Infections in End-Stage Liver Disease: Current Challenges and Future Directions. Gut, 61, 1219-1226. https://doi.org/10.1136/gutjnl-2012-302339

[7] Goel, G.A., Deshpande, A., Lopez, R., Hall, G.S., van Duin, D. and Carey, W.D. (2012) Increased Rate of Spontaneous Bacterial Peritonitis Among Cirrhotic Patients Receiving Pharmacologic Acid Suppression. Clinical Gastroenterology

and Hepatology, 10, 422-427. https://doi.org/10.1016/j.cgh.2011.11.019

[8] De Gottardi, A., Thévenot, T., Spahr, L., Morard, I., Bresson-Hadni, S., Torres, F., et al. (2009) Risk of Complications After Abdominal Paracentesis in Cirrhotic Patients: A Prospective Study. Clinical Gastroenterology and Hepatology, 7, 906-909. https://doi.org/10.1016/j.cgh.2009.05.004

[9] Ariza, X., Castellote, J., Lora-Tamayo, J., Girbau, A., Salord, S., Rota, R., et al. (2012) Risk Factors for Resistance to Ceftriaxone and its Impact on Mortality in Community, Healthcare and Nosocomial Spontaneous Bacterial Peritonitis. Journal of Hepatology, 56, 825-832. https://doi.org/10.1016/j.jhep.2011.11.010

[10] Ge, P.S. and Runyon, B.A. (2015). Preventing Future Infections in Cirrhosis: A Battle Cry for Stewardship. Clinical Gastroenterology and Hepatology, 13, 760-762. https://doi.org/10.1016/j.cgh.2014.10.025

[11] Fernández, J., Acevedo, J., Castro, M., Garcia, O., Rodríguez de Lope, C., Roca, D., et al. (2012) Prevalence and Risk Factors of Infections by Multiresistant Bacteria in Cirrhosis: A Prospective Study. Hepatology, 55, 1551-1561.

https://doi.org/10.1002/hep.25532

[12] O'Leary, J.G., Reddy, K.R., Wong, F., Kamath, P.S., Patton, H.M., Biggins, S.W., et al. (2015) Long-Term Use of Antibiotics and Proton Pump Inhibitors Predict Development of Infections in Patients with Cirrhosis. Clinical Gastroenterology and Hepatology, 13, 753-759. https://doi.org/10.1016/j.cgh.2014.07.060

[13] Tandon, P., DeLisle, A., Topal, J.E. and Garcia-Tsao, G. (2012) High Prevalence of Antibiotic-Resistant Bacterial Infections among Patients with Cirrhosis at a US Liver Center. Clinical Gastroenterology and Hepatology, 10, 1291-1298.

https://doi.org/10.1016/j.cgh.2012.08.017

[14] Karvellas, C.J., Abraldes, J.G., Arabi, Y.M. and Kumar, A. (2015) Appropriate and Timely Antimicrobial Therapy in Cirrhotic Patients with Spontaneous Bacterial Peritonitis-Associated Septic Shock: A Retrospective Cohort Study. Alimentary Pharmacology & Therapeutics, 41, 747-757. https://doi.org/10.1111/apt.13135

[15] Shah, N.L., Intagliata, N.M., Henry, Z.H., Argo, C.K. and Northup, P.G. (2016) Spontaneous Bacterial Peritonitis Prevalence in Pre-Transplant Patients and Its Effect on Survival and Graft Loss Post-Transplant. World Journal of Hepatology, 8, 1617-1622. https://doi.org/10.4254/wjh.v8.i36.1617

[16] Dam, G., Vilstrup, H., Watson, H. and Jepsen, P. (2016) Proton Pump Inhibitors as a Risk Factor for Hepatic Encephalopathy and Spontaneous Bacterial Peritonitis in Patients with Cirrhosis with Ascites. Hepatology, 64, 1265-1272.

https://doi.org/10.1002/hep.28737

[17] Bajaj, J.S., Ratliff, S.M., Heuman, D.M. and Lapane, K.L. (2012) Proton Pump Inhibitors are Associated with a High Rate of Serious Infections in Veterans with Decompensated Cirrhosis. Alimentary Pharmacology & Therapeutics, 36, 866-874. https://doi.org/10.1111/apt.12045

[18] Terg, R., Casciato, P., Garbe, C., Cartier, M., Stieben, T., Mendizabal, M., et al. (2015) Proton Pump Inhibitor Therapy Does Not Increase the Incidence of Spontaneous Bacterial Peritonitis in Cirrhosis: A Multicenter Prospective Study. Journal of Hepatology, 62, 1056-1060. https://doi.org/10.1016/j.jhep.2014.11.036

[19] Hanouneh, M.A., Hanouneh, I.A., Hashash, J.G., Law, R., Esfeh, J.M., Lopez, R., et al. (2012) The Role of Rifaximin in the Primary Prophylaxis of Spontaneous Bacterial Peritonitis in Patients with Liver Cirrhosis. Journal of Clinical Gastroenterology, 46, 709-715. https://doi.org/10.1097/MCG.0b013e3182506dbb

[20] Sengupta, S., Wroblewski, K., Aronsohn, A., Reau, N., Reddy, K.G., Jensen, D., et al. (2015) Screening for Zinc Deficiency in Patients with Cirrhosis: When Should We Start? Digestive Diseases and Sciences, 60, 3130-3135.

https://doi.org/10.1007/s10620-015-3613-0

[21] Mohammad, A.N., Yousef, L.M. and Mohamed, H.S. (2016) Prevalence and Predictors of Spontaneous Bacterial Peritonitis: Does Low Zinc Level Play Any Role? Al-Azhar Assiut Medical Journal, 14, 37. https://doi.org/10.4103/1687-1693.180461

[22] Kar, K., Dasgupta, A., Bhaskar, M.V. and Sudhakar, K. (2014) Alteration of Micronutrient Status in Compensated and Decompensated Liver Cirrhosis. Indian Journal of Clinical Biochemistry, 29, 232-237.

https:/ / doi.or g/ 10.1007/ s12291-013-0349-5

[23] Suvak, B., Torun, S., Yildiz, H., Sayilir, A., Yesil, Y., Tas, A. and Kayaçetin, E. (2013) Mean Platelet Volume Is a Useful Indicator of Systemic Inflammation in Cirrhotic Patients with Ascitic Fluid Infection. Annals of Hepatology, 12, 294-300.

[24] Lesińska, M., Hartleb, M., Gutkowski, K. and Nowakowska-Duława, E. (2014) Procalcitonin and Macrophage Inflammatory Protein-1 Beta (MIP-1β) in Serum and Peritoneal Fluid of Patients with Decompensated Cirrhosis and Spontaneous Bacterial Peritonitis. Advances in Medical Sciences, 59, 52-56.

https:/ / doi.or g/ 10.1016/ j.advm s.2013.07.006

[25] Lodato, F., Azzaroli, F., Di Girolamo, M., Feletti, V., Cecinato, P., Lisotti, A., et al. (2008) Proton Pump Inhibitors in Cirrhosis: Tradition or Evidence Based Practice? World Journal of Gastroenterology, 14, 2980-2985.

https:/ / doi.or g/ 10.3748/ wjg.14.2980

[26] Mangray, S., Zweit, J. and Puri, P. (2015) Zinc Deficiency in Cirrhosis: Micronutrient for Thought? Digestive Diseases and Sciences, 60, 2868-2870.

https:/ / doi.or g/ 10.1007/ s10620-015-3854-y

11

Liver Transient Elastography Combined to Platelet Count (Baveno VI) Predict High Esophageal Varices in Black African Patient with Compensated Hepatitis B Related Cirrhosis

Doffou Adjeka Stanislas1, Assi Constant2, Ndjitoyap Ndam Antonin Wilson2,
Kouame Hardryt Dimitri1, Bangoura Demba1, Kissi Anzouan-Kacou1, Ouattara Amadou2,
Lohoues-Kouacou Marie-Jeanne2, Attia Koffi Alain1

1 Gastrointestinal Unit, Cocody Teaching Hospital Center, Abidjan, Côte d'Ivoire 2Gastrointestinal Unit, Yopougon

Email: das_stan@yahoo.ca

Abstract

Aim: To assess the predictive value of the Baveno VI criteria for the diagnosis of large esophageal varices (EV) in Black African patient with compensated hepatitis B related cirrhosis. Methods: We carried out a cross-sectional study from January 2 to July 3 (2016), in Department of Gastroenterology at Uni-versity Hospitals of Cocody (CHUC) and Yopougon (CHUY). All the black African patients included were more than 15 years old and their liver elasticity score (LES) was carried out at Yopougon University Hospital. Hepatitis B re-lated cirrhosis was defined by LES ≥ 11 kPa (FibroScan® (Echosens, France)) with positive HBs antigen (HBsAg) and anti HBc antibody. All the patients with hepatitis B related cirrhosis performed a gastroscopy at Cocody Univer-sity Hospital and esophageal varices were ranked according to société française d'endoscopie digestive (SFED) classification. Data analysis was per-formed by SPSS model 20.0 statistics software (SPSS Inc., Chicago, IL, United States). Diagnostic performance of LES < 20 kPa and platelet count > 150,000/mmm3 (Baveno VI criteria) for the diagnosis of large EV by gastros-copy was studied (area under the ROC curve, specificity (Sp), sensitivity (Se), positive predictive value (PPV) and negative predictive value (NPV). Results: During the study period, 720 patients achieved liver FibroScan® at CHUY. Of these, 60 respondents to our inclusion criteria were prospectively included in our study. Twelve (20%) of these 60 patients met the Baveno VI criteria. EV were present in 40% of cases (n = 24) with 6.7% (n = 4), 15% (n = 9) and 18.3% (n = 11) of grade 1, 2 and 3, respectively. (66.7% (n = 40) without EV or with small EV) and 33.3% (n = 20) with large EV. The Baveno VI criteria had a Se, Sp, PPV and NPV of 100%, 41.6%, 30% and 100% respectively for the diagnosis of large EV. The area under the ROC curve of a platelet count greater than 150,000/mm3, a liver elasticity score of less than 20 kPa and com-bination of both were respectively 0.763 [0.645 - 0.880; P = 0.272]; 0.588 [0.436 - 0.739; P = 0.01] and 0.650 [0.513 - 0.787 P = 0.005]. Conclusion: The combination of liver elasticity score < 20 kPa and a blood platelet count > 150,000/mm3, allowed the exclusion of large esophageal varices at gastroscopy with a 100 % NPV in Black African patients with compensated hepatitis B re-lated cirrhosis.

Keywords

Cirrhosis, Esophageal Varice-Fibroscan®-Baveno, Negative Predictive Value, Africa

1. Introduction

Upper gastrointestinal bleeding secondary to rupture of esophageal varices (EV) in a patient with cirrhosis is fatal in more than 15% cases [1] [2] [3]. Primary prevention of this hemorrhage is based on the screening of these EV by gastroscopy. Several authors study non-invasive criteria to exclude the presence of EV at gastroscopy [4]-[11]. The Baveno VI conference experts suggested not to perform gastroscopy in a patient with compensated hepatitis B related cirrhosis if there are both: a liver elasticity score < 20 kPa and a blood platelet count > 150,000 constituents by mm3 [12] [13]. These criteria have been studied in several studies to assess their relevance [14] [15] [16] [17]. None of these studies, as far as we know, has been carried out in a black African population with hepatitis B related cirrhosis. The aim of our study was to assess the predictive value of the Baveno VI criteria for the diagnosis of large esophageal varices in patients with compensated hepatitis B related cirrhosis.

2. Patients and Method
2.1. Sample Population

We carried out a cross-sectional study from January 2 to July 31 (2016) in department of gastroenterology at University Hospitals of Cocody (CHUC) and Yopougon (CHUY). We included all the patients older than 15 years, irrespective of gender, in whom hepatitis B related cirrhosis was previously diagnosed, based on a liver elasticity score. Liver elasticity threshold for the diagnosis of cirrhosis was respectively 11 kPa [18]. Patients with at least one of the following criteria were not included: acute ethylic intoxication during the four (4) weeks before FibroScan® test, regular alcohol intake higher than 20 g per day, splenectomised ones, transaminases greater than 5-fold normal, liver tumor or portal thrombosis or significant ascites or dilatation of hepatic veins on ultrasound.

2.2. Method

In each of these patients, the following variables were gathered using a pre-established survey form: anamnestics data (age, sex, treatments in progress or received (diuretics, antiviral B and/or C treatment and alcohol), clinics data (splenomegaly, ascites, hepatic encephalopathy, jaundice), biologicals data (transaminases, blood platelets, Prothrombin, albuminemia). We established Child-Pugh-Turcott score (Class A, B and C) [19].

All the gastroscopies were carried out at Cocody University Hospital (CHUC) by two physicians (gastroenterologist) with more than 15 years of seniority in digestive endoscopy. Esophageal varices were ranked according to Société Française d'Endoscopie Digestive classification (grade 1, 2 and 3) [20]. Gastroscopy also looked for the presence of red signs on the varices, for gastric varices and for portal hypertension gastropathy.

All the liver elasticity measurements were performed at Yopougon University Hospital (CHUY) by two experienced gastroenterologists. The device used was the FibroScan® (Echosens, France). The results of FibroScan® data were expressed in kPa using a medium-size probe (M). The patient was lying on his back, his right hand raised behind his head. The doctor applied a water-based gel on the skin at the level of the liver, then takes measurements on the right side. The probe was placed in the intercostal space, perpendicularly to the skin. A minimum of 2 or 3 hours fasting was required of the patient before the test. The test consists of 10 successives measurements. The validation criteria of the measurements were: a success rate of about 60%, IQR < 1 and a rate of variability <30% of the median value.

2.3. Statistical Analyses

Data analysis was performed by SPSS model 20.0 statistics software (SPSS Inc., Chicago, IL, United States). Variables by category were presented in percentage form, the ones which are continuous by their median, standard deviation and stretch (minimum maximum). The patients were separated into two groups according to the results of the gastroscopy. Those having grade 2 or 3 varices were classified into group II (large varices) and those having grade 1 varices or not having varices were classified into group I (small varices). Comparison between these two groups was performed by the chi-square test for the variables by category and the ANOVA test for the ones which are continuous. Diagnosis performance of the liver elasticity score (LES) < 20 kPa and platelet count > 150,000/ mm3 for the diagnosis of large EV by gastroscopy was assessed with the following parameters: specificity (Sp), sensitivity (Se), positive and negative predictive value (PPV) and (NPV), positive (LR+) and negative likelihood ratio (LR-).

2.4. Ethical Clearance

A double oral informed consent was obtained: the one of the patient before his inclusion and the one of the attending physician.

3.   Results

During the study period, 720 patients performed a liver FibroScan® at the CHUY. Of these, 60 respondents to our inclusion criteria were prospectively included in our study. Twelve (20%) of these 60 patients met the Baveno VI criteria. All the patients performed a gastroscopy which found EV in 40% of cases (n= 24), with 6.7% (n = 4), 15% (n = 9) and 18.3% (n = 11) of grade 1, 2 and 3 respectively. Patients were classified into 2 categories according to the size of their EV: 66.7% (n = 40) of patients with small or no EV (group I) and 33.3 % (n = 20) with large EV (group II). None of the patients with large varices met the Baveno VI criteria. The characteristics of the two groups are shown in Table 1. There was no difference between them in age, sex, transaminases and median liver elasticity score. Patients with no large EV had no red sign and no gastric varices. Child-Pugh scores of patients with large EV were significantly higher than the ones of those who did not have EV (P < 0.001). None of the patients with platelet count > 150,000/mm3 had large EV (Table 1). Table 2 shows the diagnostic performance of these two criteria. In application of the Baveno VI score, respectively 20% and 46.7% of gastroscopies would have been needless and useless (Table 3).

4.   Discussion

Our prospective study carried out exclusively in black African patients with hepatitis B related cirrhosis shows that in these patients, Baveno VI criteria allowed to exclude the presence of large EV at gastroscopy with a 100% NPV. Comparable results (NPV 98%) have been reported by Maurice et al. but in a hepatitis C related cirrhosis predominantly [15].The prevalence of EV was 40% in our sample comparable to that reported from a sample of 1250 patients by Augustin et al. [14]. By applying the Baveno criteria, 20% of gastroscopies would have been avoided, no significant varices would have been detected, proportions also reported by Augustin et al. but with a lower percentage of unnecessary gastroscopy (38% versus 46.7%) [14].

To our knowledge, it is the only study carried out in patients with only hepatitis B related cirrhosis on the Baveno VI criteria. Our results suggest that in these patients when diagnosing compensated cirrhosis, gastroscopy screening for EV could be postponed. This alternative is interesting in Ivory Coast because few hospitals have technical platform and digestive endoscopy staff. However, there is no clear consensus on the periodicity of this non-invasive evaluation [15]. In addition, when interpreting these criteria, it is essential, as pointed out by several authors, to respect scrupulously FibroScan® conditions and contraindications to avoid misclassification of patients [16] [17]. The diagnosis of cirrhosis based on FibroScan® of the liver, notwithstanding the problems of sampling fluctuation, is difficult to currently implement in Ivory Coast because there is only one device for the whole country. On the other hand, the poor PPV of these Baveno criteria in our study as in others does not allow to base the diagnosis of significant EVs nor to start a preventive treatment of these [12]-[17]. A blood platelet count> 150,000/mm3 also had a 100% NPV to exclude large EVs. It is an easy test to obtain especially in our environment with lack of FibroScan®. The use of platelet count as a non-invasive marker for portal hypertension is the subject of many studies [9] [21]. The platelet count appears as an independent predictive factor for the diagnosis of large varice in cirrhotic patients. A retrospective Moroccan study showed that a platelet count of less than 100,000/mm3 was correlated with the presence of EVs but not that of large EVs in viral cirrhosis [9]. Another Egyptian study, like our work showed that a normal platelet level was correlated with the absence of large EVs [21]. The advantage of the combination is that a normal level of platelets is found in any healthy individual but a liver elasticity score greater than 11 kPa immediately suggests cirrhosis. On the other hand, liver elasticity score at 20 kPa alone had a lower NPV to exclude the diagnosis of large EV in our work. This result is in contradiction with the one reported by Maurice et al. where the NPV of the LES was higher than platelets count [1]. Unlike the sample of this work, in ours, we found that the platelet count and the median LES were respectively statistically significant and not significant between patients with EV and those who did not have EVs. Hua et al., like our results, observed a lack of difference between the LES between these two groups [22] in opposition to those of Sharma et al. [23]. Other authors have studied these two criteria with different cut-offs. Ding et al. also reported a 100% NPV when LES > 25 kPa and platelets count < 100,000/mm3 [24]. Augustin et al. also observed similar results, but at a higher threshold [25]. This was in contradiction to the one reported by Pas et al. In their work, liver elasticity score was higher than platelet counts to predict the presence of large EV [26]. One limitation of this study is the use of elastometric criteria (>14 kPa) for the diagnosis of cirrhosis as we know the value used was not validated in the black African population. Liver FibroScan® values were set up from a study in black Africa comparing FibroScan® performance for the diagnosis of hepatitis B related cirrhosis compared to the result of liver histology. This study showed that at the threshold of 11 kPa the LES had a Se and a Sp of 71% and 88% for the diagnosis of fibrosis F4 [18]. Another limitation of our work is to have assessed only the prediction of large EVs. Although they are responsible for nearly 80% 90% of the upper gastrointestinal hemorrhages, others lesions of portal hypertension, though rare, may be responsible. These are small EVs with red signs or occurring on severe cirrhosis, portal hypertensive gastropathy and gastric varices. They would also require measures to prevent hemorrhage [11]. In our work none of patients without large EV had gastric varices not red signs on EV. And on the other hand there was no patient with a Child Pugh C score in patient with small varice.

Table 1. Cirrhotic patients' characteristics according to the size of varices.

Variables	Small varice n = 40	High varice n = 20	P value
Sex ratio	2.3	5.7	0.172
Age median in years ± SD	49 ± 13.3	48.5 ± 12	0.720
BMI median in kg/m² ± SD	22.6 ± 3.4	22.1 ± 2.8	0.108
ASAT in UI/l ± SD	43.5 ± 42.6	47.5 ± 188.9	0.264
ALAT in UI/l ± SD	35 ± 31.5	46.5 ± 150.8	0.184
Platelets/mm³ ± SD	157500 ± 106639	59500 ± 30201	<0.0001
Platelets > 150,000/mm³	21 (52.5%)	0	<0.0001
Child Pugh score < 0.0001			
A	33 (82.5%)	7 (35%)	
B	7 (17.5%)	13 (65%)	
Median of SLE (kPa) ± SD	20.7 ± 13.4	25.9 ± 18.6	0.062
SLE < 20 kPa	19 (47.5%)	6 (30%)	0.269
SLE < 20 kPa and platelets > 150,000/mm³	12 (30%)	0	
Grade of EV			
1	4 (10%)	-	
2	-	9 (45%)	
3	-	11 (55%)	
Other endoscopic sign			
Red sign in EV	0 (0%)	9 (45%)	<0.0001
PHG	1 (2.5%)	6 (30%)	0.004
Gastric varices	0 (0%)	1 (5%)	0.333

SD = standard deviation; BMI = body mass index; ASAT = aspartat amino transferase ALAT = alanine amino transferase; EV = esophageal varice; PHG= portal hypertensive gastropathy; SLE = liver elastography score.

Table 2. Predictive negative value of the platelet count and score of the liver elastography to exclude large esophageal varices.

Criteria	PPV	NPV	Se	Sp	RV+	RV-
Platelet> 150,000/mm³	51.3%	100 %	100 %	52.5 %	2.10	0
SLE < 20 kPa	33.3 %	60 %	70 %	30 %	1	1
Platelet count > 150,000/mm³ and SLE < 20 kPa	41.6%	100%	100%	30%	1.42	0

SLE = Score of Liver Elastography; PPV = Predictive Positive Value; PNV = Predictive Negative Value; Se = Sensitivity; Sp = Specificity.

Table 3. Diagnostic performance of the Baveno VI criteria to screen for large esophageal varices.

Variables	Number	Percentage
Large varices non diagnosed	0	0
Needless gastroscopies	12/60	20 %
Useless gastroscopies	28 /60	46.7 %

5. Conclusion

The combination of liver elasticity score <20 kPa in association with a count of platelet >150,000/mm3 could allow to exclude the presence of large esophageal varices at gastroscopy with a 100% NPV in hepatitis B related cirrhosis. These parameters could allow us to identify patients who could be delayed gastroscopy screening for large EVs, knowing that those patients do not have EVs. Prospective internal and external validation on large samples is necessary to confirm these cut-off values.

References

[1] Cabrera, L., Tandon, P. and Abraldes, J.G. (2016) An Update on the Management of Acute Esophageal Variceal Bleeding. Gastroenterologia y Hepatologia, 40, 34-40.

[2] Zaman, A. (2003) Current Management of Esophageal Varices. Current Treatment Options in Gastroenterology, 6, 499-507.

https://doi.org/10.1007/s11938-003-0052-3

[3] Bouglouga, O., Bagny, A., Lawson-Ananissoh, L. and Djibril, M. (2014) Hospital Mortality Associated with Upper Gastrointestinal Hemorrhage Due to Ruptured Esophageal Varices at the Lome Campus Hospital in Togo. Medecine et Sante Tropicales, 24, 388-391.

[4] Elalfy, H., Elsherbiny, W., Abdel Rahman, A., Elhammady, D., Shaltout, S.W., Elsamanoudy, A.Z. and El Deek, B. (2016) Diagnostic Non-Invasive Model of Large Risky Esophageal Varices in Cirrhotic Hepatitis C Virus Patients. World Journal of Hepatology, 8, 1028-1037. https://doi.org/10.4254/wjh.v8.i24.1028

[5] Mahassadi, A.K., Bathaix, F.Y., Assi, C., Bangoura, A.D., Allah-Kouadio, E., Kissi, H.Y., Touré, A., Doffou, S., Konaté, I., Attia, A.K., Camara, M.B. and Ndri-Yoman, T.A. (2012) Usefulness of Noninvasive Predictors of Oesophageal Varices in Black African Cirrhotic Patients in Côte d'Ivoire (West Africa). Gastroenterology Research and Practice, 2012, 216390. https://doi.org/10.1155/2012/216390

[6] Abraldes, J.G., Bureau, C., Stefanescu, H., Augustin, S., Ney, M., Blasco, H., Procopet, B., Bosch, J., Genesca, J., Berzigotti, A. and Anticipate Investigators (2016) Noninvasive Tools and Risk of Clinically Significant Portal Hypertension and Varices in Compensated Cirrhosis: The "Anticipate" Study. Hepatology, 64, 2173-2184. https://doi.org/10.1002/hep.28824

[7]   de Franchis, R. and Dell'Era, A. (2014) Invasive and Noninvasive Methods to Diagnose Portal Hypertension and Esophageal Varices. Clinical Liver Disease, 18, 293-302.

[8]   European Association for Study of Liver (2015) Asociacion Latinoamericana para el Estudio del Higado. EASL-ALEH Clinical Practice Guidelines: Non-Invasive Tests for Evaluation of Liver Disease Severity and Prognosis. Journal of Hepatology, 63, 237-264.

[9]   Nada, L., Samira el, F., Bahija, B., Adil, I. and Nourdine, A. (2015) Noninvasive Predictors of Presence and Grade of Esophageal Varices in Viral Cirrhotic Patients. The Pan African Medical Journal, 20, 145. https://doi.org/10.11604/pamj.2015.20.145.4320

[10] Castera, L. (2007) Use of Elastometry (FibroScan) for the Non-Invasive Staging of Liver Fibrosis. Gastroenterologie Clinique Et Biologique, 31, 524-530; quiz 500, 531-532.

[11] Shi, K.Q., Fan, Y.C., Pan, Z.Z., Lin, X.F., Liu, W.Y., Chen, Y.P. and Zheng, M.H. (2013) Transient Elastography: A Meta-analysis of Diagnostic Accuracy in Evaluation of Portal Hypertension in Chronic Liver Disease. Liver International, 33, 6271.

[12] De Franchis, R. and Baveno VI Faculty Collaborators (2015) Expanding Consensus in Portal Hypertension: Report of the Baveno VI Consensus Workshop: Stratifying Risk and Individualizing Care for Portal Hypertension. Journal of Hepatology, 63, 743-752. https://doi.org/10.1016/j.jhep.2015.05.022

[13] Cardenas, A. and Mendez-Bocanegra, A. (2016) Report of the Baveno VI Consensus Workshop. Annals of Hepatology, 15, 289-290.

[14] Augustin, S., Pons, M. and Genesca, J. (2016) Validating the Baveno VI Recommendations for Screening Varices. Journal of Hepatology, 66, 459-460. https://doi.org/10.1016/j.jhep.2016.09.027

[15] Maurice, J.B., Brodkin, E., Arnold, F., Navaratnam, A., Paine, H., Khawar, S., Dhar, A., Patch, D., O'Beirne, J., Mookerjee, R., Pinzani, M., Tsochatzis, E. and Westbrook, R.H. (2016) Validation of the Baveno VI Criteria to Identify Low Risk Cirrhotic Patients Not Requiring Endoscopic Surveillance for Varices. Journal of Hepatology, 65, 899-905. https://doi.org/10.1016/j.jhep.2016.06.021

[16] Berzigotti, A., Boyer, T.D., Castéra, L., de Franchis, R., Genescà, J. and Pinzani, M. (2015) Reply to "Points to Be Considered When Using Transient Elastography for Diagnosis of Portal Hypertension According to the Baveno's VI Consensus. Journal of Hepatology, 63, 1049-1050. https://doi.org/10.1016/j.jhep.2015.07.012

[17] Perazzo, H., Fernandes, F.F., Castro Filho, E.C. and Perez, R.M. (2015) Points to Be Considered When Using Transient Elastography for Diagnosis of Portal Hypertension According to the Baveno's VI Consensus. Journal of Hepatology, 63, 1048-1049.

[18] Bonnard, P., Sombie, R., Lescure, F.X., Bougouma, A., Guiard-Schmid, J.B., Poynard, T., Cales, P., Housset, C., Callard, P., Le Pendeven, C., Drabo, J., Carrat, F. and Pialoux, G. (2010) Comparison of Elastography, Serum Marker Scores, and Histology for the Assessment of Liver Fibrosis in Hepatitis B Virus (HBV)-Infected Patients in Burkina Faso. The American Journal of Tropical Medicine and Hygiene, 82, 454-458. https://doi.org/10.4269/ajtmh.2010.09-0088

[19] Pugh, R.N., Murray-Lyon, I.M., Dawson, J.L., Pietroni, M.C. and Williams, R. (1973) Transection of the Oesophagus for Bleeding Oesophageal Varices and Performed Ultrasound for Each Patient (Size of the Spleen, Splenomegaly If the Diameter of the Spleen >130 mm) and Possible Presence of an Ascite. British Journal of Surgery, 60, 646-649.

[20] Lebrec, D., Vinel, J.P. and Dupas, J.L. (2005) Complications of Portal Hypertension in Adults: A French Consensus. European Journal of Gastroenterology & Hepatology, 17, 403-410. https://doi.org/10.1097/00042737-200504000-00003

[21] Abd-Elsalam, S., Habba, E., Elkhalawany, W., Tawfeek, S., Elbatea, H., El-Kalla, F., Soliman, H., Soliman, S., Yousef, M., Kobtan, A., El Nawasany, S., Awny, S., Amer, I., Mansour, L. and Rizk, F. (2016) Correlation of Platelets Count with Endoscopic Findings in a Cohort of Egyptian Patients with Liver Cirrhosis. Medicine (Baltimore), 95, e3853.

[22] Hua, J., Liu, G.Q., Bao, H., Sheng, L., Guo, C.J., Li, H., Ma, X. and Shen, J.L. (2015) The Role of Liver Stiffness Measurement in the Evaluation of Liver Function and Esophageal Varices in Cirrhotic Patients. Journal of Digestive Diseases, 16, 98-103.

[23] Sharma, P., Kirnake, V., Tyagi, P., Bansal, N., Singla, V., Kumar, A. and Arora, A. (2013) Spleen Stiffness in Patients with Cirrhosis in Predicting Esophageal Varices. The American Journal of Gastroenterology, 108, 1101-1107. https://doi.

org/10.1038/ajg.2013.119

[24] Ding, N.S., Nguyen, T., Iser, D.M., Hong, T., Flanagan, E., Wong, A., Luiz, L., Tan, J.Y., Fulforth, J., Holmes, J., Ryan, M., Bell, S.J., Desmond, P.V., Roberts, S.K., Lubel, J., Kemp, W. and Thompson, A.J. (2016) Liver Stiffness plus Platelet Count Can Be Used to Exclude Highrisk Oesophageal Varices. Liver International, 36, 240-245. https://doi.org/10.1111/liv.12916

[25] Augustin, S., Millán, L., González, A., Martell, M., Gelabert, A., Segarra, A., Serres, X., Esteban, R. and Genescà (2014) Detection of Early Portal Hypertension with Routine Data and Liver Stiffness in Patients with Asymptomatic Liver Disease: A Prospective Study. Journal of Hepatology, 60, 561-569. https://doi.org/10.1016/j.jhep.2013.10.027

[26] Pritchett, S., Cardenas, A., Manning, D., Curry, M. and Afdhal, N.H. (2011) The Optimal Cutoff for Predicting Large Oesophageal Varices Using Transient Elastography Is Disease Specific. Journal of Viral Hepatitis, 18, e7580.

12

Serum Total Triiodothyronine versus Free Tetraiodothyronine and TSH in Patients with HCV Related Cirrhosis and Their Correlation to the Severity of Cirrhosis

Ashraf A. Hammam[1], Amal A. Jouda[2]*, Mona E. Hashem[3]

[1]Internal Medicine Department, Zagazig University, Zagazig, Egypt
[2]Tropical Medicine Department, Zagazig University, Zagazig, Egypt
[3]Clinical Pathology Department, Zagazig University, Zagazig, Egypt
Email: *ashraf_hammam2003@yahoo.com, dr.amaljouda@zu.edu.eg, mm_hashem2001@yahoo.com

Abstract

Background and Aim: The levels of thyroid hormones and their binding proteins are altered in patient with cirrhosis. We aim to study the changes in triiodothyronine level in HCV related cirrhosis and its correlation to the severity of liver decompensation. **Patients and Methods:** This study included seventy two patients with HCV related cirrhosis in three groups Group I: 24 patients with Child A class Group II: 24 patients with Child B and C classes without hepatic encephalopathy Group III: 24 patients with Child B and C classes with hepatic encephalopathy. **Results:** T3 level was significantly lower in group III than group I and II (0.74 ng/ml vs 1 and 1.3 ng/ml in group II and I in succession). The correlation between Child's score and T3 level was highly significant (r = −0.64, P < 0.001). **Conclusion:** Triiodothyronine level is lower in cirrhosis and its level is correlated to the severity of decompensation.

Keywords

Free T4, TSH, Clinical Euthyroid, Cirrhosis, HCV

1. Introduction

Cirrhosis is the ninth leading cause of death in the United States and is responsible for 1.2% of all US deaths;

---

*Corresponding author.

about 35,000 deaths each year. Cirrhosis is often preceded by hepatitis and fatty liver (steatosis), independent of the cause. If the cause is treated at this stage, the changes are reversible. The pathological hallmark of cirrhosis is the development of fibrosis that replaces normal parenchyma. Damage to the hepatic parenchyma (due to inflammation) leads to activation of the stellate cell, which increases fibrosis (through production of myofibroblasts) and obstructs blood flow in the circulation [1]. In addition, it secretes TGF-β1, which leads to a fibrotic response and proliferation of connective tissue. Furthermore, it secretes TIMP 1 and 2, naturally occurring inhibitors of matrix metalloproteinases, which prevents them from breaking down fibrotic material in the extracellular matrix. The fibrous tissue bands (septa) separate hepatocyte nodules, which eventually replace the entire liver architecture [2].

The thyroid status depends not only on thyroxine secretion but also on normal thyroid hormone metabolism, delivery of T_3 to nuclear receptors and on receptor distribution and function. Normal thyroid function, which is essential for normal growth, development and the regulation of energy metabolism within cells, is dependent on a normally functioning thyroid and liver axis [3].

The liver performs important functions in the process of thyroid hormone transport and metabolism. The liver extracts 5% - 10% of plasma T_4 during a single passage, as shown by studies using $[I^{131}]$ T_4 [4]. The liver also synthesizes a number of plasma proteins that bind the lipophilic thyroid hormones. More than ninety nine percent of the thyroid hormones are bound to thyroxine-binding globulin, thyroxine-binding prealbumin and albumin in plasma. This bound portion of the hormone works like a large reservoir of free circulating hormone. The free hormone component within plasma is in equilibrium with the protein-bound hormone, and it is this free fraction which accounts for the hormone's biological activities. The plasma concentrations of free T_4 and T_3 are at a steady concentration, so that the tissues are exposed to the same concentrations of the free hormone [5].

There are evidences showing an association between chronic liver diseases and changes in thyroid gland. Furthermore, it is demonstrated that levels of thyroid hormones and their binding proteins are altered in patient with hepatic disorders, especially cirrhosis [6]. Some studies even say that there's a significant increase in thyroid glandular volume in cirrhotic patients when compared with controls. However, almost all of them are clinically euthyroid [7].

On the other hand, some authors believe that the changes in thyroid hormones levels may be regarded as an adaptive hypothyroid state that helps to decrease the basal metabolic rate within hepatocytes and preserve liver function and total body protein stores [8]. This hypothesis is based on a study in cirrhotic patients which showed that the onset of hypothyroidism due to intrinsic thyroid disease during cirrhosis resulted in a biochemical improvement in liver function as compared to cirrhotic controls [9]. Hypothyroidism has also been associated with lesser degrees of decompensation in cirrhosis [10].

Egypt has the highest prevalence of hepatitis C virus (HCV) in the world it is estimated that about 15% of population are infected with hepatitis C. About 85% of patients infected with HCV develop chronic hepatitis C (CHC) and are at risk for fibrosis progression. About 20% - 30% of CHC patients will develop cirrhosis of the liver within years. Once cirrhosis is established the rate of HCC development is 1% - 4% per year [11]. Being a major health problem in Egypt, and due to its role in development of cirrhosis we chose post-HCV cirrhosis to evaluate the relation between thyroid functions and liver disease.

2. Aim of the Work

This study aims at exploring the changes in thyroid functions in patients with HCV related cirrhosis and the correlation between them and the severity of liver dysfunction.

3. Patients and Methods

This study had been carried out in Zagazig University Hospitals in the period between August 2014 and August 2015. The study design was approved by the Institutional Review Board (IRB) of Faculty of Medicine, Zagazig University. The study included 72 patients with liver cirrhosis due to chronic HCV infection, 40 males and 32 females and their ages ranged from 40 to 70 years old.

3.1. Inclusion Criteria

Patients with liver cirrhosis diagnosed by combination of clinical, ultrasonic and laboratory assessments due to chronic HCV hepatitis diagnosed by positive anti-HCVAb and PCR.

3.2. Exclusion Criteria

- Patients < 18 and > 60 years old
- Patients who refused to give written consent to be included in the study (the consent included description of the study and subsequent editing and publication)
- Patients suffering from previously known thyroid dysfunction
- Patients on medications that could affect thyroid functions e.g. Carbamazepine, Phenobarbitone, Phenytoin, Salicylates and Nonsteroidal anti-inflammatory drugs (NSAIDs)
- Patients with other etiology for liver cirrhosis e.g. history of significant alcohol consumption > 60 g/day, patients with positive HbsAg, patients with evidence of metabolic or autoimmune disease
- Patients under interferon therapy
- Patients admitted to the hospital in an acute event such as upper GI bleeding or sepsis.
- Patients with acute hepatitis and fulminant liver failure.

Patients were allocated to three groups

Group I: included 24 patients with compensated post-hepatitis C cirrhosis (Child's grade A)

Group II: included 24 Patients with decompensated post-hepatitis C cirrhosis (Child's grade B or C) without encephalopathy

Group III: included 24 Patients with decompensated post-hepatitis C cirrhosis (Child's grade B or C) with encephalopathy

All patients were subjected to:

- Thorough history taking regarding the duration of liver disease, history of alcohol intake
- Physical examination with special stress on manifestation of liver cirrhosis.
- Laboratory investigations including: Complete blood count (CBC) by Dyn 1700, Liver and kidney function tests including: Serum albumin, Total and direct bilirubin, Aspartate aminotransferase (AST), Alanine aminotransferase (ALT), serum creatinine and blood urea nitrogen by integra 400 analyser, coagulation profile, Prothrombin Time (PT) and (INR) by sysmex CA 1500, Serum antibodies to HCV and HbsAg by ELISA, auto immune markers (ANA, ALKM Ab, SMA) by slide immunoflorescence.
- Thyroid parameters including total T3, free T4 and TSH serum levels:
- T3 was measured using (Calbiotech, Inc. (CBI) total T3 ELISA kit). The assay was run fully automated. Normal level = (0.52 - 1.58 ng/ml).
- FT4 was measured using (Chemux BioScience, INC kit). The assay was run fully automated. Normal level = (0.65 - 1.97 ng/dl).
- TSH test done using (Chemux BioScience, INC kit). The assay was run fully automated. Normal level = (0.4 - 7.0 uIU/ml).
- Abdominal ultrasonography: examine patients for manifestations of liver cirrhosis and portal hypertension e.g. increased liver echogenicity, coarse echotexture, irregular borders, splenomegally, ascites, dilated portal vein.
- **The severity of the liver dysfunction** was graded according to Child-Pugh classification [12].

	1	2	3
Bilirubin Total	<2 mg/dl	2 - 3 mg/dl	>3 mg/dl
Serum albumin	>3.5 g/dl	2.8 - 3.5 g/dl	<2.8 g/dl
INR	<1.7	1.71 - 2.20	>2.20
Ascites	None	Suppressed with medication	Refractory
Hepatic encephalopathy	None	Grade I-II (or suppressed with medication)	Grade III-IV (or refractory)

Points	Class
5 - 6	A
7 - 9	B
10 - 15	C

4. Statistical Analysis

Data were analyzed using SPSS (Statistical Package for the Social Sciences) version 15 for data processing and statistics. Numbers and percentages were used for qualitative data while mean ± standard deviation (SD) was used for quantitative ones. Chai square test X^2 was used to compare categorical data and ANOVA was used to compare numerical data. KW test was used for numerical data when normal distribution is lacking. Linear regression was used to test the correlation between variables. P value < 0.05 was considered significant.

5. Results

Comparison between the three studied groups as regards their demographic data shows that there were no significant differences as regards age and gender distribution as shown in **Table 1**. **Table 2** represents comparison between the three studied groups as regards all the laboratory parameters. It shows that group I had significantly higher hemoglobin concentration, WBC's count and albumin concentration than the other two groups. Group I also had significantly lower bilirubin level, liver enzymes, PT, INR, urea and creatinine than the other two groups. When it comes to thyroid functions **Table 2** shows that group I had significantly higher level of serum total T3 than the other two groups and group II had significantly higher level of serum total T3 than group III. Comparing the level of free T4 revealed no significant difference between the studied groups while the level of TSH was significantly higher in group II. **Table 3** shows that most of the patients included in the study had normal thyroid function tests. Comparing the three studied groups as regards the prevalence of subnormal levels of the studied thyroid parameters shows that group III had statistically higher incidence of subnormal total T3 level than the other two groups and group II shows significantly higher incidence of subnormal total T3 than group I as shown in **Table 3**. **Table 3** also shows that the prevalence of subnormal serum level of free T4 is not significantly different among the three groups. Also the level of TSH was within normal range in all patients of the three groups. **Table 4** shows that there's highly significant negative correlation between total T3 and the Child's score while TSH has significant positive correlation with it.

6. Discussion

This work aims to study the relation between serum levels of some thyroid parameters specially total T3 and HCV related cirrhosis and their relation to the severity of decompensation. The three groups of patients in this study represented three degrees of hepatic decompensation to evaluate the relation between thyroid and liver functions. There were no significant differences between the three studied groups in this study as regard age and gender distribution.

There was a statistically significant difference between different studied groups regarding the mean value of serum total T3 and serum TSH. We found that, the mean value of serum total T3 was the lowest among decompensated cirrhotic patients with hepatic encephalopathy (group III) (0.74 ± 0.3 ng/ml) and Kayacetin *et al.*, 2003 agrees with this finding in is study on patients with hepatic encephalopathy, followed by decompensated cirrhotic patients without hepatic encephalopathy (group II) (1.0 ± 0.3 ng/ml) but the highest mean value was seen among compensated cirrhotic patients (group I) (1.33 ± 0.3 ng/ml) [13]. This means that mean level of total T3 in groups II and III was higher than in group I. These results come in agreement with most of the previous studies dealing with this topic (Yamanaka *et al.*, 1980, Brozio *et al.*, 1983, Shimada *et al.*, 1988, Agha *et al.*, 1989, Antonelli *et al.*, 2006, Spadaro *et al.*, 2004, Elkabbany *et al.*, 2012, Mansour-Ghanaei *et al.*, 2012 and Dehghani *et al.*, 2013) [6] [14]-[21]. this finding is explained by the fact that liver plays a central role in the peripheral

Table 1. Demographic data.

		Group I No = 24		Group II No = 24		Group III No = 24		Test value	P	Signif.
Age (years) Mean ± SD		53.2 ± 5.1		58.5 ± 8.1		61.5 ± 7.2		8.89*	<0.001	HS
		No	%	No	%	No	%			
Gender	**Male**	14	58.3	9	37.5	17	70.8	5.51#	0.06	NS
	Female	10	41.7	15	62.5	7	29.2			

*ANOVA, #X².

Table 2. Comparison between the three groups as regards all laboratory parameters.

	Group I No = 24	Group II No = 24	Group III No = 24	F	P	Signif.
HB(g/dl) Mean ± SD	12.1 ± 2.0	10.1 ± 2.4	11.1 ± 2.2	4.9	0.009	S
WBC's(cells x10³/ml) Mean ± SD	6.1 ± 2.1	7.1 ± 3.5	9.9 ± 5.8	5.29	0.007	S
Platelet(cellsx10³/ml) Mean ± SD	94.6 ± 28.2	91.3 ± 30	110.3 ± 38.9	2.3	0.1	NS
Total Bilirubin (mg/dl) Mean ± SD	0.9 ± 0.4	2.6 ± 1.4	3.7 ± 2.5	16.7	<0.001	HS
Direct Bilirubin(mg/dl) Mean ± SD	0.3 ± 0.2	1.1 ± 0.8	1.8 ± 1.5	12.5	<0.001	HS
Albumin (g/dl) Mean ± SD	3.6 ± 0.5	2.4 ± 0.3	2.5 ± 0.5	53.2	<0.001	HS
ALT (IU/ml) Mean ± SD	47.8 ± 32.5	46.2 ± 28.3	80.4 ± 126	KW = 1.16	0.55	NS
AST(IU/ml) Mean ± SD	51.4 ± 25.4	63 ± 45.9	131.7 ± 214.7	KW = 5.2	0.07	NS
INR Mean ± SD	1.16 ± 0.14	1.7 ± 0.5	1.6 ± 0.4	12.3	<0.001	HS
PT(sec) Mean ± SD	13.5 ± 0.7	16.3 ± 3.8	16.1 ± 2.7	7.68	<0.001	HS
Blood Urea (mg/dl) Mean ± SD	33.8 ± 17.6	59.9 ± 43.8	102 ± 37.1	11.2	<0.001	HS
Creatinine (mg/dl) Mean ± SD	0.8 ± 0.2	1.1 ± 0.6	1.6 ± 1.2	5.6	0.005	S
Child's Score(Points) Mean ± SD	5.4 ± 0.5	10.2 ± 1.5	11.2 ± 2.1	99.03	<0.001	HS
Total T3(ng/ml) Mean ± SD	1.33 ± 0.3	1.0 ± 0.3	0.74 ± 0.3	19.56	<0.001	HS
Free T4(ng/dl) Mean ± SD	1.46 ± 0.5	1.47 ± 0.8	1.69 ± 1.1	0.56	0.58	NS
TSH (uIU/ml) Mean ± SD	0.86 ± 0.57	1.36 ± 1.0	0.81 ± 0.6	KW = 6.5	0.03	S

Table 3. Comparison between the studied groups as regards the prevalence of hypothyroidism.

	<Lower Limit N = 72(%)	Normal Range N = 72(%)	>Upper Limit N = 72(%)			
Total T3	14 (19.4%)	57 (79.2%)	1 (1.4%)			
Free T4	10 (13.9%)	56 (77.8%)	6 (8.3%)			
TSH	8 (11.1%)	64 (88.9%)	0 (0.0%)			
	Group I No = 24	Group II No = 24	Group III No = 24	X^2	P	Signif.
Total T3 — Normal 0.52 - 1.58 ng/ml	24 (100%)	20 (83.3%)	14 (58.3%)	13.48	0.001	HS
Total T3 — <Lower Limit	0 (0%)	4 (16.7%)	10 (41.7%)			
Free T4 — Normal 0.65 - 1.97 ng/dl	22 (91.7%)	19 (79.2%)	21 (87.5%)	1.63	0.4	NS
Free T4 — <Lower Limit	2 (8.3%)	5 (20.8%)	3 (12.5%)			
TSH — Normal 0.4 - 7.0 uIU/ml	24 (100%)	24 (100%)	24 (100%)	0.0	1	NS
TSH — >Upper Limit	0 (0%)	0 (0%)	0 (0%)			

Table 4. Correlation between thyroid function tests and Child's score.

	R	P	Signif.
Total T3	−0.64	<0.001	HS
Free T4	0.15	>0.05	NS
TSH	0.25	<0.05	S

conversion of T4 to T3 and that in conditions of liver cirrhosis the enzyme system of the diseased liver converts T4 to reverse T3b which is less active than T3 [8]. This finding was also explained by Novis et al., 2001 that found that the lower T3 level was accompanied by a higher reverse T3 level which is a less active form of the hormone that appear due to peripheral conversion of T4 [22]. This finding is also emphasized by Tas et al., 2012 in a study that linked this lower T3 levels to higher mortality in critically ill cirrhotic patients [23]. The level of free T4 showed no significant differences between the three groups, this agrees with Spadaro et al, 2004 and Elkabbany et al., 2012 [19] [20]. This finding disagrees with Yamanaka et al., 1980, Shimada et al., 1988, Agha et al., 1989, Kayacetin et al., 2003, Antonelli et al., 2006 and Dehghani et al., 2013 who said that the level of F T4 was also lower than non-cirrhotic patients [13] [14] [16]-[18] [21]. This also disagrees with Brozio et al., 1983 who found that free T4 was higher than non-cirrhotic patients [15]. The authors who say that the free T4 is lower in cirrhotic patients though they are clinically euthyroid explain this by the fact that there are some changes in the thyroid gland itself [19]. Comparing the three studied groups as regards TSH level revealed that group II had significantly higher TSH, this agrees with Aizawa et al., 1980, Schlienger et al., 1980 and Antonelli et al., 2006 [18] [24] [25]. This finding agrees also with Atalav et al., 2015 that said that cirrhotic patients have lower levels of TSH especially at night [26]. This finding disagrees with Spadaro et al., 2004 and Elkabbany et al., 2012 that found that there was no change in TSH level [19] [20].

A highly significant negative correlation was observed, between total T3 and Child's score and a significant positive correlation between TSH and Child's score and non significant correlation as regard serum free T4. This relatively comes in the same line with Mansour-Ghanaei et al. (2012) who found that only for serum total T3, its level decreased with advancing Child-Pugh score [6].

Comparing the three groups as regards prevalence of subnormal thyroid function levels, we found that the incidence of low serum total T3 (below the lower limit) was the highest in decompensated cirrhotic patients with hepatic encephalopathy (group III) (41.7%), followed by decompensated cirrhotic patients without hepatic encephalopathy (group II) (16.7%) versus none of the compensated cirrhotic patients (group I). This difference was statistically highly significant when compare compensated group I with other two decompensated groups (II, III), which means, in other words that the frequency of patients with serum total T3 level below the lower limit significantly increased with severity of liver cirrhosis. This finding agrees with Tas et al., 2012 who found that T3 level is correlated to MELD and Child's score of the patients. As regard serum FT4 level, in our study, we found that, There is no statistically significant correlation with the Child's score this also agrees with Tas et al., 2012. The study by Tas et al. disagrees with our study as regards the fact that the level of TSH is also correlated to the Child's score. Tas et al. believe that TSH isn't significantly correlated to the severity of liver disease [23].

In our study, it was found that 11.1% (8 patients) of all patients with liver cirrhosis had serum TSH level below the lower limit, serum TSH level was normal in most cases. With a closer look at these abnormalities, patients with low serum TSH level included 3 patients with compensated liver cirrhosis (group I) and 5 patients with decompensated liver cirrhosis with hepatic encephalopathy (group III). This can be explained by the fact that late alteration in thyroid metabolism is a decrease in the pituitary secretion of TSH. Such changes may be a self-protective adaptation to illness, as the body attempts to conserve energy [27].

Such thyroid function derangements sought in this study may be attributed either to a true thyroid dysfunction associated with liver disease or the well-established entity of non-thyroidal illness syndrome (NTIS) formerly known as sick euthyroid syndrome. These findings agree with previous studies that analyzed thyroid dysfunction during critical illness as Fliers et al. study that reported a significant inverse correlation between serum total T3 concentrations and the severity of liver dysfunction. Also, Borzio et al. study that compared cirrhotic with normal subjects and chronic hepatitis patients. They found that serum total and free T3 levels inversely paralleled severity of liver dysfunction. TSH levels are described to be commonly within the normal range in NTIS but may decrease in prolonged illness [15] [28].

7. Conclusion

Patients with chronic liver disease may have lower serum total T3 level than normal though clinically euthyroid. The decline in thyroid function is correlated to the severity of liver disease.

Conflict of Interests

The research was conducted in the absence of any commercial or financial relationships that could be construed as a potential conflict of interest.

References

[1]   Iredale, J.P. (2003) Cirrhosis: New Research Provides a Basis for Rational and Targeted Treatments. *BMJ*, **327**, 143-147. http://dx.doi.org/10.1136/bmj.327.7407.143

[2]   Puche, J.E., Saiman, Y. and Friedman, S.L. (2013) Hepatic Stellate Cells and Liver Fibrosis. *Comprehensive Physiology*, **3**, 1473-1492. http://dx.doi.org/10.1002/cphy.c120035

[3]   Hennemann, G., Docter, R., Friesema, E.C., de Jong, M., Krenning, E.P. and Visser, T.J. (2001) Plasma Membrane Transport of Thyroid Hormones and Its Role in Thyroid Hormone Metabolism and Bioavailability. *Endocrine Reviews*, **22**, 451-476. http://dx.doi.org/10.1210/edrv.22.4.0435

[4]   Mendel, C.M., Cavalieri, R.R. and Weisiger, R.A. (1988) Uptake of Thyroxine by the Perfused Rat Liver: Implications for the Free Hormone Hypothesis. *The American Journal of Physiology*, **255**, E110-E119.

[5]   Bianco, A.C., Salvatore, D., Gereben, B., Berry, M.J. and Larsen, P.R. (2002) Biochemistry, Cellular and Molecular Biology, and Physiological Roles of the Iodothyronine Selenodeiodinases. *Endocrine Reviews*, **23**, 38-89. http://dx.doi.org/10.1210/edrv.23.1.0455

[6]   Mansour-Ghanaei, F., Mehrdad, M., Mortazavi, S., Joukar, F., Khak, M. and Atrkar-Roushan, Z. (2012) Decreased Serum Total T3 Level in Hepatitis B and C Related Cirrhosis by Severity of Liver Damage. *Annals of Hepatology*, **11**, 667-671.

[7]   Bianchi, G.P., Zoli, M., Marchesini, G., Volta, U., Vecchi, F., Iervese, T., Bonazzi, C. and Pisi, E. (1991) Thyroid Gland Size and Function in Patients with Cirrhosis of the Liver. *Liver*, **11**, 71-77. http://dx.doi.org/10.1111/j.1600-0676.1991.tb00495.x

[8]   Malik, R. and Hodgson, H. (2002) The Relationship between the Thyroid Gland and the Liver. *QJM*, **95**, 559-569. http://dx.doi.org/10.1093/qjmed/95.9.559

[9]   Oren, R., Sikuler, E., Wong, F., Blendis, L.M. and Halpern, Z. (2000) The Effects of Hypothyroidism on Liver Status of Cirrhotic Patients. *Journal of Clinical Gastroenterology*, **31**, 162-163. http://dx.doi.org/10.1097/00004836-200009000-00016

[10]   Oren, R., Brill, S., Dotan, I. and Halpern, Z. (1998) Liver Function in Cirrhotic Patients in the Euthyroid versus the Hypothyroid State. *Journal of Clinical Gastroenterology*, **27**, 339-341. http://dx.doi.org/10.1097/00004836-199812000-00012

[11]   Mohamed, M.K. (2005) Epidemiology of HCV in Egypt. *The Afro-Arab Liver Journal*, **3**, 41-52.

[12]   Child, C.G. and Turcotte, J.G. (1964) Surgery and Portal Hypertension. In: Child. C.G., Ed., *The Liver and Portal Hypertension*, Saunders, Philadelphia, 50-64.

[13]   Kayacetin, E., Kisakol, G. and Kaya, A. (2003) Low Serum Total Thyroxine and Free Triiodothyronine in Patients with Hepatic Encephalopathy Due to Non-Alcoholic Cirrhosis. *Swiss Medical Weekly*, **133**, 210-213.

[14]   Yamanaka, T., Ido, K., Kimura, K. and Saito, T. (1980) Serum Levels of Thyroid Hormones in Liver Diseases. *Clinica Chimica Acta*, **101**, 45-55.

[15]   Borzio, M., Caldara, R., Borzio, F., Piepoli, V., Rampini, P. and Ferrari, C. (1983) Thyroid Function Tests in Chronic Liver Disease: Evidence for Multiple Abnormalities Despite Clinical Euthyroidism. *Gut*, **24**, 631-636. http://dx.doi.org/10.1136/gut.24.7.631

[16]   Shimada, T., Higashi, K., Umeda, T. and Sato, T. (1988) Thyroid Functions in Patients with Various Chronic Liver Diseases. *Endocrinologia Japonica*, **35**, 357-369. http://dx.doi.org/10.1507/endocrj1954.35.357

[17]   Agha, F., Qureshi, H. and Khan, R.A. (1989) Serum Thyroid Hormone Levels in Liver Cirrhosis. *Journal of Pakistan Medical Association*, **39**, 179-183.

[18]   Antonelli, A., Ferri, C., Fallahi, P., Ferrari, S.M., Ghinoi, A., Rotondi, M. and Ferrannini, E. (2006) Thyroid Disorders in Chronic Hepatitis C. *The American Journal of Medicine*, **117**, 10-13.

[19] Spadaro, L., Bolognesi, M., Pierobon, A., Bombonato, G., Gatta, A. and Sacerdoti, D. (2004) Alterations in Thyroid Doppler Arterial Resistance Indices, Volume and Hormones in Cirrhosis: Relationships with Splanchnic Haemodynamics. *Ultrasound in Medicine and Biology*, **30**, 19-25. http://dx.doi.org/10.1016/j.ultrasmedbio.2003.10.008

[20] El-Kabbany, Z.A., Hamza, R.T., Abd El Hakim, A.S. and Tawfik, L.M. (2012) Thyroid and Hepatic Haemodynamic Alterations among Egyptian Children with Liver Cirrhosis. *ISRN Gastroenterol*, **2012**, 595734. http://dx.doi.org/10.5402/2012/595734

[21] Dehghani, S.M., Haghighat, M., Eghbali, F., Karamifar, H., Malekpour, A., Imanieh, M.H. and Malek-Hoseini, S.A. (2013) Thyroid Hormone Levels in Children with Liver Cirrhosis Awaiting a Liver Transplant. *Experimental and Clinical Transplantation*, **11**, 150-153. http://dx.doi.org/10.6002/ect.2012.0182

[22] Novis, M., Vaisman, M. and Coelho, H.S. (2001) Thyroid Function Tests in Viral Chronic Hepatitis. *Arquivos De Gastroenterologia*, **38**, 254-260.

[23] Taş, A., Köklü, S., Beyazit, Y., Kurt, M., Sayilir, A., Yeşil, Y. and Çelik, H. (2012) Thyroid Hormone Levels Predict Mortality in Intensive Care Patients with Cirrhosis. *The American Journal of the Medical Sciences*, **344**, 175-179. http://dx.doi.org/10.1097/MAJ.0b013e318239a666

[24] Aizawa, T., Yamada, T., Tawata, M., Shimizu, T., Furuta, S., Kiyosawa, K. and Yakata, M. (1980) Thyroid Hormone Metabolism in Patients with Liver Cirrhosis, as Judged by Urinary Excretion of Triiodothyronine. *Journal of the American Geriatrics Society*, **28**, 485-491. http://dx.doi.org/10.1111/j.1532-5415.1980.tb01126.x

[25] Schlienger, J.L., Jacques, C., Sapin, R. and Stephan, F. (1980) Thyroid Function in Patients with Alcoholic Cirrhosis. *Ann Endocrinol* (*Paris*), **41**, 81-94.

[26] Atalay, R., Ersoy, R., Demirezer, A.B., Akın, F.E., Polat, S.B., Cakir, B. and Ersoy, O. (2015) Day-Night Variations in Thyroid Stimulating Hormone and Its Relation with Clinical Status and Metabolic Parameters in Patients with Cirrhosis of the Liver. *Endocrine*, **48**, 942-948. http://dx.doi.org/10.1007/s12020-014-0364-1

[27] Hamblin, P.S., Dyer, S.A., Mohr, V.S., Le Grand, B.A., Lim, C.F., Tuxen, D.V., Topliss, D.J. and Stockigt, J.R. (1986) Relationship between Thyrotropin and Thyroxine Changes during Recovery from Severe Hypothyroxinemia of Critical Illness. *The Journal of Clinical Endocrinology & Metabolism*, **62**, 717-722. http://dx.doi.org/10.1210/jcem-62-4-717

[28] Fliers, E., Bianco, A.C., Langouche, L. and Boelen, A. (2015) Thyroid Function in Critically Ill Patients. *The Lancet Diabetes & Endocrinology*, **3**, 816-825. http://dx.doi.org/10.1016/S2213-8587(15)00225-9

13

Partial Splenic Artery Embolization in Cirrhosis Is a Safe and Useful Procedure

Fakhar Ali Qazi Arisar1*, Syed Hasnain Ali Shah1, Tanveer Ul Haq2

1Section of Gastroenterology, Department of Medicine, The Aga Khan University Hospital, Karachi, Pakistan
2Section of Interventional Radiology, Department of Radiology, The Aga Khan University Hospital, Karachi, Pakistan

Email: *fakhar.qazi@aku.edu, hasnain.alishah@aku.edu, tanveer.haq@aku.edu

Abstract

Background: Portal Hypertension is a common complication of cirrhosis. It leads to splenomegaly which manifests with features of hypersplenism. This results in leucopenia which increases the likelihood of sepsis and prevents treatment with interferon. Thrombocytopenia increases the risk of bleeding including variceal bleeds which make the anemia worse. This study was done to determine the usefulness and safety of partial splenic artery embolization (PSAE) in portal hypertension due to cirrhosis. Methods: Patients with PSAE were identified by using International Classification of Diseases (ICD)-10 coding from medical records and their charts were reviewed retrospectively. 25 patients underwent splenic artery embolization at The Aga Khan University Hospital Karachi from November 2000 to December 2016. 18 patients who underwent PSAE for disabling hypersplenism caused by cirrhosis were included. Patients who were under 18 year of age, or in whom PSAE were performed for reasons other than cirrhosis and those with missing records/incomplete data were excluded (n = 7). Information was collected regarding demographic details, procedure indications, nature, technique, clinical efficacy, repeat embolization and complications along with laboratory and radiological investigations. Results: Eighteen patients of cirrhosis with a mean age of 43.47 ± 10.926 years, of which 14 were males, underwent PSAE (19 procedures). Indications were severe hypersplenism which precluded treatment with interferon and ribavirin (n = 8) and recurrent Gastro-oesophageal variceal (GOV) bleeds due to advanced Child-Pugh grade and thrombocytopenia (n = 10). Hematological parameters improved significantly following PSAE. Three out of eight patients successfully completed interferon + ribavirin treatment for hepatitis C (HCV) infection post PSAE, and GOV bleeds stopped in eight out of 10 patients. Complications included mild Left upper quadrant (LUQ) abdominal pain n = 9 (47.3%), post-embolization syndrome n = 4 (21%), and clinically insignificant pleural effusion n = 4 (21%). One patient developed spontaneous bacterial peritonitis (SBP) which was appropriately managed. One patient needed re-emobilization after 6 months. Conclusion: PSAE is a safe and effective procedure in the treatment of hypersplenism due to cirrhosis.

Keywords

Partial Splenic Artery Embolization, Chronic Liver Disease, Cirrhosis, Hypersplenism, Safety, Outcome

1. Introduction

Hypersplenism occurs in the majority of patients suffering from cirrhosis and portal hypertension [1]. Splenic pooling and sequestration of blood cells lead to cytopenias, which precludes them from treatment options for chronic viral hepatitis or any invasive procedure/surgeries and also makes them prone to recurrent GI bleed.

Currently partial splenic artery embolization (PSAE) is an approved minimally invasive intervention for management of portal hypertension, gastric variceal hemorrhage, hypersplenism, control of splenic hemorrhage from trauma or prior to surgical resection, idiopathic thrombocytopenic purpura, splenic artery aneurysms or pseudo-aneurysms, and splenic steal syndrome (nonocclusive hepatic artery hypoperfusion syndrome [NOHAH]) in liver transplant recipients [2] [3] [4].

Splenic artery embolization interrupts arterial flow to the splenic artery or to one of its branches; it may be considered as an alternative, either to splenectomy or to ligation of the splenic artery. Total splenectomy may be a useful cure for hypersplenism, but it is associated with perioperative complications and also impairs the capability of the body to create antibodies against encapsulated microorganisms, thus making the individual prone to sepsis. Many authors have advocated incomplete or partial embolization (PSAE), in which a segment of the splenic parenchyma is left viable to maintain the immunologic function [3] [5].

Splenic artery embolization was introduced in 1973 as a nonsurgical treatment for variceal hemorrhage and hypersplenism [6]. Six years after this initial report, Spigos et al. treated patients with partial splenic artery embolization using antibiotic coverage and post-embolization pain control [7]. The process helps occlude the arterial supply of the spleen more peripherally, which results in ischemic necrosis of much of the functional spleen leading to a reduction in splenic size and consequently counters hypersplenism. Initially the application of procedure remained limited because of the high rate of morbidity and mortality, but later on, the outcomes improved once partial embolization and antibiotic coverage were adopted [2] [3]. This technique can be used safely when total embolization volume is ~50% and the procedural and peri-procedural time periods are covered with antibiotics [2]. Also, it is a simple, and rapid procedure that can be performed easily under local anesthesia and incurs less morbidity, and there is no need for blood transfusion [5].

The current study is planned with the purpose of providing a detailed assessment of the safety, effectiveness, and clinical outcome in cirrhotic patients who underwent PSAE. To the best of our knowledge, no such study has yet been carried out in Pakistan and it will definitely be a good addition in the upcoming literature from our country.

2. Material and Methods

2.1. Patients

Patients with PSAE were identified by using International Classification of Diseases (ICD)-10 coding from medical records and their charts were reviewed retrospectively. 25 patients underwent splenic artery embolization at The Aga Khan University Hospital Karachi from November 2000 to December 2016. 18 patients who underwent PSAE for disabling hypersplenism caused by cirrhosis were included. Patients who were under 18 year of age, or in whom PSAE was performed for reasons other than cirrhosis, and those with missing records/incomplete data were excluded (n = 7).

Information was collected regarding age, gender, primary and secondary outcomes, length of hospital stay, comorbid conditions, procedure indications, nature, technique, clinical efficacy, repeat embolization and complications along with laboratory and radiological investigations. This study was granted exemption from ethical approval by the institutional ethical review committee of The Aga Khan University Hospital, Pakistan.

2.2. Pre-Procedure Preparation

The day before examination the patients were admitted to the hospital. Their laboratory data were revised and they were started on antibiotics (ceftriaxone 2 g IV OD). Six units of platelets were given the night before the examination. Prophylactic vaccination was not given to any patient.

2.3. Embolization Protocol

The PSAE procedures were performed in the interventional radiology suite. After standard preparation, under local anesthesia, the percutaneous arterial access was obtained with Seldinger's technique (4F introducer). The celiac trunk and the splenic artery were selectively catheterized with a 4F C1 catheter. Embolization was then done by administration of polyvinyl alcohol (PVA) particles. The size of the particles of embolization ranged from 255 to 710 micron. The infarction volume was estimated by a selective angiogram showing the reduction in the splenic vascularization and the residual spleen parenchyma.

2.4. Follow-Up

The post-procedure follow-up was analyzed retrospectively from the medical case notes. Post-PSAE supportive care included appropriate hydro-electrolytic infusion, systemic prophylaxis with antibiotics, using ceftriaxone within hospital and cefixime (400 mg/day) on discharge for at least 5 days, and analgesic treatment using paracetamol or tramadol. All patients stayed in hospital after PSAE until they became clinically stable, and were then followed up at the outpatient clinic. Peripheral blood cell parameters including hemoglobin (Hb), white blood cell count (WBC), and platelet count were monitored on the 3rd, 14th, 30th day after PSAE, and subsequently at 3-month intervals during the 1-year follow-up period.

2.5. Outcome

- Primary outcome: technically and clinically successful embolization, defined by the ability to resume or begin treatment that induces pancytopenia (such as interferon), and the prevention or reduction of recurrent bleeding.

- Secondary outcome: morbidity and mortality related to PSAE.

2.6. Statistical Analysis

Data were entered in a commercial statistical software package (IBM SPSS Statistics for Windows, version 19.0). The results were expressed as means $\pm$ standard deviations or median $\pm$ range for quantitative variables, and compared according to the volume of embolized parenchyma. Proportions were calculated for qualitative variables. Hb, platelet count and WBC count pre and post-PSAE were compared using a pairedt-test. The association between the splenic infarction rate and the rate of increase in blood cell counts was analyzed by the Spearman rank correlation coefficient (rsp). Significance was established at $P < 0.05$.

3. Results

A total of 19 procedures were performed on 18 cirrhotic patients. 14 (77.7%) were male, age between 15 to 64 years (mean age: 43.47 $\pm$ 10.926 years). The indication of parenchymal reduction for eight patients was cytopenias too severe to initiate or continue treatment with interferon and ribavirin (Hb < 10 g/dl or platelets < 90 × 109/L) [8] [9]; 10 cases presented with recurrent gastrointestinal bleeding with severe thrombocytopenia.

The baseline characteristics of the total cohort of PSAE-treated patients are shown in Table 1. Nine patients each were classified as Child-Pugh Class A and B [10] [11]. The etiologic factors of cirrhosis were hepatitis C in 15 patients and hepatitis B in 2 patients, while one patient had Non-B, Non-C related cirrhosis. 13 patients had esophageal varices. No HCC was detected in the study subjects before PSAE.

All procedures were done electively except for one who underwent emergency embolization due to failure to control upper GI bleed by endoscopic variceal band ligation (EVBL). PSAE was effectively performed in all patients with percentage embolization volume ranging from 30% to 50%. Patients were followed up for a median period of 304 (Range: 07 - 1126) days after the procedure. Repeat procedure was only necessary for one patient due to persistent thrombocytopenia.

Table 1. Baseline characteristics of the total cohort of psae-treated patients.

Variables		n = 18 (%)
Gender	Male	14 (77.7%)
	Female	4 (22.3%)
Etiology	HCV	15 (83.3%)
	HBV	1 (5.5%)
	HBV + HDV	1 (5.5%)
	NBNC	1 (5.5%)
Co-morbid	DM	4 (22.3%)
	HTN	5 (27.7%)
	IHD	1 (5.5%)
	Hereditary spherocytosis	1 (5.5%)
Child Class	A	9 (50%)
	B	9 (50%)
Esophageal varices	Absent	5 (27.7%)
	Small	3 (16.6%)
	Large	10 (55.5%)
Indication	Upper GI Bleed	10 (55.5%)
	HCV treatment	8 (44.5%)
Nature of procedure	Urgent	1 (5.5%)
	Elective	17 (94.5%)
PVA particles (micron)	255 - 300	1 (5.5%)
	355 - 500	6 (33.3%)
	500 - 710	11 (61.1%)
	Both 355 - 500 & 500 - 710	1 (5.5%)
Percentage embolization volume	(mean ± SD)	37.22 ± 7.321
Length of stay in days	(Median + Range)	02 (01 - 25)
Follow up in days	(Median + Range)	304 (07 - 1126)

HBV: Hepatitis B virus; HCV: Hepatitis C virus; HDV: Hepatitis D virus; NBNC: Non-Hepatitis B, Non-HepatitisC; DM: Diabetes mellitus; HTN: Hypertension; IHD: Ischemic heart disease; PVA: Polyvinyl alcohol.

Irrespective of the percentage embolization volume, all cell lineages improved significantly immediately after PSAE until 1 week. After which a progressive decline was noted in all cell counts as shown in Table 2. However, despite a decrease in counts, leukocytes and platelets remained persistently improved at 1 month, 6 months and until 1 year of follow up with a significant difference from baseline (P-value of 0.003 and 0.003 at 1 month, 0.017 and <0.001 at 6 months and 0.006 and 0.03 at 1 year respectively). The average maximum spleen length before PSAE was 150.34 ± 67.36 cm, while after 4 weeks of PSAE, it was 136.0 ± 65.18 cm.

Antiviral treatment with interferon was started after an adequate rise in cell counts in five patients, of which 3 were successfully able to complete the full duration of treatment as described in Table 3. Of the other 2 patients, 1 got decompensated with ascites while other was lost to follow-up. In three patients, planned antiviral treatment was not started due to persistent thrombocytopenia (n = 2) and anemia (n = 1). The overallprimary outcome of PSAE was achieved in 13 (72%) patients (upper GI bleed control: n = 8, resume/begin interferon therapy: n = 5).

Table 2. Comparison of pre- and post-procedure spleen size and hematologic parameters.

	Pre PSAE (Mean ± SD)	Post PSAE (Mean ± SD)				
		1 Day	1 Week	4 Weeks	26 Weeks	52 Weeks
n	18	17	17	17	15	10
Spleen size (mm)	150.34 ± 67.37			136.00 ± 65.18 P = 0.079		
Hemoglobin (g/dl)	10.47 ± 2.13	11.33 ± 2.06	11.62 ± 1.49	11.18 ± 1.75 P = 0.113	10.68 ± 1.73 P = 0.311	12.4 ± 1.74 P = 0.08
White blood cells (×10⁹/L)	3.29 ± 0.93	8.02 ± 5.57	9.12 ± 6.99	5.46 ± 2.39 P = 0.003	4.8 ± 2.23 P = 0.017	5.2 ± 1.42 P = 0.006
Platelets (×10⁹/L)	38.11 ± 15.41	54 ± 23.39	116 ± 79.87	107.93 ± 74.71 P = 0.003	63.92 ± 14.94 P ≤ 0.001	91.33 ± 61.21 P = 0.03

PSAE: Partial splenic artery embolization.

The median hospital stay after PSAE was 2 days (range from 1 to 25 days). The pain was the most common complication occurring in 9 (47.3%) patients followed by post-embolization syndrome [12] characterized by fever and abdominal pain in 4 (21%) patients (Table 4). Mild pleural effusion was noticed in 4 (21%) patients, which was picked up on follow-up images without any clinical symptom and sign. One patient had spontaneous bacterial peritonitis (SBP) and sepsis post procedure leading to prolong hospital stay. None of the patients developed splenic abscess. No immediate or late procedure-related mortality occurred in our patients on follow-up. No relation of percentage volume embolization with the development of complications was found.

4. Discussion

Pakistan has been labeled as a cirrhotic state long ago [13], with hepatitis C being the commonest cause (60% 90%) according to recent data [14] [15] [16]. According to a study, 68% of the cirrhotic patients in Pakistan were found to have hypersplenism [1].

The published efficacy rates for splenic artery embolization are fairly high; a recent meta-analysis of splenic artery embolization in the non-operative management of blunt splenic trauma found an overall failure rate of 15.7% [17], while success in the treatment of splenic arterial aneurysms and pseudoaneurysms is approximately 90% [18] [19] [20]. In patients with portal hypertension and hypersplenism, splenic artery embolization has been shown to produce significant and sustained improvements in both liver function and hematologic indices, as well as an 80% reduction in annual bleeding episodes in patients with recurrent variceal hemorrhage [21] [22].

Table 3. Primary outcomes achieved after PSAE.

Outcomes	n	%
Upper GI Bleed control		
For recurrent bleed	7/9	77.7%
For acute bleed	1/1	100%
Interferon started	5/8	62.5%
Interferon completed	3/5	60%
Re-embolization (After 6 months)	1/17	5.8%

Table 4. Morbidity after PSAE.

Complications	n	Percentage of total procedures (out of 19)
Pain	09	47.4%
Post-embolization syndrome	04	21.1%
Clinically insignificant pleural effusion	04	21.1%
Initial thrombocytopenia	02	21.1%
SBP, sepsis and prolong hospital stay	01	11.1%

PVA particles were used for embolization in our patients, ranging from 250 to 710 micron in size. Although gelfoam particles are also reported to be used by other researchers and found equivalent to PVA in terms of outcomes [23], the embolization volume was determined by confirming a reduction in the splenic vascularization using selective angiogram. The extent of embolization varied between 30% - 50% (mean 37.22 ± 7.321) which according to the literature is sufficient to cause infarction of around 70% - 80% of splenic parenchyma [24].

The highest mean values of WBC and platelets were recorded at 1-week post-PSAE, after which a gradual decline was noted for next 6 month followed by steady levels. The initial rapid rise could be either due to the release of platelets sequestered by spleen and splenic regeneration could have explained the gradual decline in platelets [24]. Initial leukocytosis could be due to an inflammatory process against infarcted spleen [25] [26]. Neither the percent embolized volume nor any other factor was found to be related with an increase in platelet count or WBC. Hayashi et al. also did not find any factor related to increase cell counts except infarcted splenic volume [27].

Our study supports the utilization of PSAE in controlling GI bleed. We demonstrated control of recurrent GI bleed in 7 out of 9 patients (77%) which is consistent with previous reports [4] [22].

Although serum albumin concentration was reported to rise after PSAE by some authors, no major change in liver enzymes is described [28]. We also did not find any significant impact on liver functions in patients undergoing PSAE consistent with the previous reports [24] [29] [30].

Post-embolization syndrome occurred in only four patients (22%) which is much lower than what was noted in previous studies (95% - 100%) [5] [23] [29] [31]. The reported mortality rate of PSAE is around 4% [32], while major complications are reported up to 17% [33]. In our cohort, no mortality was reported; however, one patient had sepsis and SBP requiring to prolong hospital stay.

Child C disease has been associated with complications after PSAE [31] [33]. However, none of the patients in our cohort belonged to child class C. Advanced age and post-procedure hydrothorax are the other factors related to mortality [4]. However, we were unable to identify any significant relationship between mortality and these factors.

Retrospective nature is the major weaknesses of this study. Other limitations include a small sample size, single center data collection, and non-randomized design. The long duration of enrollment and prolonged follow-up of patients are the major strengths of our study.

5. Conclusion

PSAE is a safe and effective procedure in the treatment of hypersplenism of cirrhosis.

Conflicts of Interest

No conflict of interest to disclose from all authors.

Supported Foundation

Not applicable.

Author's Contributions

Qazi Arisar FA designed and performed the research and wrote the manuscript. Shah SHA designed the research and supervised the manuscript. All authors agreed with the content of the manuscript. Haq TU designed the research, performed PSAE procedures and supervised the manuscript.

References

[1]   Ashraf, S. and Naeem, S. (2010) Frequency of Hypersplenism in Chronic Liver Disease Patients Presenting with Pancytopenia. Annals of King Edward Medical University, 16, 108-110.

[2]   Smith, M. and Ray, C.E. (2012) Splenic Artery Embolization as an Adjunctive Procedure for Portal Hypertension. Seminars in Interventional Radiology, 29, 135-139. https:/ / doi.or g/ 10.1055/ s-0032-1312575

[3]   Madoff, D.C., Denys, A., Wallace, M.J., Murthy, R., Gupta, S., Pillsbury, E.P., et al. (2005) Splenic Arterial Interventions: Anatomy, Indications, Technical Considerations, and Potential Complications. RadioGraphics, 25, S191-S211.

https:/ / doi.or g/ 10.1148/ r g.25si055504

[4]   Gaba, R.C., Katz, J.R., Parvinian, A., Reich, S., Omene, B.O., Yap, F.Y., et al. (2013) Splenic Artery Embolization: A Single Center Experience on the Safety, Efficacy, and Clinical Outcomes. Diagnostic and Interventional Radiology, 19, 49-55.

[5]   Amin, M.A., El Gendy, M.M., Dawoud, I.E., Shoma, A., Negm, A.M. and Amer, T.A. (2009) Partial Splenic Embolization versus Splenectomy for the Management of Hypersplenism in Cirrhotic Patients. World Journal of Surgery, 33, 1702-1710. https:/ / doi.or g/ 10.1007/ s00268-009-0095-2

[6]   Maddison, F.E. (1973) Embolic Therapy of Hypersplenism. Investigative Radiology, 8, 280-281. https:/ / doi.or g/ 10.1097/ 00004424-197307000-00054

[7]   Spigos, D., Jonasson, O., Mozes, M. and Capek, V. (1979) Partial Splenic Embolization in the Treatment of Hypersplenism. American Journal of Roentgenology, 132, 777-782. https:/ / doi.or g/ 10.2214/ajr .132.5.777

[8]   (2011) EASL Clinical Practice Guidelines: Management of hepatitis C Virus Infection. Journal of Hepatology, 55, 245-264.

https:/ / doi.or g/ 10.1016/ j.jhep.2011.02.023

[9]   AASLD-IDSA Recommendations for Testing, Managing, and Treating Hepatitis C. http:/ / www.hcvguidelin es.or g

[10] Child, C.G. and Turcotte, J.G. (1964) Surgery and Portal Hypertension. Major Problems in Clinical Surgery, 1, 1-85.

[11] Pugh, R.N., Murray-Lyon, I.M., Dawson, J.L., Pietroni, M.C. and Williams, R. (1973) Transection of the Oesophagus for

Bleeding Oesophageal Varices. British Journal of Surgery, 60, 646-649.

[12] Sakai, T., Shiraki, K., Inoue, H., Sugimoto, K., Ohmori, S., Murata, K., et al. (2002) Complications of Partial Splenic Embolization in Cirrhotic Patients. Digestive Diseases and Sciences, 47, 388-391. https:/ / doi.or g/ 10.1023/ A:1013786509418

[13] Ahmad, K. (2004) Pakistan: A Cirrhotic State? The Lancet, 364, 1843-1844. https:/ / doi.or g/ 10.1016/ S0140-6736(04)17458-8

[14] Fahim, U., Khan, S., Afridi, A.K. and ur Rahman, S. (2012) Frequency of Different Causes of Cirrhosis of Liver in Local Population. Gomal Journal of Medical Sciences, 10, 178-181.

[15] Arisar, F.A.Q., Khan, S.B. and Umar, A. (2014) Hepatic Encephalopathy in Chronic Liver Disease: Predisposing Factors in a Developing Country. Asian Journal of Medical Sciences, 6, 35-42.

[16] Umar, A., Arisar, F.A.Q., Sattar, R.A. and Umar, B. (2014) Non-Invasive Parameters for the Detection of Variceal Bleed in Patients of Liver Cirrhosis, an Experience of a Tertiary Care Hospital in Pakistan. Asian Journal of Medical Sciences, 6, 61-66. https:/ / doi.or g/ 10.3126/ ajm s.v6i1.9624

[17] Requarth, J.A., D'Agostino Jr., R.B. and Miller, P.R. (2011) Nonoperative Management of Adult Blunt Splenic Injury with and without Splenic Artery Embolotherapy: A Meta-Analysis. Journal of Trauma and Acute Care Surgery, 71, 898-903. https:/ / doi.or g/ 10.1097/ TA.0b013e318227ea50

[18] Loffroy, R., Guiu, B., Cercueil, J.-P., Lepage, C., Cheynel, N., Steinmetz, E., et al. (2008) Transcatheter Arterial Embolization of Splenic Artery Aneurysms and Pseudoaneurysms: Shortand Long-Term Results. Annals of Vascular Surgery, 22, 618-626. https:/ / doi.or g/ 10.1016/ j.avsg.2008.02.018

[19] Belli, A.-M., Markose, G. and Morgan, R. (2012) The Role of Interventional Radiology in the Management of Abdominal Visceral Artery Aneurysms. CardioVascular and Interventional Radiology, 35, 234-243.

https:/ / doi.or g/ 10.1007/ s00270-011-0201-3

[20] McDermott, V.G., Shlansky-Goldberg, R. and Cope, C. (1994) Endovascular Management of Splenic Artery Aneurysms and Pseudoaneurysms. CardioVascular and Interventional Radiology, 17, 179-184. https:/ / doi.or g/ 10.1007/ BF00571531

[21] Hirai, K., Kawazoe, Y., Yamashita, K., Kumagai, M., Tanaka, M., Sakai, T., et al. (1986) Transcatheter Partial Splenic Arterial Embolization in Patients with Hypersplenism: A Clinical Evaluation as Supporting Therapy for Hepatocellular Carcinoma and Liver Cirrhosis. Hepatogastroenterology, 33, 105-108.

[22] Koconis, K.G., Singh, H. and Soares, G. (2007) Partial Splenic Embolization in the Treatment of Patients with Portal Hypertension: A Review of the English Language Literature. Journal of Vascular and Interventional Radiology, 18, 463-481.

https:/ / doi.or g/ 10.1016/ j.jvir.2006.12.734

[23] Albadry, A., Elbatea, H.E. and Elfert, A.A. (2010) Long-Term Outcome of Angiographic Partial Splenectomy in Patients with Decompensated Liver Cirrhosis and Hypersplenism. Arab Journal of Gastroenterology, 11, 202-205.

https:/ / doi.or g/ 10.1016/ j.ajg.2010.09.004

[24] Nassef, A.A., Zakaria, A.A. and ElBary, M.S.A. (2013) Partial Splenic Artery Embolization in Portal Hypertension Patients with Hypersplenism: Two Interval-Spaced Sessions' Technique. The Egyptian Journal of Radiology and Nuclear Medicine, 44, 531-537. https:/ / doi.or g/ 10.1016/ j.ejr nm.2013.04.004

[25] Sundaresan, J.B., Dutta, T., Badrinath, S., Jagdish, S. and Basu, D. (2005) Study of Hypersplenism and Effect of Splenectomy on Patients with Hypersplenism. Journal of Indian Academy of Clinical Medicine, 6, 291-296.

[26] McCormick, P.A. (2007) Hypersplenism. Textbook of Hepatology: From Basic Science to Clinical Practice, 3rd Edition, 771-778.

[27] Hayashi, H., Beppu, T., Masuda, T., Mizumoto, T., Takahashi, M., Ishiko, T., et al. (2007) Predictive Factors for Platelet Increase after Partial Splenic Embolization in Liver Cirrhosis Patients. Journal of Gastroenterology and Hepatology, 22, 1638-1642. https:/ / doi.or g/ 10.1111/ j.1440-1746.2007.05090.x

[28] Tajiri, T., Onda, M., Yoshida, H., Mamada, Y., Taniai, N. and Kumazaki, T. (2002) Long-Term Hematological and Bio-

chemical Effects of Partial Splenic Embolization in Hepatic Cirrhosis. Hepatogastroenterology, 49, 1445-1448.

[29] Zhu, K., Meng, X., Qian, J., Huang, M., Li, Z., Guan, S., et al. (2009) Partial Splenic Embolization for Hypersplenism in Cirrhosis: A Long-Term Outcome in 62 Patients. Digestive and Liver Disease, 41, 411-416.

https:/ / doi.or g/ 10.1016/ j.dld.2008.10.005

[30] Pålsson, B., Hallén, M., Forsberg, A.M. and Alwmark, A. (2003) Partial Splenic Embolization: Long-Term Outcome. Langenbeck's Archives of Surgery, 387, 421-426.

[31] Hussein, W.M., Ahmed, A.T., Magdy, M., Amer, T.A. and Habba, M.R. (2017) Predictive Factors of Platelet Increase and Complications after Percutaneous Trans-Arterial Partial Splenic Embolization for Hypersplenism in Chronic Liver Disease Patients. The Egyptian Journal of Radiology and Nuclear Medicine, 48, 393-401. https:/ / doi.or g/ 10.1016/ j.ejr nm.2017.01.010

[32] Kumar, A., Yoon, J., Thakur, V. and Contractor, S. (2013) Abstract No. 317 Safety and Efficacy of Partial Splenic Embolization for Hypersplenism: A Meta-Analysis. Journal of Vascular and Interventional Radiology, 24, S138.

https:/ / doi.or g/ 10.1016/ j.jvir.2013.01.342

[33] Hayashi, H., Beppu, T., Okabe, K., Masuda, T., Okabe, H. and Baba, H. (2008) Risk Factors for Complications after Partial Splenic Embolization for Liver Cirrhosis. British Journal of Surgery, 95, 744-750. https:/ / doi.or g/ 10.1002/ bjs.6081

14

Epidemiological, Clinical and Paraclinical Aspects of Cirrhosis at Borgou Departmental University Hospital Center (Benin)

Comlan Albert Dovonou1,2, Cossi Adébayo Alassani1*, Kadidjatou Sake1,
Cossi Angelo Attinsounon1,2, Angèle Azon-Kouanou3, Agossou Romaric Tandjiekpon1,
Djimon Marcel Zannou3, Fabien Houngbe3

1Medicine Department and Medical Specialities, Medical Faculty, Parakou University, Parakou, Benin
2Internal Medicine Department, Departmental Hospital Center of Borgou, Parakou, Benin
3Medicine Department and Medical Specialities of Cotonou Health and Science Faculty, Cotonou, Benin

Email: *adebayoalassani@gmail.com

Abstract

Objective: To study the epidemiological, clinical and paraclinical aspects of cirrhosis at Borgou Departmental University hospital Center. Methods: This is a retrospective study for descriptive purpose conducted in the Internal Medicine Department. The study population consists of patients hospitalized in the Internal Medicine Department during the period from 1st January 2009 to 31st December 2016. Results: The frequency of cirrhosis was 1.35%. The sex ratio was 3.76. The average age of patients was 45.22 ± 15.23 years old, with a range from 15 to 82 years. There is a post hepatitis Bcirrhosis predominance in 87.5% of cases, followed by alcoholic cirrhosis in 21.59% of cases. The complications of cirrhosis are dominated by ascites (78.4%) and jaundice (52%). Conclusion: Cirrhosis is a condition that is wide spread. The hepatitis B virusis the main cause of liver cirrhosis followed by alcoholism. Ascites is the most encountered complication. It's very important now to educate the populations for a behaviour change and to promote vaccination against viral hepatitis.

Keywords

Epidemiology, Hepatic Cirrhosis, Parakou, Benin

1. Introduction

Hepatic cirrhosis is a pathology characterized by the disorganization of liver architecture attributable to hepatocytes destruction since the hepatocellular regeneration remains in form of regeneration nodules with the presence of fibrosis.

Nowadays, it's a public health problem [1]. The real world prevalence of hepatic cirrhosis is not well known but can exceed 1% if a liver biopsy and anatomopathological examination are realized for the benefit of the general population. In United State, Cirrhosis prevalence is 0.15% and represents respectively the tenth and the twelfth cause of death in men and in women, killing about 35,000 people every year [2]. In Europe, almost 170,000 of death related to cirrhosis are registered per year and represent 1.8% of all causes of death [3]. In Africa, the hospital prevalence of cirrhosis is at 7.02% in 2012 in Togo [4] and 2.35% in 2008 in Mali [5]. In Cotonou (Benin), Sèhonou et al. have reported 22.6% as prevalence in 2006; Viral hepatitis Band chronic alcoholism are the most frequent reported causes [6]. No studies on hepatic cirrhosis have been performed in northern Benin. The present study was initiated to identify the different and the most common cause of hepatic cirrhosis. The results of this study will help prevent liver cirrhosis. The goal of this study is to describe the epidemiological, clinical and paraclinical aspects of cirrhosis in Borgou Departmental University hospital Center.

2.   Patients and Methods

This is a retrospective study for a descriptive purpose conducted in the Internal Medicine Department. The study population consists of the patients hospitalized in the Internal Medicine Department from 1 St January 2009 to 31 St December 2016. We included only the patients with liver cirrhosis. Diagnostic criteria used for the diagnosis of liver cirrhosis were as follows:

-   Signs in accordance with cirrhosis in the laboratory analyses: AST/ALT ratio > 1, presence of thrombocytopenia, and prolonged prothrombin time.

-   Imaging findings (ultrasonography and/or tomography): decrease in the liver size, parenchymal heterogeneity, superficial nodular changes, hypertrophy of the left lobe, splenomegaly and dilatation of portal vein.

-   Clinical and endoscopic signs suggestive of cirrhosis; splenomegaly, esophageal varices, ascites, hepatic encephalopathy.

The study variables are:

-   Epidemiological Data

    o   Frequency of liver cirrhosis. The diagnosis of cirrhosis is decided based on clinical and paraclinical arguments. Liver biopsy and anatomopathological examination were not done due to the lack of technical capacity means.

    o   Age of the patients

    o   The sex

-   Clinical Data

    o   Functional signs

    o   History

    o   General signs

    o   Physical signs

-   Paraclinical Data

    o   Liver function tests

    o   Liver morphology explorations

The analysis has been done using the software Epi Data, version 3.1.

This study has used the data of a retrospective cohort. Neither the names nor others characteristics allowing the patients recognition were collected. The agreement of the National Ethics Committee for Health Research (http:/ / www.ethique-san te.or g/) according to the recommendations was not requireddue to its retrospective character.

3. Results

Epidemiological Data

Out of 9260 patients who consulted or have been hospitalized in the Internal Medicine Department, 125 suffered from hepatic cirrhosis or a frequency at 1.35%.

Among the cirrhotic patients, a male predominance has been noticed: 99 men for 26 women. The sex ratio was 3.76.

Figure 1 shows cirrhotic patients distribution by age. The average age of the patients was 45.22 ± 15.23 years, with the extremes 15 and 82 years of age. About 8 out 10 patients were between the ages of 25 and 64 years old.

Clinical characteristics

> Consultation period: All the patients were seen at a late phase with complications. The average period of consultation was 4.45 ± 1.2 months with the extremes 1 and 48 months; 60% of patients consulted three months after the first signs of decompensation.

> Reason for consultation: dominated by abdominal pain (56.8%), fever (43.2%) and right hypochondrium lump (40%).

> Patient's history: Dominated by jaundice (36.8%), alcoholism (29.6%) and smoking (29.6%).

> The examination of the general condition found asthenia (83.2%), weight loss (78.4%), anorexia (70.4%) and fever (37.6%).

> The physical signs were dominated by ascites (78.4%), hepatomegaly (66.88%) lower limb oedema (63.2%). Table 1 shows the different signs found during the physical exam.

Ascitic fluid examination

> Ascitic fluid was macroscopically yellow citrine (89.6%), hemorrhagic (8.6%) or cloudy (1.6%).

> Cytobacteriological examination of the fluid showed a leukocyte count lower than 250/mm3 in 96.8% of the cases.

Table 2 points out the results of the ascitic fluid examination.

Paraclinical characteristics

> Liver function tests

o   The average level of prothrombin was 51.85% ± 5.7%.

o   The average level of AST (Aspartate Aminotransferase) was 189.13 ± 24.02 UI/L and the one of ALT (Alanine Aminotransferase) was 87.98 ± 17.15 UI/L. The ratio AST/ALT ≥ 1 in 92.1% of cases.

o   The average value of totalbilirubinemia was 77.82 ± 14.6 mg/L and varied from 1 to 498 mg/L.

o   The level of Gamma Glutamyl Transferase (GGT) varied from 39 to 3024 UI/L with an average level at 413 ± 33 UI/L.

o   The average level of alkaline phosphatase was 473.44 ± 24.89 UI/L and varied from 138 to 1000 UI/L.

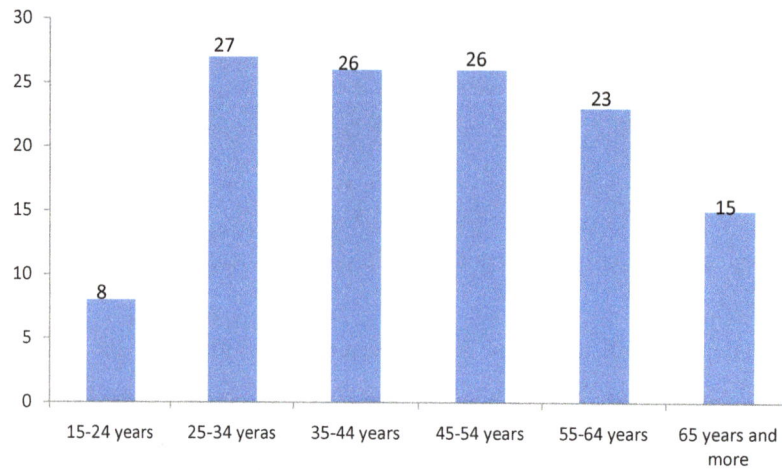

Figure 1. Distribution of the 125 cirrhotic patients by age groups.

Table 1. Physical signs found in 125 cirrhotic patients.

	n	%
Ascites	98	78.40
Lower limbs oedema	79	63.20
Clinical anemia	74	59.20
Jaundice	65	52
Hepatomegaly	86	68.88
Normal or atrophic liver	39	31.12
Splenomegaly	70	56
Collateral venous circulation	50	40
Pleurisy	6	4.80
Scarification	11	3.20
Flapping tremor	4	8.80

Table 2. Results of the ascites fluid examination in 125 cirrhotic patients.

	n	%
Macroscopic aspects		
Citrine	112	89.6
Hemorrhagic	11	8.8
Cloudy	2	1.6
White blood cells count		
<250/mm^3	121	96.8
≥250/mm^3	4	3.2

> ➤ The abdominal ultrasound focused on liver realized in 105 patients showed:

  o The average size of the liver was 150.47 ± 46.15 mm and varied between 50 and 250 mm.

  o The average size of the spleen was 156.10 ± 41.11 mm and varied from 80 to 251 mm. 56.38% of patients had a splenomegaly.

  o The average diameter of the portal vein was 15.87 ± 2.27 mm with 21 mm as maximum.

Table 3. Results of abdominal ultrasonography in 105 cirrhotic patients.

	N	%
Size		
Hepatomegaly	69	65.71
Normal or atrophic liver	36	34.29
Contour		
Regular	0	0
Irregular	105	100
Homogeneity		
Heterogeneous	105	100
Homogeneous	0	0
Nodular aspect		
Micronodular	50	47.61
Macronodular	47	44.76
Mixed	8	7.61
Dilatation of portal vein		
<13 mm	34	32.38
≥13 mm	71	67.62
Splenomegaly		
Yes	58	55.23
No	47	44.77
Ascites		
Yes	84	80
No	21	20

> ➤ The etiology of cirrhosis has been identified in 87 cases with a predominance of post hepatitis B cirrhosis in 68 cases (87.5%) followed by alcoholic liver cirrhosis in 11 cases (21.59%). Figure 2 shows the etiologies of cirrhosis.

> ➤ The complications of cirrhosis are dominated by ascites (78.4%) and jaundice (52%). Table 4 shows the complications of cirrhosis.

4. Discussion

The frequency of cirrhosis in Internal Medicine Department of Borgou Departmental University hospital Center was 1.35%. The prevalence of cirrhosis is extremely variable from one country to another and from one continent to another. In West Africa Sèhonou et al. in 2006 [6] have reported 22.60% as frequency in Internal Medicine Department of National Teaching Hospital in Cotonou; Bouglouga et al. in 2012 [4] have reported a prevalence at 7.02% in Togo. The frequency of cirrhosis at 1.35% in Borgou Departmental University hospital Center is lower than the one at 22.60% reported by Sèhonou et al. at National Teaching Hospitalin Cotonou. In reality, the prevalence reported by Sèhonou et al. was the one of hepatitis cirrhosis at the hepatogastroenterology department that usually receives patients suffering from digestive pathologies like chronic hepatopathies. The Internal Medicine Department of Borgou Departmental University hospital Centerreceives patients coming from all the medical specialities.

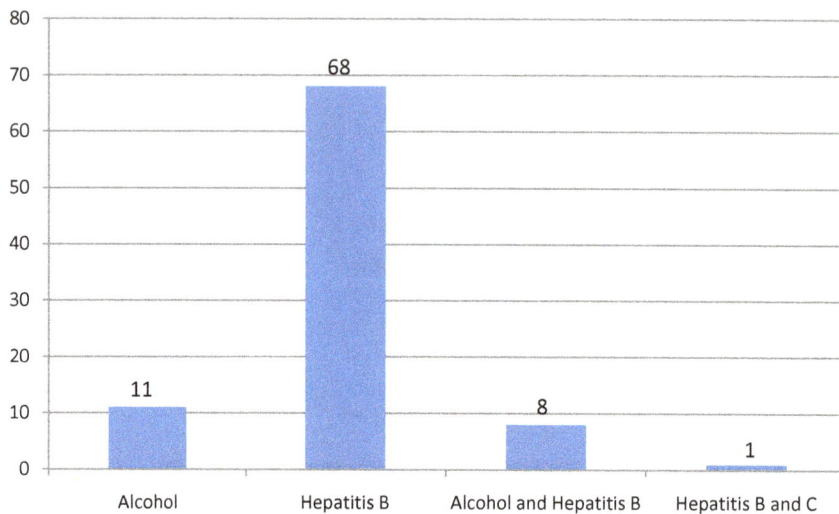

Figure 2. Patient distribution by etiologies of cirrhosis.

Table 4. Complications encountered in 125 cirrhotic patients.

	N	%
Ascites	98	78.40
Lower limbs oedema	79	63.20
Ascitesandoedema	72	57.60
Jaundice	65	52
Carcinoma	64	51.20
Biological anemia	54	43.20
Infection	51	40.80
Digestive hemorrhage	25	20
Kidney failure	24	19.20
Hepatic encephalopathy	13	10.40

Among the 125 cirrhotic patients of our series, there were 99 men and 26 women with a sex ratio at 3.76. Many studies have reported a male predominance; the sex ratio varies between 1.38 and 4.6 [7] [8] [9]. Where as Houissa et al., reported a female predominance with a sex ratio at 0.83 [10].

The average age of our patients was 45.22 ± 15.23 years. That was similar to the one reported by many authors notably Sèhonou et al. (49 years of age) and Ouavene et al. (45 years) [6] [11]. In Turkey, en Greece and Romania, the average age of the cirrhotic patients was higher. Those authors have respectively reported 55.3 years, 56 years and 59 years [12] [13] [14]. That higher average age could be explained by the exposure to the hepatitis B virus (HBV) which was the main cause of cirrhosis in childhood in developing countries like Benin and also the absence of vaccination against that virus. Mother-to-child transmission is a real source of infection. This is augmented by unprotected sex mostly for teenagers.

All the patients were seen at a phase of decompensation, the average period for consultation was 4.45 months. In Togo, that has been observed by Bouglouga et al. who reported that 78.1% of patients consulted one year after the progression of the disease [4].

Among the patients history, jaundice and alcoholism account for respectively 36.80% and 29.60% of cases. According to the study of Ouavene et al., 63.5% had jaundice as history and 34% are chronic alcoholics [11].

Most of the patients had a poor general condition with asthenia (83%) and loss of weight (78%). The cirrhotic patients were undernourished. The prevalence of under nutrition during cirrhosis is approximately 50% at all the stages of the disease. That under nutrition is due to the reduction of food intake, a malabsorption and a hypercatabolism. The cirrhotic patient doesn't eat much since he is still anorexic, nauseous, with an altered sense of taste and a feeling of gastric fullness. According to the study of Xie et al., 56.6% of the patients were asthenic and 34.1% were anorexic [15]. Hepatomegaly was found in 68.8% of cases, ascites (78.40%) andsplenomegaly (56%), collateral venous circulation (40%) and abdominal pain (56.80%). Hepatomegaly was the main morphological change reported by the authors [4] [6]. Aboutsplenomegaly and collateral venous, Bouglouga et al. [4] have respectively reported 13% and 5.2% and Ouavene et al. reported [11] 77.50% and 95.50%.

Hepatic ultrasonography realized showed hepatomegaly in 65.71% of cases. In the series of Ouavene et al. [11], hepatomegaly accounted for only 30% whereas the liver size was normal in 67.5% of cases.

The analysis of the ascitic fluid revealed a leukocyte count inferior to 250/mm3 in 96.8% of cases, or an absence of infection encountered in most of noncomplicated hepatic cirrhosis [16]. Ouavene et al. have reported a yellow citrineascitic fluid (82%) with a negative Rivalta's test results (76.7%) [11]. In this study, the average level of prothrombin was 51.85% reflecting a hepatocellular insufficiency. In Togo, Bouglouga et al. have reported a decrease in the level of prothrombin at 70% in 85.5% of cases [4].

The average value of AST was 189.13 UI/L and the one of ALT was 87.98 UI/L with the ratio ASAT/ALAT ≥ 1 in 92.1% of cases. Ouavene et al. reported a cytolysis in 71.50% of cases for the AST and 68% for the ALT [11]. Xie et al. have also reported a predominant hepatic cytolysis for the AST with a ratio AST/ALT at 2.00 ± 1.20 [15].

The average level of total bilirubinemia was 77.82 mg/L, the one of Gamma GT was 413 UI/L and 473.44 UI/L for alkaline phosphatase witness of cholestasis. In their series, over half of the patients suffered from cholestasis according to Ouavene et al. [11].

In this study, hepatitis B virus is the main cause of hepatic cirrhosis (87.5%) followed by chronic alcoholism (21.59%). That predominance of post hepatitis B cirrhosis has been pointed out by Topdagi et al. who reported hepatitis B as the major cause of hepatic cirrhosis in the developing countries [12]. Post hepatitis cirrhosis are the most represented (54.5%) followed by alcoholic cirrhosis (32.5%) according to Ouavene et al. [11]. In his study conducted in Turkey, 52% of hepatic cirrhosis were due to hepatitis B virus; Alcohol were found in 2% of cases. The low proportion of alcoholic cirrhosis could be explained by the low alcohol consumption in Muslim countries like Turkey. In Tunisia, viral causes (75%) and mostly hepatitis C virus (62.5%) have been reported as the main causes of hepatic cirrhosis. In developed countries, it's rather alcohol and hepatitis C virus the main causes of hepatic cirrhosis [17]. In Germany, alcohol is responsible of hepatic cirrhosis for 52% followed by hepatitis C virus (28%) and hepatitis B virus (14%) [18]. Hepatic cirrhosis inescapably progresses to the complications [11]. In this study, many complications have been identified. Among these, ascites was observed in 78.40%, hepatocellular carcinoma 51.20%, jaundice 52%, and infections 40.8%.Similar results have been reported by many authors. Bouglouga et al. [4] have reported some complications like ascites, hepatocellular carcinoma and hepatic encephalopathy in respectively 60%, 26.3% and 7.5% of cases. Among the complications reported by Topdagi et al., there is an ascites predominance (83%) followed by digestive hemorrhage (56%), Peritonitis (42%), hepatic encephalopathy (26%) [12]. Xie et al. [15] and Bruno et al. [18] have reported an ascites predominance in respectively 60% and 76.65% of cases.

Esophageal varices have been observed in 60% of cases and ascites in 49% according to Kittner et al. [19].

This study has limits cause of limit number of patients and the type of study. Another study on a grand scale and cross sectional study would be necessary.

5. Conclusion

Hepatic cirrhosis is a growing pathology. Hepatitis B virus is the main cause of hepatic cirrhosis followed by chronic alcoholism. Vaccination against hepatitis B virus and a decrease in alcoholic consumption are necessary in order to reduce the incidence of hepatic cirrhosis. That vaccination against hepatitis B virus is necessary from childhood especially in our developing countries.

References

[1] Ouavene, J.O., Koffi, B., Mobima, T., Bekondji, C., Massengue, A. and Guenebem,

A.K. (2014) Cirrhoses du foie à l'hôpital de l'amitié de Bangui aspects épidémiologiques, cliniques, échographiques et problèmes de diagnostic. Journal African d'Imagerie Médicale, 5, 1-12.

[2] Mokdad, A.A., Lopez, A.D., Shahraz, S., Lozano, R., Mokdad, A.H., Stanaway, J., et al. (2014) Liver Cirrhosis Mortality in 187 Countries between 1980 and 2010: A Systematic Analysis. BMC Medicine, 12, 145.

[3] Balkan, A., Alkan, S., Barutçu, S., Konduk, B.T., Yildirim, A.E. and Erdem, R. (2016) Etiological Distribution and Clinical Features of Cirrhotic Patients: Single Tertiary Referral Center Experience. Acta Medica Mediterranea, 32, 669-675.

[4] Verhelst, X., Geerts, A. and Vlierberghe, H.V. (2017) Cirrhosis: Reviewing the Literature and the Future Perspectives. European Medical Journal, 1, 111-117.

[5] Bouglouga, O., Bagny, A. and Djibril, M. (2012) Aspects épidémiologiques, diagnostiques et évolutifs de la cirrhose hépatique dans le service d'hépatogastroentérologie du chu campus de Lomé. J. Rech. Sci. Univ. Lomé (Togo), Série D, 14, 1-7.

[6] Diarra, M., Konaté, A. and Soukho (2010) Aspects évolutifs de la maladie cirrhotique dans un service d'hépato-gastro-entérologie au mali. Mali Médical, 14, 42-46.

[7] Sèhonou, J., Kodjoh, N., Saké, K. and Mouala, C. (2010) Cirrhose hépatique à Cotonou: Facteurs liés au décès. Médecinetropicale, 70, 375-378.

[8] Mohammed, S., Abdo, A. and Mudawi, H. (2016) Mortality and Rebleeding Following Variceal Haemorrhage in Liver Cirrhosis and Periportal Fibrosis. World Journal of Hepatology, 8, 1336-1342. https:// doi.or g/ 10.4254/ wjh.v8.i31.1336

[9] Mathur, A., Chakrabarti, A., Mellinger, J., Volk, M., Day, R. and Singer, A. (2017) Hospital Resource Intensity and Cirrhosis Mortality in United States. World Journal of Gastroenterology, 23, 1857-1865. https:// doi.or g/ 10.3748/ wjg.v23.i10.1857

[10] Zhang, X., Qi, X., de Stefano, V., Hou, F., Ning, Z. and Zhao, J. (2016) Epidemiology, Risk Factors, and in Hospital Mortality of Venous Thromboembolisme in Liver Cirrhosis: A Single Center Retrospective Observational Study. Medical Science Monitor, 22, 969-976. https:// doi.or g/ 10.12659/ MSM.896153

[11] Houissa, F., Mouelhi, L., Amouri, N., Salem, M., Bouzaidi, S., Debbeche, R., et al. (2012) Factors Predicting Mortality in Infected Hospitalized Cirrhotics Patients: About 97 Cases. La Tunisie medicale, 90, 807-811.

[12] Topdagi, O., Okcu, N. and Bilen, N. (2014) The Frequency of Complications and the Etiology of Disease in Patients with

Liver Cirrhosis in Erzurum. The Eurasian Journal of Medicine, 46, 110-114. https:/ / doi.or g/ 10.5152/ eajm .2014.25

[13] Goulis, C.E., Arsos, J., Birtsou, G., Nakouti, C., Papadopoulou, S. and Akriviadis, E. (2013) Association between Ratio of Sodium to Potassium in Random Urine Samples and Renal Dysfunction and Mortality in Patients with Decompensated Cirrhosis. Clinical Gastroenterology and Hepatology, 11, 862-867.

https:/ / doi.or g/ 10.1016/ j.cgh.2013.02.005

[14] Onuigbo, W. (2016) The Prevention of Liver Cancer and the Epidemiology of Cirrhosis. Journal of Cancer Prevention and Current Research, 4, 132-133.

https:/ / doi.or g/ 10.15406/ jcpcr .2016.04.00132

[15] Xie, Y., Feng, B., Gao, Y. and Wei, L. (2013) Characteristics of Alcoholic Liver Disease and Predictive Factors for Mortality of Patients with Alcoholic Cirrhosis. Hepatobiliary & Pancreatic Diseases International, 12, 594-601.

https:/ / doi.or g/ 10.1016/ S1499-3872(13)60094-6

[16] Ascione, T., Flumeri, G., Boccia, G. and de Caro, F. (2017) Infections in Patients Affected by Liver Cirrhosis: An Update. Le Infezioni in Medicina, 1, 91-97.

[17] Da Silva, M., Miozzo, S., Dossin, I., Tovo, C., Branco, F. and de Mattos, A. (2016) Incidence of Hepatocellular Carcinoma in Outpatients with Cirrhosis in Brazil: A 10 Year Retrospective Cohort Study. World Journal of Gastroenterology, 22, 10219-10225. https:/ / doi.or g/ 10.3748/ wjg.v22.i46.10219

[18] Bruno, S., Saibeni, S., Bagnardi, V., Vandelli, C., de Luca, M. and Felder, M. (2013) Mortality Risk According to Different Clinical Characteristics of First Episode of Liver Decompensation in Cirrhotic Patients: A Nationwide, Prospective, 3-Year Follow-Up Study in Italy. The American Journal of Gastroenterology, 108, 1112-1122. https:/ / doi.or g/ 10.1038/ ajg.2013.110

[19] Kittner, S.V., Sprinzl, J.M., Weinmann, M.F., Wiltink, K.S. and Schattenberg, J.M. (2014) Etiology and Complications of Liver Cirrhosis: Data from a German Centre. Deutsche Medizinische Wochenschrift, 139, 1758-1762.

15

Prognostic Factors for Cirrhosis Hospital in Abidjan (Côte d'Ivoire)

Mamert Fulgence Yao Bathaix[*], Akelesso Bagny, Kouamé Alassane Mahassadi, Anassé Jean-Baptiste Okon, Ya Henriette Kissi-Anzouan, Stanislas Doffou, Aboubacar Demba Bangoura, Hatrydt Dimitri Kouamé, Kadiatou Diallo, Antonin N'Dam, Aoudi Ousman De, Koffi Alain Attia, Aya Thérèse N'dri Yoman

Department of Hepatology and Gastroenterology of the Centre Hospitalier Universitaire de Yopougon (CHU-Y), Abidjan, Côte d'Ivoire
Email: [*]bathaixful@yahoo.fr

Abstract

Cirrhosis is the cause of a high rate of death in hospitals. The aim of this research was to estimate the incidence of mortality and identify the risk factors associated with cirrhosis patients in hospital in Côte d'Ivoire. Methodology: It is a retrospective study covering from January 1st, 2002 to December 31st, 2011 at Centre Hospitalier et Universitaire de Yopougon in Abidjan. We concerned the cirrhosis patients that have been followed at the hepatology and gastroenterology department. Survival was estimated by the Kaplan-Meier curve and comparison of survival curves by the log-rank test. The multi-varied analysis of the survivals has been achieved with the Cox proportional Hazard regression. A p value < 0.05 was taken as significant. Results: We recruited, 221 patients (135 men) of whom the medium age was 59 ± 15.12 years. Among those patients, 34.5% were classified as Child Pugh C and 52.94% Child Pugh B, 19.45% suffered from digestive hemorrhage, 26.5% suffered from renal deficiency, 47% suffered from hepatic encephalopathy and 10.7% from hyponatremia. The median overall survival of patients was 0.50 person-months. The variables that were significantly associated to a reduction of survival were hepatic encephalopathy ($p = 0.0029$), spontaneous ascitesfluid infection ($p = 0.0208$), hyponatremia ($p = 0.0434$) and stage Cof Child-Pugh score ($p = 0.046$). Conclusion: The incidence of mortality in cirrhotic patients hospitalized in Abidjan is high. Pejorative prognostic factors were essentially hepatic encephalopathy, spontaneous ascites fluid infection, hyponatremia and stage C of Child-Pugh score.

Keywords

Cirrhosis, Portal Hypertension, Child-Pugh, Encephalopathy, Prognostic, Côte d'Ivoire

---

[*]Corresponding author.

1. Introduction

Cirrhosis is the final stage of development of liver fibrosis induced more by chronic liver disease. It is defined by the existence of an architectural modification diffuse hepatic parenchyma characterized by the existence of extensive fibrosis, dissecting and delimiting annular hepatocyte nodules so-called regeneration [1]. The complications of cirrhosis are frequent and potentially serious: portal hypertension, the cause of bleeding from ruptured gastroesophageal varices, hepatic encephalopathy, infection of ascites fluid, hepatorenal syndrome and hepatocellular carcinoma [2].

Cirrhosis is an important hepatobiliary disorder under our tropics [3] and is the leading cause of hospitalization for chronic liver disease in gastro enterology hospital in Côte d'Ivoire [4].

The management of portal hypertension with cirrhotic patients is a major concern for hospital practitioners. It is the cause of a high rate of mortality [5] [6]. In fact, many factors predispose cirrhosis patient to the risk of unexpected arrival of a multi deep-rooted default likely to imperil the vital prognostic: immune dysregulation that will promote the risk of infection, pulmonary involvement by the hepatopulmonary syndrome, a decrease of coagulation factors and renal hypoperfusion caused by the hepatorenal syndrome [7]. The evaluation of the prognosis of cirrhotic patients followed in hospitalization would improve their care. We conducted this study in cirrhotic patients in order to estimate the death incidence and to identify risk factors to which cirrhotic patients in Ivorian hospitals are exposed.

2. Methodology

2.1. Study Oversight

From January 2002 through December 2011, we consecutively enrolled patients with cirrhosis and portal hypertension who were admitted to Centre Hospitalier et Universitaire de Yopougon in Abidjan. No commercial support was involved in the study. All the authors vouch for the integrity and the accuracy of the analysis and for the fidelity of the study. No one who is not an author contributed to the manuscript.

2.2. Selection of Patients

Patients who had cirrhosis and portal hypertension were considered for the inclusion. The diagnosis of cirrhosis was based on the combination of clinical, biological, echographical and/or endoscopical criteria. Exclusion criteria were a lost sight below12-months-follow-up and hepatocellular carcinoma (HCC) at the inclusion, whose diagnosis was based on the Barcelona criteria.

2.3. Studied Variables

The sociodemo graphic variables were the age, the gender and the socioeconomic status. The clinic variables studied were ascites, limbs edema, gastrointestinal bleeding, hepatic encephalopathy. The endoscopic variables were the esophageal varices, the gastric varices and the portal hypertensive gastropathy. The biological variables were the prothrombin rate, the bilirubin, the albuminemia, the natremia, the creatinine and the platelets. The etiology of cirrhosis has also been studied.

These data were collected from the files of hospitalized patients on pre-established survey forms.

2.4. Statistical Analysis

The results are expressed as frequency, percentage, or mean ± standard deviation and median. The survival is estimated by the Kaplan-Meier Curve and the comparison of survival curbs by the Log-rank test. The multivariable analysis was conducted with Cox proportional Hazard regression. A p value < 0.05 was taken as significant.

3. Results

Over the period of our study we recruited 221 patients (135 men) with an average of 59 ± 15.12 years. The middle socio-economic status was the most represented with 56.15% of the patients. The main etiology of cirrhosis was HVB (76.04%) followed by alcohol. The edema and ascites syndrome was present in almost all patients. Also 19.45% of patients had gastrointestinal bleeding and hepatic encephalopathy (47.05%). On average, patients had a high rate of creatinine (16.18 mg/L ± 19.58) and a decreasing of platelet ($131.27 \times 10^9 \, \text{L}^{-1} \pm 96.82$).

Patients were classified as Child-Pugh B and C in the respective proportions of 52.92% and 32.76%. The epidemiological, clinical, biological and endoscopic characteristics of the patients during their hospitalization are shown in **Table 1**.

The overall survival of patients was between 0.03 and 16.66 person-months, with a median of 0.50 person-months (**Figure 1**).In univariate analysis, hepatic encephalopathy, Child-Pugh C, renal failure, spontaneous ascites fluid infection (SAI) and hyponatremia were associated with a significant survival decrease. However, the diuretic outlet would significantly improve the survival (**Table 2**).

Stage C of the Child-Pugh score was associated with a significant increase ($p = 0.046$) of 3.351 times of death

Table 1. Demographics, clinical, biological and endoscopic characteristics of patients.

Number of patients	221
Socio-demography	
-Age	14 years - 86 years (59 ± 15.12 years)
-Sex ratio (M/F)	2.06
-Socioeconomic status	high (8), middle (117), low (80), unspecified (16)
Etiology	HBV[*] (76.04%), alcohol (11.06%) not identified (9%), HCV[**] (3%)
Clinic	
-Ascites ± edema in the lowerextremities	ascite (136), ascite + ankle edema (91)
-gastrointestinal bleeding	43
- hepatic encephalopathy	104
Portal hypertension endoscopicsigns (113 patients)	OVs[***] stage I (20), OVs Stage II (58), OVs stage III (35), bleeding gastric varices (04) Gastropathy http (71)
Biologicalparameters	
-Prothrombinrate (N ≥ 70%)	16% - 100% (mean 57.27 ± 23.21; median 55)
- Bilirubin (N < 10 mg/l)	0.28 - 6.2 mg/l (mean30.66 ± 23.64; median 22.5)
-Albumin (N = 35 - 50 g/l)	6.1 - 63.7 g/l (mean 25.35 ± 9.98; median 25)
-Natremia (N = 135 - 145 meq/l)	96 - 172 meq/l (mean 135.08 ± 9.52; median 135)
-Creatinine (N < 15)	3 - 176 mg/l (mean 16.18 ± 19.58; median 10.25)
-platelets (N = 150 - 450 GIGA/L)	14 - 670 GIGA/L (mean 131.27 ± 96.82; median 108)
Child Pugh score	
Stage A/Stage B/Stage C	28/117/76

HBV[*]: hepatitis B; HCV[**]: Hepatitis C; OVs[***]: Bleeding esophageal varices.

Table 2. Prognostic factors in univariate analysis.

Variables	n	p
Male	149	0.3029
Ascite decompensation	163	0.4457
hepatic encephalopathy	104	0.0029[*]
Infection of ascites	50	0.0208[*]
Child Pugh C	76	0.0460[*]
gastrointestinal bleeding	43	0.0992
Thrombocytopenia	124	0.5693
Renal failure	38	0.0858
Diuretics	60	0.0458[*]
Administration of beta blockers	84	0.3010
Hyponatremia	35	0.0434[*]

[*]$p < 0.05$.

Hazard Ratio [CI: 1.023 to 10.977] as compared with stage A (**Figure 2**). In multivariate analysis, hepatic encephalopathy and gastrointestinal bleeding were associated with the survival decrease and the use of diuretic would improve the prognostic (**Table 3**).

4. Discussion

This study confirms the high frequency of viral B cirrhosis in our country in the light of our literary review [4].

Table 3. Prognostic factors of survival in multi varied analysis.

Variables	Hazard Ratio	p	Interval of confidence
Child-Pugh score C	1.281	0.497	0.627 - 2.618
Added bacterial infection	1.794	0.169	0.779 - 4.131
Diuretics	0.387	0.002*	0.209 - 0.714
Hyponatremia	1.093	0.788	0.571 - 2.094
Gastrointestinal bleeding	0.402	0.030*	0.177 - 0.917
Encephalopathy	2.223	0.035*	1.056 - 4.699

*$p < 0.05$.

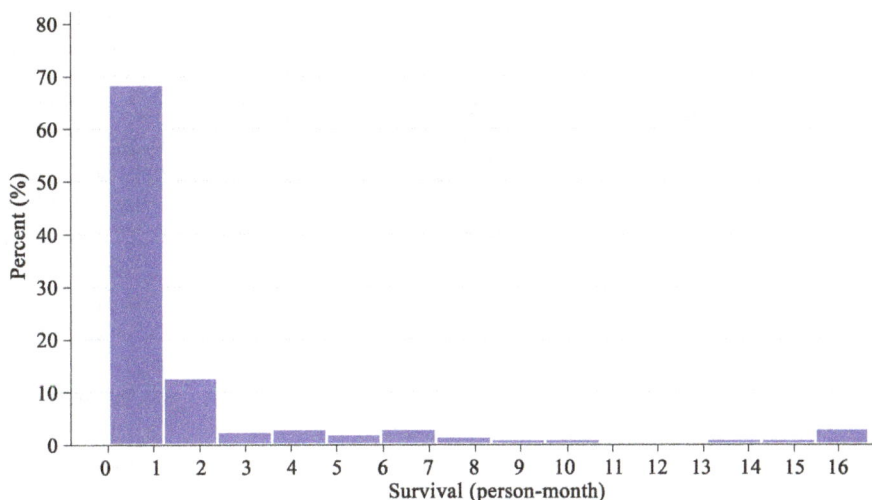

Figure 1. Distribution of patients according to survival time.

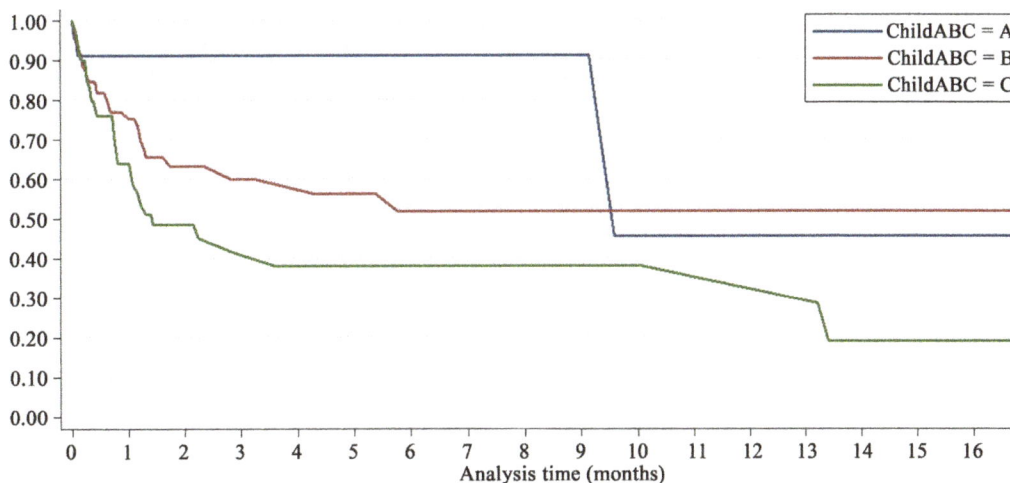

Figure 2. Survival in patients with cirrhosis according to the child-pugh score.

Besides, we have also established a weak proportion of alcoholic cirrhosis which is quite probably under estimated because of the consumption of alcohol that was not confessed. On the contrary, the etiology of the cirrhosis has not been established within 10% of the cases. This considerable proportion is due to many factors: insufficiency of investigations because of lack of means, undeclared consumption of alcohol.

The duration of the overall survival of the patients has run from 0.03 to 16.66 per month with a median of 0.50 patient-month. This means that we have registered 0.5 deaths out of 221 patients-months. Also, this is a proof that the patients were at an advanced stage of cirrhosis before their hospitalization with a reserved prognostic. In fact, in our study 34.39% of the cirrhosis patients were classified at stage C where as 52.94% of them were classified in stage B. It is not easy to establish the direct causes of cirrhosis patient's death [7]. In univariate analysis, 5 parameters (encephalopathy, spontaneous ascetic fluid infection, hyponatremia, renal failure and the score Child-Pugh C) were significantly associated with bad prognosis and the taking of diuretic was significantly associated with improved survival.

The occurrence of hepatic encephalopathy significantly has decreased the survival of our patients (p = 0.0029). This is consistent with several studies that have shown that hepatic encephalopathy was responsible for a high rate of death with cirrhotic patients [8].

In our study, ascites was the most common complication of cirrhosis [9] and is associated with a bad prognosis with a considerable reduction of survival-rate of patients [10] [11].

Moreover, with regard to our test sample, the spontaneous ascites fluid infection with our patients significantly reduced their chances of survival (p = 0.0208). This massive mortality can be likely associated with the development of a systematic inflammatory reaction [12]. The spontaneous ascites fluid infection is a frequent factor of complication with cirrhotic patients [13].

The survival of patients who have recovered from a prior episode of spontaneous ascites fluid infection is 30% over a period of 1 year with a prognostic that depends on the gravity of cirrhosis [14].

The risk factors of the spontaneous ascites fluid infection are manifold: antecedents of ascites fluid infection in the absence of antibiotic prophylaxis [15], gastrointestinal bleeding [16], albuminemiea < 28 gr/l, alcoholic etiology and Child-Pugh C [17]. In Ivory Coast, the ascites fluid infection is the first cause of death with cirrhotic patients [18]. The prognosis has improved considerably in recent years with the early use of antibiotic, a support of renal failure when it exists and a primary or secondary antibiotic prophylaxis [14].

The use of diuretic has improved significantly the survival of patients (p = 0.0458). Research studies should be urged on in order to better assess the added value of diuretics in the survival of cirrhosis patients with edema and ascites syndrome. However, their side effects and unfitness with cirrhosis patients minimize their use [19].

The presence of an hyponatremiea was forcibly associated with a decrease of a survival among the study sample (p = 0.0434). Many studies [20] [21] have revealed that a natremia ≤ 130 m·mol/l was a negative prognostic factor to which an elevated rate of death is associated with cirrhotic patients.

Renal failure has decreased survival without statistical significance with our patients (p = 0.0858) in contrast to the literature [21] where they are significantly associated. This lack of significance of our results can be explained by the fact that the inclusion criteria of hepato-renal syndrome were not clearly established during our inquiry, which means that it was under estimated and that the retrospective character of our study did not allow us to reach that stage. Renal function is frequently impaired in cirrhosis [7] [21] [22]. Hospital mortality due to renal failure despite adequate support is around 30% [23].

Stage C of Child-Pugh score was associated with an important increase (p = 0.046) by 3.351 time of death hazard-ratio in our sample.

Stage C of the Child-Pugh score is an independent factor that predisposes the patient to death [24] [25].

By a multivariate analysis, the survival of a hepatic encephalopathy in the course of cirrhosis was an independent parameter that has significantly reduced the patients chances of survival. In fact, the hepatic encephalopathy was associated with a significant increase in mortality (p = 0.035). Despite a better understanding of hepatic encephalopathy, of its symptomatic factors and its care taking, its mortality both in hospitalization as well as in the reanimation units remains high [26]. The development of hepatic encephalopathy with cirrhosis patients is associated with high lethality: 41% to 80% at one (1) year and 77% to 85% at three (3) years [26] [27].

The use of diuretics was associated with a significant decrease (p = 0.002) of 0.386 time (more than half) of the death Hazard Ratio. The gastrointestinal bleeding was associated with a significant decrease of mortality (p = 0.030). Indeed, the death cases that are caused by the first episodes of gastrointestinal bleeding of portal hypertension on cirrhosis is high but is also dependent on the area where bleeding occurs [28]. Our results could be explained by a proper application of resuscitation measures especially with patients having crossed the transfusion threshold although the specific charge made is not optimal [29] and the high rate of primary and secondary prophylaxis with beta blockers [22] [25].

The score of Child-Pugh is not independently associated to morality (p = 0.497) as the literary review seems to indicate [7]. Indeed, it does not take into account some factors that may have a significant impact on prognosis as renal function. This has led to the creation of other scores such as the MELD score (model for end-stage liver disease) [30] which is used in the assessment of medium prognostic of cirrhotic patients and the indication of hepatic transplantation. However, the prognostic value of the Child-Pugh score at one (1) or two (2) years is clearly known [6] [24].

The limitations to our study are the relatively small number of patients, due to the retrospective nature of the study with several missing data. Indeed, some hospitalized patients have not been able to do their para clinical examinations. Therefore, the survival time associated with the various parameter studied could be less than the duration of survival in subjects with cirrhosis and portal hypertension in the general population. Insufficient etiological research because of the cost of viral markers and the lack of quantitative and qualitative assessment of the consumption of alcohol is prior to the study.

5. Conclusion

Cirrhotic patients with portal hypertension mortality are high in Abidjan. Pejorative prognostic factors are hyponatremia, ascites fluid infection and hepatic encephalopathy. The use of diuretics and beta blockers were indicated when associated with improved survival.

References

[1] Friedman, S.L. (2003) Liver Fibrosis—From Bench to Bedside. *Journal of Hepatology*, **38**, S38-S53. http://dx.doi.org/10.1016/S0168-8278(02)00429-4

[2] Sawadogo, A., Diba, N., and Calès, P. (2007) Physiopathologie de la cirrhose et de ses complications. *Réanimation*, **16**, 557-562. http://dx.doi.org/10.1016/j.reaurg.2007.09.001

[3] Sehonou, J., Kodjoh, N., Sake, K. and Mouala, C. (2010) Cirrhose hépatique à Cotonou (République du Bénin): Aspects cliniques et facteurs liés au décès. *Médecine Tropicale*, **70**, 375-378.

[4] Attia, K.A., N'dri Yoman, A.T., Talla, P., Bathaix, Y., Mahassadi, A., Kissi, H., Brou, I., Touré, A. and Diomandé, I. (2003) Facteurs prédictifs des signes endoscopiques d'hypertension portale sévère chez le cirrhotique en milieu Africain: A propos de 131 cas. *Médecine d' Afrique Noire*, 50, 109-114.

[5] Anderson, R.N. (2002) Deaths. Leading Causes for 2000. *National Vital Statistics Reports*, **50**.

[6] Attia, K.A., Ackoundou, K.C., N'dri, A.T., *et al.* (2008) Child-Pugh-Turcott versus MELD Score for Predicting Survival in a Retrospective Cohort of Black African Cirrhotic Patients. *World Journal of Gastroenterology*, **14**, 286-291. http://dx.doi.org/10.3748/wjg.14.286

[7] Robert, R. and Veinstein, A. (2003) Pronostic du malade atteint de cirrhose en réanimation. *Gastroentérologie Clinique et Biologique*, **27**, 877-881.

[8] Bustamante, J., Rimola, A., Ventura, P.J., Nassava, M., Cirera, I., Reggiardo, V. and Rhodes, J. (1999) Prognostic Significance of Hepatic Encephalopathy in Patients with Cirrhosis. *Journal of Pathology*, **30**, 890-895. http://dx.doi.org/10.1016/s0168-8278(99)80144-5

[9] Runyon, B.A. (2013) American Association for the Study of Liver Diseases Introduction to the Revised American Association for the Study of Liver Diseases Practice Guideline Management of Adult Patients with Ascites Due to Cirrhosis 2012. *Hepatology*, 57, 1651-1653. http://dx.doi.org/10.1002/hep.26359

[10] Ginès, P., Cardenas, A., Arroyo, V. and Rodes, J. (2004) Management of Cirrhosis and Ascites. *NEJM*, **350**, 1646-1654. http://dx.doi.org/10.1056/NEJMra035021

[11] Singhal, S., Baikati, K.K., Jabbour, I.I. and Anand, S. (2012) Management of Refractory Ascites. *American Journal of Therapeutics*, **19**, 121-132. http://dx.doi.org/10.1097/MJT.0b013e3181ff7a8b

[12] de Guibert, B. (2012) Caractéristiques des patients présentant une hyponatrémie sévère en service de médecine: Epidémiologie, clinique, étiologies et évolution. *Human Health and Pathology*, HAL ID: dumas-00769783.

[13] Grangé, J.D. and Amiot, X. (1992) La prophylaxie des complications infectieuses par décontamination bactérienne digestive sélective chez les malades atteints de cirrhose. *Gastroentérologie Clinique et Biologique*, **16**, 692-700.

[14] Chagneau, C. (2004) Traitement et prévention de l'infection du liquide d'ascite. *Gastroentérologie Clinique et Biolo-*

gique, **28**, 138-145. http://dx.doi.org/10.1016/S0399-8320(04)95249-9

[15] Gines, P., Rimola, A., Flessing, R., Vargas, V., Marco, F., Almela, M., *et al.* (1990) Norfloxacin Prevents Spontaneous Bacterial Peritonitis Recurrence in Cirrhosis: Result of a Double-Blind, Placebo-Controled Trial. *Hepatology*, **12**, 716-724. http://dx.doi.org/10.1002/hep.1840120416

[16] Fernández-Esparrach, G., Sanchez-Fucip, A., Gines, P., Uriz, J., Quintó, L., Ventura, P.-J., *et al.* (2001) A Prognostic Model for Predicting Survival in Cirrhosis with Ascites. *Journal of Hepatology*, **34**, 46-52. http://dx.doi.org/10.1016/S0168-8278(00)00011-8

[17] Sandhu, B.S., Gupta, R., Sharma, J., Singh, J., Murthy, N.S. and Sarin, S.K. (2005) Norfloxacin and Cisapride Combination Decreases the Incidence of Spontaneous Bacterial Peritonitis in Cirrhotic Ascites. *Journal of Gastroenterology and Hepatology*, **20**, 599-605. http://dx.doi.org/10.1111/j.1440-1746.2005.03796.x

[18] Attia, K.A., N'driYoman, T., Sawadogo, A., Mahassadi, A., Bathaix-Yao, F., Sermé, K. and Kassi, L.M. (2001) L'infection spontanée du liquide d'ascite chez le cirrhotique africain. Etude descriptive à propos de 12 cas. *Bulletin de la Société de Pathologie Exotique*, **94**, 319-321.

[19] Perri, G.-A. (2013) L'ascite chez les patients atteints de cirrhose. *Canadian Family Physician*, **59**, e538-e540.

[20] Taoufik, R. (2011) Profil épidémiologique de l'hypertension portale au chu hassan II de fes. Thèse N 014/11, P.2.

[21] Attia, K.A., N'driYoman, A.T., Mahassadi, A.K., Ackoundou-Nguessan, K.C., Kissi, H.Y. and Bathaix, Y.F. (2008) Impact of Renal Failure on Survival of African Patients with Cirrhosis. *Saudi Journal of Kidney Diseases and Transplantation*, **19**, 587-592.

[22] Ouakaa-Kchaou, A., Belhadj, N., Abdelli, N., Azzouz, M., Mami, N.B., Dougui, M.H., *et al.* (2010) Survie chez Le cirrhotique Tunisien. *La Tunisie Medicale*, **88**, 804-808.

[23] Follo, A., Liovet, J.M., Navara, M., Planas, R., Forns, X., Francitorra, A., *et al.* (1994) Renal Impairment after Spontaneous Bacterial Peritonitis in Cirrhosis: Incidence, Clinical Course, Predictive Factors and Prognosis. *Hepatology*, **20**, 1495-1501.

[24] Gex, L., Bernard, C. and Spahr, L. (2010) Scores en hépatologie: Child-Pugh, MELD et Maddrey. *Revue Médicale Suisse*, 1803-1808.

[25] Castera, L., Pauwels, A. and Lévy, V. (1996) Indicateurs pronostiques chez les malades atteints de cirrhose admis en service de réanimation. *Gastroentérologie Clinique et Biologique*, **20**, 263-268.

[26] Bustamante, J., Rimola, A., Ventura, P.J., Navasa, M. and Cirera, V. (1999) Prognostic Significance of Hepatic Encephalopathy in Patients with Cirrhosis. *Journal of Hepatology*, **30**, 890-895. http://dx.doi.org/10.1016/S0168-8278(99)80144-5

[27] Benhaddouch, Z., Abidi, K., Naoufel, M., Abouqal, R. and Zeggwagh, A.A. (2007) Mortalité et facteurs pronostiques des patients cirrhotiques en encéphalopathie hépatique admis en réanimation. *Annales Françaises d'Anesthésie et de Réanimation*, **26**, 490-495. http://dx.doi.org/10.1016/j.annfar.2007.04.005

[28] Oberti, F. (1998) Pronostic de l'hypertension portale: Hemorragie digestive par rupture de varices œsophagiennes. *Hépato-Gastro & Oncologie Digestive*, **5**, 371-377.

[29] K.A. Mahassadi, Ndri Yoman, T., Kissi, Y.H., Bathaix-Yao, M.F., Doffou, S., Toualy, W. and Attia, K.A. (2007) Evaluation de la qualité de la prise en charge de l'hémorragie digestive liée à l'hypertension portale dans un pays en développement: Exemple du CHU de Yopougon (Abidjan-Côte d'Ivoire). *Revue Internationale des Sciences Médicales d'Abidjan*, **9**, 35-42.

[30] Botta, F., Giannini, E., Romagnoli, P., Fasoli, A., Malfatti, B. and Testa, R. (2003) MELD Scoring System Is Useful for Predicting Prognosis in Patients with Liver Cirrhosis and Is Correlated with Residual Liver Function: A European Study. *Gut*, **52**, 134-139. http://dx.doi.org/10.1136/gut.52.1.134

16

Spontaneous *Streptococcus mitis* Meningitis in a Patient with Liver Cirrhosis

Andrew Villion[1], Michael Lishner[1,2], Michal Chowers[1,3], Sharon Reisfeld[1,2*]

[1]Sackler School of Medicine, Tel Aviv University, Tel Aviv, Israel
[2]Department of Medicine A, Meir Medical Center, Kfar Saba, Israel
[3]Infectious Diseases Unit, Meir Medical Center, Kfar Saba, Israel
Email: *sharon.reisseld@clalit.org.il

Abstract

Streptococcus mitis is a component of the normal oropharynx, skin, gastrointestinal system, and genital tract florae. It is generally considered as a relatively benign bacterium. We present a case of spontaneous *Streptococcus mitis* meningitis in a patient with liver cirrhosis and no known risk factors for invasive infectious diseases.

Keywords

Streptococcus mitis, Cirrhosis, Meningitis

1. Introduction

Streptococcus mitis (*S. mitis*) is an alpha-hemolytic species belonging to the family of viridans streptococci. *S. mitis* is a component of the normal oropharynx, skin, gastrointestinal system, and genital tract floras [1].

Patients with cirrhosis who have not developed major complications are classified as having compensated cirrhosis.

Patients who have developed complications of cirrhosis, such as variceal hemorrhage, ascites, spontaneous bacterial peritonitis, hepatocellular carcinoma, hepatorenal syndrome, or hepatopulmonary syndrome are considered to have decompensated cirrhosis and have a worse prognosis than those with compensated cirrhosis [2]. A few cases of *S. mitis meningitis* were described in the literature; most of them were associated with risk factors like invasive procedures or a recent upper respiratory tract infection [3] [4]. We present a rare case of spontane-

---

*Corresponding author.

ous *S. mitis* meningitis in a patient with liver cirrhosis and no other risk factors for such an invasive infection.

2. Case Report

A 58-year-old male presented with complaints of headache, photophobia, and vomiting that started two days prior to admission and worsened on the admission day. The patient had no fever at home or signs of altered mental status. He had no history of invasive procedures or upper respiratory tract infection. His medical history included Type II diabetes mellitus, controlled with oral medications with current glycated hemoglobin level of 6%, cryptogenic liver cirrhosis with thrombocytopenia, splenomegaly, elevated liver enzymes (5-fold increase) and negative serological and autoimmune workup. On physical examination, temperature was 37°C (98.6°F). He appeared somnolent, with nuchal rigidity and no focal neurological deficits. There was no rash or any skin lesions. The rest of the physical examination was normal. Noncontrast head computed tomography scan did not reveal signs of hemorrhage or increased intracranial pressure. Laboratory values are shown in **Table 1**. Lumbar puncture was performed and yielded turbid cerebrospinal fluid (CSF) with xantochromia. CSF results showed white blood cell count of 5500 cells/μl with 96% segmented neutrophils, red blood cell count of 1300 cells/μl with normal morphology, glucose level of 2 mg/dl and protein level 703 mg/dl (normal range 15 - 50 mg/dl). Gram positive diplococci were seen on gram stain. Empirical treatment with dexamethasone, ampicillin plus ceftriaxone was started.

The patient was admitted to the intensive care unit. Two days after admission blood and CSF cultures were found positive for penicillin sensitive *Streptococcus mitis*. Therefore, treatment was changed to high dose penicillin. Transesophageal echocardiogram did not reveal vegetations or any other findings suggestive of infective endocarditis.

On the fifth day of admission, the patient developed mild right side hemiparesis. Head magnetic resonance imaging (MRI) revealed high signal intensity in the globus pallidum on T1-weighted images. On the same day, the patient developed massive variceal bleeding that was treated with terlipressin and endoscopic band ligation. Treatment for hepatic encephalopathy was also started. The patient stabilized and continued to improve. He was discharged two weeks after admission with no neurological sequelae. Unfortunately the patient was lost to follow up and we have no information about his current condition.

Table 1. Laboratory data at admission.

Component	Patient Value	Normal Value
White blood cells	13.93 Kc/μl	4.8 - 10.8 Kc/μl
Hemoglobin	15.00 g/dl	13.5 - 17.5 g/dl
Platelets	85 Kc/μl	150 - 450 Kc/μl
Glucose	162 mg/dl	70 - 100 mg/dl
Sodium	137 mEq/L	135 - 145 mEq/L
Potassium	3.3 mEq/L	3.5 - 5.1 mEq/L
Aspartate aminotransferase	50 U/L	0 - 35 U/L
Alanine transaminase	30 U/L	0 - 45 U/L
Gammaglutamyl transpeptidase	138 U/L	7 - 49 U/L
Total bilirubin	2.8 mg/dl	0.2 - 1.5 mg/dl
Albumin	3.4 gr/dl	3.5 - 5.5 gr/dl
C reactive protein	5.01 mg/dl	0.00 - 0.5 mg/dl
International normalized ratio	1.16	
Creatinine	0.9 mg/dl	0.5 - 1.2 mg/dl
Urea	44 mg/dl	10 - 50 mg/dl

3. Discussion

Streptococcus mitis (*S. mitis*) is an alpha-hemolytic species belonging to the family of viridans streptococci. *S. mitis* is a component of the normal oropharynx, skin, gastrointestinal system, and genital tract floras [1], and has generally been considered a relatively benign bacterium. Nevertheless, *S. mitis* can cause a range of invasive diseases in humans. Endocarditis caused by *S. mitis* was reported both in adult and pediatric patients [5]. Liver, lung and even myocardial abscesses caused by *S. mitis* were described in a number of case reports in recent years [6]-[9]. *S. mitis* is an important and underestimated cause of bacteremia and sepsis in neutropenic patients and in cancer patients receiving chemotherapy [10]-[11].

S. mitis has been reported in conjunction with meningitis, but in the majority of the cases, patients had previously undergone invasive procedures such as spinal anesthesia [3] [4]. A case of spontaneous *S. mitis* meningitis was reported in an adult patient with a history of alcoholism, without chronic liver disease, who also had very poor oral hygiene and maxillary sinusitis [12].

Our patient was diagnosed with *S. mitis* meningitis; therefore he was evaluated for endocarditis by transesophageal echocardiogram that was negative. The antibiotic treatment was changed to penicillin and gentamycin according to endocarditis guidelines since there are no specific guidelines for *S. mitis* meningitis [13]. The patient completed 2 weeks of antibiotic treatment as recommended in cases of highly susceptibe bacteria as in our case.

To the best of our knowledge, this is the first case of spontaneous *S. mitis* meningitis in an adult with no risk factors for such an invasive infection, like poor oral hygiene, recent invasive procedures, a recent upper respiratory infection or a history of alcoholism.

Chronic liver disease and especially liver cirrhosis is considered an immunocompromised state that leads to a variety of infections [14]. The patient reported in our case had decompensated cirrhosis and most probably this was the only risk factor for such an invasive infection from a usually benign bacterium.

We are not aware of any reported cases of meningitis caused by *S. mitis* in cirrhotic patients.

4. Conclusions

We described herein, the first case of spontaneous *Streptococcus mitis* meningitis in a patient with liver cirrhosis and no known risk factors for such an invasive disease.

We emphasize the need to broaden the differential diagnosis of cirrhotic patients with altered mental status. It is important to remember that they are prone to infections from avirulent pathogens due to their liver disease and impaired immune state, including uncommon central nervous system infections similar to that presented here.

References

[1] Patterson, M.J. (1996) Streptococcus Pyodenes, Other Streptococci, and Enterococcus. In: Baron, S., Ed., *Streptococcus*, *Medical Microbiology*, *Chapter* 13, 4th Edition, University of Texas Medical Branch at Galveston, Galveston.

[2] Goldberg, E. and Chopra, S. (2014) Cirrhosis in Adults: Overview of Complications, General Management, and Prognosis. In: Post, T.W., Ed., *UpToDate*, Waltham.

[3] Janssen, M. (1996) Alpha-Hemolytic Streptococci: A Major Pathogen of Iatrogenic Meningitis Following Lumbar Puncture. Case Reports and a Review of the Literature. *Infection*, **24**, 29-33. http://dx.doi.org/10.1007/BF01780647

[4] Villevieille, T., Vincenti-Rouquette, I., Petitjeans, F., *et al.* (2000) *Streptococcus mitis*-Induced Meningitis after Spinal Anesthesia. *Anesthesia Analgesia*, **90**, 500-501.

[5] Rapeport, K.B., Giron, J.A. and Rosner, F. (1986) *Streptococcus mitis* Endocarditis: Report of 17 Cases. *Archives of Internal Medicine*, **146**, 2361-2363.

[6] Sarthy, J. and DiBardino, D. (2013) Pyogenic Liver Abscess Caused by *Streptococcus mitis*. *The Lancet Infectious Diseases*, **13**, 822. http://dx.doi.org/10.1016/S1473-3099(13)70166-X

[7] Takayanagi, N., Kagiyama, N., Ishiguro, T., Tokunaga, D. and Sugita Y. (2010) Etiology and Outcome of Community-Acquired Lung Abscess. *Respiration*, **80**, 98-105. http://dx.doi.org/10.1159/000312404

[8] Basil, A., Schoch, P.E. and Cunha, B.A. (2012) Viridans Streptococcal Biosynthetic Aortic Prosthetic Valve Endocarditis (PVE) Complicated by Complete Heart Block and Paravalvular Abscess. *Heart & Lung: The Journal of Acute and Critical Care*, **41**, 610-612. http://dx.doi.org/10.1016/j.hrtlng.2012.05.002

[9] Lo, R., Rae, J., Noack, D., Curnutte, J.T. and Avila, P.C. (2005) Recurrent Streptococcal Hepatic Abscesses in a 46-Year-Old Woman. *Annals of Allergy, Asthma Immunology*, **95**, 325-329.

http://dx.doi.org/10.1016/S1081-1206(10)61149-0

[10] Marron, A., Carratala, J., Gonzalez-Barca, E., Fernandez-Sevilla, A., Alcaide, F. and Gudiol, F. (2010) Serious Complications of Bacteremia Caused by Viridans Streptococci in Neutropenic Patients with Cancer. *Clinical Infectious Diseases*, **31**, 1126-1130. http://dx.doi.org/10.1086/317460

[11] Ahmed, R., Hassall, T., Morland, B. and Gray, J. (2003) Viridans Streptococcus Bacteremia in Children on Chemotherapy for Cancer: An Underestimated Problem. *Journal of Pediatric Hematology/Oncology*, **20**, 439-444. http://dx.doi.org/10.1080/08880010390220144

[12] Kutlu, S.S., Sacar, S., Cevahir, N. and Turgut, H. (2008) Community-Acquired *Streptococcus mitis* Meningitis: A Case Report. *International Journal of Infectious Diseases*, **12**, 107-109. http://dx.doi.org/10.1016/j.ijid.2008.01.003

[13] Baddour, L.M., Wilson, W.R., Bayer, A.S., Fowler Jr., V.G., Bolger, A.F., Levison, M.E., Ferrieri, P., Gerber, M.A., Tani, L.Y., Gewitz, M.H., Tong, D.C., Steckelberg, J.M., Baltimore, R.S., Shulman, S.T., Burns, J.C., Falace, D.A., Newburger, J.W., Pallasch, T.J., Takahashi, M. and Taubert, K.A. (2005) Infective Endocarditis: Diagnosis, Antimicrobial Therapy, and Management of Complications; A Statement for Healthcare Professionals from the Committee on Rheumatic Fever, Endocarditis, and Kawasaki Disease, Council on Cardiovascular Disease in the Young, and the Councils on Clinical Cardiology, Stroke, and Cardiovascular Surgery and Anesthesia, American Heart Association. *Circulation*, **111**, e394-e433. http://dx.doi.org/10.1161/CIRCULATIONAHA.105.165564

[14] Tandon, P. and Garcia-Tsao, G. (2008) Bacterial Infections, Sepsis, and Multiorgan Failure in Cirrhosis. *Seminars in Liver Disease*, **28**, 26-42. http://dx.doi.org/10.1055/s-2008-1040319

17

New Non-Invasive Index for Detecting Esophageal Varices in Patients with Liver Cirrhosis

Mona A. Amin1, Ahmed E. El-Badry1, May M. Fawzi1, Dalia A. Muhammed2, Shorouk M. Moussa1

1Internal Medicine, Cairo University, Cairo, Egypt
2Community Medicine, Cairo University, Cairo, Egypt

Email: monasleman@hotmail.com

Abstract

Introduction: Many studies have shown that clinical, biochemical and ultrasonographic parameter are predictive of the presence and grading of esophageal varices. Aim of Study: Validation of a non-invasive test called P2/MS and its comparison with other noninvasive tests for the detection of high risk esophageal varices. Patients and Methods: We prospectively enrolled 125 consecutive patients with liver cirrhosis. Complete blood count [CBC], Platelet count by direct method, Liver functions [serum bilirubin, AST, ALT, prothrombin time and concentration and serum albumin], kidney functions, hepatitis markers for B & C, abdominal ultrasonography and upper gastrointestinal endoscopy were done for each patient. Calculation of P2/MS [Platelet count)2/{monocyte fraction (%) × segmented neutrophil fraction (%)], API [age-platelet index], APRI [AST-to-platelet ratio index], SPRI [spleen-toplatelet ratio index], ASPRI [age-spleen-to-platelet ratio index] scores and correlating the different scores with the grade of esophageal varices found on upper endoscopy. Results: During processing of our patient's data, we found certain relation between segmented neutrophils, monocytes, platelet count, total bilirubin and the degree of esophageal varices for the detection of high risk varices and a new equation was formulated and we called it P2/MS-B. In predicting high risk esophageal varices HREV, the area under the curve for this new variable was [0.909, 95% confidence interval 0.858-0.961, p = 0.000] which was significantly higher than all the other variables including P2/MS for the detection of HREV. The sensitivity of the new equation for the detection of HREV is 85.3%, the specificity is 83.1%, the positive predictive value is 87.9%, the negative predicative value is 86.0 % and the overall accuracy of the test is 85.6%. Conclusion: A newly detected noninvasive variable for detecting HREV may reliably screen liver cirrhosis patients for HREV and avoid unnecessary endoscopy in low risk patients.

Keywords

Esophageal Varices, Noninvasive Indices

1. Introduction

Portal hypertension is a progressive complication of liver cirrhosis and it is the cause of high morbidity and mortality. Gastroesophageal varices are present in approximately 50% of patients with cirrhosis. The management of cirrhotic patients with varices differs according to the grade of varices or the presence of acute variceal bleeding. While varices are found in 40% of Child A patients, they can be present in up to 85% of Child C patients [1]. Cirrhotic patients develop varices at a rate of 8% per year and the strongest predictor for their development in those who have no varices at the time of initial endoscopic screening is a portal-hepatic venous pressure gradient (HVPG) more than 10 mmHg [2] [3]. Variceal hemorrhage occurs at a yearly rate of 5%-15%, and its most important predictor is the size of varices, with the highest risk of first hemorrhage occurring in patients with large varices [4].

The gold standard for the diagnosis of varices is esophagogastroduodenoscopy (EGD). It is recommended that patients with cirrhosis undergo endoscopic screening for varices at the time of diagnosis [5] [6]. Since the point prevalence of medium/large varices is approximately 15%-25% [1], the majority of subjects undergoing screening EGD either do not have varices or have varices that do not require prophylactic therapy. Thus, several models have been proposed to predict the presence of high risk varices by nonendoscopic methods and have excited considerable interest among researchers. Multiple studies have evaluated possible noninvasive markers of esophageal varices in patients with cirrhosis such as: the platelet count, Fibrotest, spleen size, portal vein diameter, and transient elastography [7] [8]. Lee and coworkers recently proposed a simple noninvasive test, P2/MS, which they developed in a study of patients with virusrelated chronic liver disease (CLD) [9]. They used the following formula: (platelet count)2/[monocyte fraction (%) − segmented neutrophil fraction (%)]. However, P2/MS has received little external validation of its diagnostic accuracy and cut-off values for detection of esophageal varices [10]. We, therefore, conducted the current study to externally validate P2/MS, to determine optimal thresholds to predict high risk esophageal varices (HREV) in patients with liver cirrhosis, and to compare results of the P2/MS index with those from other noninvasive tests.

2. Patients and Methods

2.1. Patients

Between August 2010 and May 2011, we prospectively enrolled 125 consecutive patients with liver cirrhosis presenting for routine follow up of their condition at Internal Medicine Department Kasr El-Aini Hospital. Cirrhosis was diagnosed clinically by history and physical examination, as well as by standard laboratory and sonographic data. The exclusion criteria included the following: the presence of infection or fever; alcohol ingestion in excess of 30 g/day for more than 45 years; previous variceal bleeding; beta-blocker therapy; previous endoscopic treatments (bandligation or sclerotherapy); previous surgery forportal hypertension or Transjugular Intrahepatic Portosystemicstent shunt placement; portal vein orsplenic vein thrombosis and Hepatocellular Carcinoma. All subjects received complete biochemical evaluations, ultrasonography and endoscopy within 2 days of admission. The study protocol followed the ethical guidelines of the 1975 Declaration of Helsinki. We obtained written, informed consent from each participant or a responsible family member after fully explaining the possible complications of the diagnostic procedures. The Institutional ethical committee approved this study.

2.2. P2/MS, and Other Noninvasive Tests

For the calculation of noninvasive tests including P2/MS, the laboratory data obtained on the same day as the endoscopic examination were used. Within one day following or preceding the endoscopy, all patients underwent an ultrasonographic examination of the upper abdomen, performed by an experienced operator blinded to the patients' clinical and laboratory data. A spleen bipolar diameter was defined as the greatest longitudinal dimension at the level of splenic hilum on the image monitor using electronic calipers [11].

The values for P2/MS and other noninvasive tests were calculated automatically, using previously published data (Table 1) [12] [13] [14] [15].

2.3. Endoscopic Evaluation

An experienced gastroenterologist blinded to the patients' clinical and laboratory data confirmed all endoscopic findings. Esophageal varices were classified as: small [veins minimally elevated above the esophageal mucosal surface], medium [tortuous veins occupying less than one third of the esophageal lumen], or large [those occupying more than one-third of the esophageal lumen]. In this study, patients with high risk esophageal varices (HREV) were defined as those with medium or large esophageal varices and those with small varices but with red signs [No. = 66 patients] and represent 52.8% of all cases.

Table 1. Simple fibrosis tests composed of clinical and laboratory parameters.

Fibrosis test	Calculation
P2/MS	[Platelet count $(109/L)]^2$/[monocyte fraction (%) _segmented neutrophil fraction (%)]
AAR	AST/ALT
API	Age (years): <30 = 0; 30 - 39 = 1; 40 - 49 = 2; 50 - 59 = 3; 60 - 69 = 4; ≥70 = 5 Platelet count (109/L): ≥225 = 0; 200 - 224 = 1; 175 - 199 = 2; 150 - 174 = 3; 125 - 149 = 4; <125 = 5 AP index is the sum of the above (possible value 0 - 10)
APRI	[(AST/ULN)/platelet count (109/L)]_100
SPRI	Spleen size (cm)/platelet count (109/L)_100
ASPRI	Age (years): <30 = 0; 30 - 39 = 1; 40 - 49 = 2; 50 - 59 = 3; 60 - 69 = 4; ≥70 = 5 ASPRI is the sum of age and SPRI
Formula by Berzigotti *et al.* (16)	Risk score = [−0.193+ (−0.359 × albumin) + (16.456 × INR) + (−0.016 × ALT)].
Our discovered formula	Final equation= e*/1 + e* Where * = −5.192 + (0.086 × segmented neutrophils) + (0.381 × monocytes) − (0.04 × platelet) + (0.637 × T. bilirubin) And e = exponential

2.4. Statistical Analysis

The goals of this study were to validate the diagnostic value of P2/MS for the detection of esophageal varices and to estimate optimal P2/MS cut-off points to indicate when a patient with liver cirrhosis should undergo prophylactic treatment. To assess the diagnostic accuracy of each noninvasive index, receiver operating characteristic [ROC] curves were constructed and the corresponding areas under the ROC curve [AUROC] were computed. The data was coded and entered using the statistical package SPSS version 15. The data was summarized using descriptive statistics: median and range, minimal and maximum values for quantitative variables and number and percentage for qualitative values. Statistical differences between groups were tested using Chi Square test for qualitative variables, independent sample [T] test for quantitative normally distributed variables while Nonparametric Mann Whitney test was used for quantitative variables which aren't normally distributed. Correlations were done to test for linear relations between variables. Logistic regression analysis was done to test for significant predictors of outcome variable.ROC curve was used to test the validity of different scores in diagnosing high risk esophageal varices. P-values less than or equal to 0.05 were considered statistically significant.Sensitivity, Specificity, positive predictive value PPV and negative predictive value NPV of different tests were calculated.

3.   Results

3.1. Patient Characteristics

The mean age of the patients [86 males, 39 females] was 55.17 ± 7.73 years (Table 2). Of these 125 patients, 115 [92%] had esophageal varices [61 classified as small, 36 as medium and 18 as large] and 66 had high risk esophageal varices [52.8%]. The median platelet count was 100 [109]/L [interquartile (IQR) 20-440], the median segmented neutrophil fraction 65% [IQR 42-89] and the median monocyte fraction 8% [IQR 1 - 15].

3.2. Comparisons of the P2/MS Index and Our New Index with Other Noninvasive Tests

Table 2. Main clinical characteristics and laboratory results of the patients.

Variables	Median	Minimum	Maximum
Age years	55	33	70
Gender (M/F)		86/39	
Body mass index (kg/m^2)	23	18	28
Child-Pugh (A/B/C)		1/74/50	
White cell count (/µl)	5400	1700	9800
Segmented neutrophil fraction (%)	65	42	89
Monocyte fraction (%)	8	1	15
Haemoglobin (g/dl)	10	7	12
Platelet count (10^9/L)	100	20	440
Prothrombin time (INR)	1.50	1.06	2.60
Total bilirubin (mg/dl)	1.30	0.10	5
Albumin (g/dl)	2.50	1.40	3.30
AST (IU/L)	57	16	200
ALT (IU/L)	38	8	170
BUN (mg/dl)	10	3	18
Creatinine (mg/dl)	0.90	0.40	1.20
Spleen diameter (cm)	15	9.50	26.60
Varices present (yes/no)		115/10	
esophageal varices size (1/2/3)		61/36/18	
High-risk esophageal varices (yes/no)		66/59	
Portal hypertensive gastropathy (yes/no)		70/55	

AST: aspartate aminotransferase, ALT: alanine aminotransferase, BUN: blood urea nitrogen.

P2MS was calculated [mean 67.67, median 20.74 (IQR 3.00-849.06)]. The patients without esophageal varices [median 46.94, IQR 9.04-849.06] had a higher P2/MS value than those with esophageal varices [median 12.48, IQR (3.00-107.53), P = 0.000], suggesting that the higher the score, the lower the likelihood of esophageal varices. In predicting high risk esophageal varices, area under the curve for P2/MS was [0.897, 95% confidence interval (CI) 0.841-0.953] which showed values better than those of AAR [0.511, 95% CI 0.405-0.618; P = 0.828] , API [0.757, 95% CI 0.669-0.845; P = 0.000], SPRI [0.767, 95% CI 0.684-0.850; P = 0.000], ASPRI [0.771, 95% CI 0.688-0.853; P = 0.000] and APRI [0.697, 95% CI 0.605-0.788; P = 0.000] and the formula by Berzigotti et al. [16] were 0.573 [95% CI 0.471-0.675], all of which were significantly lower than that of P2/MS (Table 3).

During processing of our patients' data and on doing bivariate analysis, we found a certain relation between segmented neutrophil, monocytes, platelet count and total bilirubin in detection of HREV, so we entered those variables in a logistic regression model and found data depicted in Table 4 from which we obtained a new formula for predicting HREV. We named this formula P2/MS-B. Accordingly, a probability score was calculated for each patient, and then we analyzed this score to the ROC curve to validate it. In predicting high risk esophageal varices, area under the curve for this new variable was [0.909, 95% confidence interval (CI) 0.858-0.961, P = 0.000]. This means it showed better values than all other variables including P2/MS.

3.3. Determination of the Optimal Cut-Off Values

As the central goal of this study, we sought to validate the noninvasive P2/MS test as a predictor of HREV and use it to determine which patients should undergo prophylactic treatment.

Table 3. Correlation of the noninvasive scores and the degree of esophageal varices.

| | High risk esophageal varices | | P value |
	No	Yes	
Number	59	66	
Spleen diameter (cm)	14.80 (10 - 21.7)	15.95 (9.5 - 26.6)	0.080
P2MS	46.94 (9.04 - 849.06)	12.48 (3 - 107.53)	0.000
AAR	1.39 (0.54 - 5.38)	1.61 (0.64 - 3.46)	0.828
API	7 (2 - 10)	8 (3 - 10)	0.000
SPRI	11.55 (3 - 46.44)	18.79 (5.83 - 80)	0.000
ASPRI	14.67 (6.67 - 50.44)	22.11 (7.83 - 83)	0.000
Berzigotti	23.25 (16.40 - 38.28)	24.23 (16.48 - 41.77)	0.178
APRI	1.31 (0.29 - 4.78)	1.98 (0.40 - 10.26)	0.000

Details of abbreviations above are outlined in **Table 1.**

Table 4. Analysis of the variables of the new equation.

| | B | S.E | P-Value | EXP [B] | 95.0% C.I for EXP [B] | |
					Lower	Upper
Segmented neutrophils	0.086	0.029	0.003	1.089	1.030	1.152
Monocytes	0.381	0.085	0.000	1.463	1.238	1.729
Platelet count 10^9/L	−0.040	0.009	0.000	0.961	0.943	0.979
T. bilirubin mg/dl	0.637	0.289	0.027	1.981	1.074	3.328
Constant	−5.192	2.099	0.013	0.006		

At a P2/MS cut-off value of 28.84 the test achieved a PPV of 79.7%, sensitivity 89.4%, specificity 74.6%, NPV 86.3% and total accuracy of 82.4%. Thus, P2/MS reliably predicted HREV if the result was equal to or less than 28.84 with high accuracy. Above this number, HREV may be excluded with high accuracy and low-risk patients may avoid endoscopy (Table 5, Figure 1 & Figure 2).

The cut off points for the other variables showed lower sensitivity and specificity compared to P2/MS or the new test variable P2/MS-B. Compared to other cut off values, the new variable was the only test that showed better accuracy in detecting HREV. At a cut-off value of 0.5743, high risk esophageal varices were found when the numbers were greater than or equal to this number. The new test achieved a PPV of 87.9 %, a sensitivity of 85.3%, a specificity of 83.1%, a NPV of 86.0% and a total accuracy of 85.6%.

4. Discussion

Current guidelines recommend periodic endoscopic screening to all cirrhotic patients and prophylactic treatment for patients with HREV. But universal screening will lead to many unnecessary endoscopies [17]. Thus, various noninvasive tests based on biochemical and imaging studies have been proposed [12] [13] [14] [15]. This is particularly important in nations whose healthcare budget is low and the availability of endoscopic units is limited.

Table 5. Suggested cut-off values all test variables for prediction of high risk esophageal varices.

Test result variable	Cut off point.	Sensitivity %	Specificity %	PPV %	NPV %	Total Accuracy %
AAR	1.4170	65.2	54.2	59.7	75.1	60.2
API	7.5000	71.2	69.5	72.2	81.3	79.5
SPRI	11.8561	89.4	50.8	56.8	86.2	80.2
ASPRI	14.7917	89.4	50.8	56.8	86.2	80.2
Berzigotti	23.3458	62.1	50.8	56.8	72.1	59.5
APRI	1.3575	72.7	52.5	57.4	80.2	76.1
P2MS	28.85[a]	89.4	74.6	79.7	86.3	82.4
New test	0.5743[b]	85.3	83.1	87.9	86.0	85.6

[a]High risk esophageal varices positive if the result is less than or equal to this number. [b]High risk esophageal varices positive if the result is greater than or equal to this number.

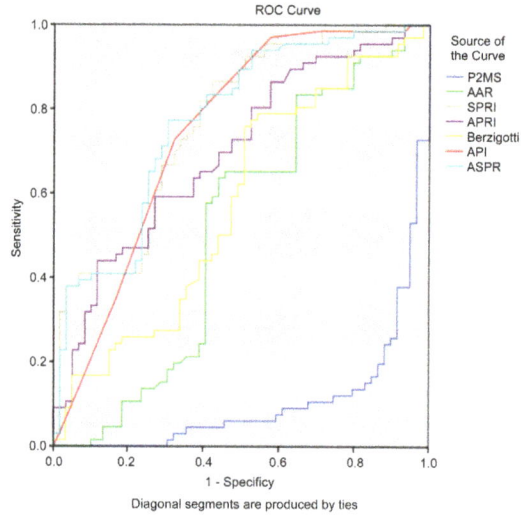

Figure 1. ROC curve of various test variables.

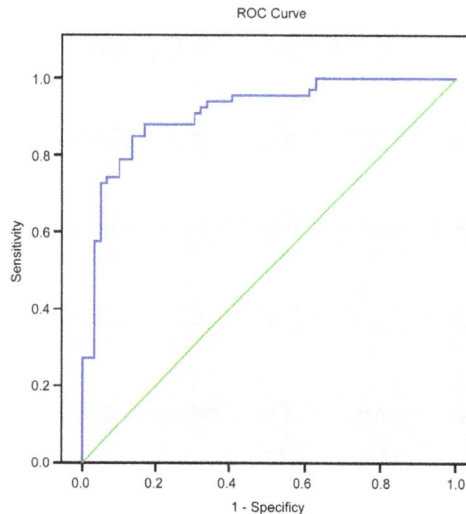

Figure 2. ROC curve of our new test variable.

Indeed, selective screening endoscopy becomes cost-effective with respect to universal screening endoscopy when non-invasive tests are sufficiently reliable to rule-in or rule-out the presence of esophageal varices.

A new index, P2/MS, based on a complete blood count, is specifically designed to predict esophageal varices in chronic liver disease. We conducted validation of the P2/MS index, and can now suggest optimal cut-off points to predict the presence of HREVs in patients with liver cirrhosis. Our study, has shown that a combination of simple, non-invasive serum markers could avoid performing unnecessary endoscopies, with only a small number of misdiagnosed cases.

In terms of the AUROC, P2/MS showed a high likelihood of reliably identifying patients with HREV [0.897], with values slightly lower than those seen in the other study by Beom Kyung et al. [0.941] [18]. In predicting HREV, P2/MS showed a higher accuracy than all variables except for our new test variable. We have suggested one cut off point for detection of HREV, which differ slightly from those of Beom Kyung et al. who used two cut off values so patients may be in the zone between the two cut off values. Above a cut-off value for P2/MS of 28.85, HREV could be excluded, with a negative predictive value [NPV] of 86.3%. Based on this value, patients could avoid unnecessary endoscopy. These patients have a low risk of bleeding and periodic follow up using this formula could be considered adequate. In contrast to other studies, our study aimed primarily to predict the presence of HREV rather than varices of any size, with the aim of selecting these patients for prophylactic endoscopic ligation. Empirical Beta blocker

therapy for primary prophylaxis can no longer be recommended for all cirrhotic patients without diagnostic endoscopy; it was not found to incur long term benefit. The formula P2/MS has several clinical advantages. First of all, one can easily calculate P2/MS at the bedside or in the outpatient clinic, as it does not require standardization and is free of intra-/interobserver variability. This make it different from other noninvasive tests that use ultrasonographic parameters such as portal vein velocity, portal vein diameter, hepatic impedance indexes, splenic impedance indexes and splenic diameter [19] [20]. We were able to detect a new test variable for detection of HREV that shows better results than P2/MS [AUROC = 0.909]. The sensitivity of the new equation for the detection of high risk esophageal varices is 85.3%, the specificity is 83.1%, the positive predictive value is 87.9% the negative predicative value is 86.0% and the overall accuracy of the test is 85.6% compared to P2/MS test which has a sensitivity of 89.4% a specificity of 74.6%, a positive predictive value of 79.7%, a negative predicative value of 86.3% and an overall accuracy of 82.4%. This new formula is a preliminary effort; its strength lies in the incorporation of a parameter that is affected by the degree of cirrhosis but it requires further assessment and validation over a large scale of patients.

5. Conclusion

P2/MS as well as P2/MS-B formulae are reliable means for detecting HREV. They are noninvasive, exhibit a high rate of accuracy and are cost effective.

Recommendation

Due to the small numbers of patients included in our study, reassessment of our new variable on a larger number of patients before validation is recommended.

Conflict of Interest

No conflict of interest of any of the authors.

Consent

The study protocol followed the ethical guidelines of the 1975 Declaration of Helsinki. We obtained written, informed consent from each participant or a responsible family member after fully explaining the possible complications of the diagnostic procedures.

References

[1] Pagliaro, L., D'Amico, G., Pasta, L., Politi, F., Vizzini, G., Traina, M., et al. (1994) Portal Hypertension in Cirrhosis: Natural History. In: Bosch, J. and Groszmann, R.J., Eds., Portal Hypertension. Pathophysiology and Treatment, Blackwell Scientific, Oxford, 72-92.

[2] Groszmann, R.J., Garcia-Tsao, G., Bosch, J., Grace, N.D., Burroughs, A.K., Planas, R., et al., The Portal Hypertension Collaborative Group (2005) Betablockers to Prevent Gastroesophageal Varices in Patients with Cirrhosis. New England Journal of Medicine, 353, 22542261. http://dx.doi.org/10.1056/NEJMoa044456

[3]   Merli, M., Nicolini, G., Angeloni, S., Rinaldi, V., De Santis, A., Merkel, C., et al. (2003) Incidence and Natural History of Small Esophageal Varices in Cirrhotic Patients. Journal of Hepatology, 38, 266-272. http://dx.doi.org/10.1016/S0168-8278(02)00420-8

[4] The North Italian Endoscopic Club for the Study and Treatment of Esophageal Varices (1988) Prediction of the First Variceal Hemorrhage in Patients with Cirrhosis of the Liver and Esophageal Varices. A Prospective Multicenter Study. New England Journal of Medicine, 319, 983-989. http://dx.doi.org/10.1056/NEJM198810133191505

[5] Grace, N.D., Groszmann, R.J., Garcia-Tsao, G., Burroughs, A.K., Pagliaro, L., Makuch, R.W., et al. (1998) Portal Hypertension and Variceal Bleeding: An AASLD Single Topic Symposium. Hepatology, 28, 868-880. http://dx.doi.org/10.1002/hep.510280339

[6] D'Amico, G., Garcia-Tsao, G., Cales, P., Escorsell, A., Nevens, F., Cestari, R., et al. (2001) Diagnosis of Portal Hypertension: How and When. In: de Franchis, R., Ed., Portal Hypertension III. Proceedings of the Third Baveno International Consensus Workshop on Definitions, Methodology and Therapeutic Strategies, Blackwell Science, Oxford, 36-64.

[7] D'Amico, G. and Morabito, A. (2004) Noninvasive Markers of Esophageal Varices: Another Round, Not the Last. Hepatology, 39, 30-34. http://dx.doi.org/10.1002/hep.20018

[8] Garcia-Tsao, G., D'Amico, G., Abraldes, J.G., Schepis, F., Merli, M., Kim, W.R., et al. (2006) Predictive Models in Portal Hypertension. In: de Franchis, R., Ed., Portal Hypertension IV. Proceedings of the Fourth Baveno International Consensus Workshop on Methodology of Diagnosis and Treatment, Blackwell, Oxford, 47-100.

[9] Lee, J.H., Yoon, J.H., Lee, C.H., et al. (2009) Complete Blood Count Reflects the Degree of Oesophageal Varices and Liver Fibrosis in Virus-Related Chronic Liver Disease Patients.

Journal of Viral Hepatitis, 16, 444-452. http://dx.doi.org/10.1111/j.1365-2893.2009.01091.x [10] Kim, B.K., Han, K.H., Park, J.Y., et al. (2009) External Validation of P2/MS and Comparison with Other Simple Non-Invasive Indices for Predicting Liver Fibrosis in HBV-Infected Patients. Digestive Diseases and Sciences.

[11] Dittrich, M, Milde, S., Dinkel, E., Baumann, W. and Weitzel, D. (1983) Sonographic Biometry of Liver and Spleen Size in Childhood. Pediatric Radiology, 13, 206-211. http://dx.doi.org/10.1007/BF00973157

[12] Kim, B.K., Kim, S.A., Park, Y.N., et al. (2007) Noninvasive Models to Predict Liver Cirrhosis in Patients with Chronic Hepatitis B. Liver International, 27, 969-976. http://dx.doi.org/10.1111/j.1478-3231.2007.01519.x

[13] Sheth, S.G., Flamm, S.L., Gordon, F.D. and Chopra, S. (1998) AST/ALT Ratio Predicts Cirrhosis in Patients with Chronic Hepatitis C Virus Infection. American Journal of Gastroenterology, 93, 44-48. http://dx.doi.org/10.1111/j.1572-0241.1998.044_c.x

[14] Chan, H.L., Wong, G.L., Choi, P.C., et al. (2009) Alanine Aminotransferase-Based Algorithms of Liver Stiffness Measurement by Transient Elastography (Fibroscan) for Liver Fibrosis in Chronic Hepatitis B. Journal of Viral Hepatitis, 16, 36-44. http://dx.doi.org/10.1111/j.1365-2893.2008.01037.x

[15] Poynard, T. and Bedossa, P. (1997) Age and Platelet Count: A Simple Index for Predicting the Presence of Histological Lesions in Patients with Antibodies to Hepatitis C Virus. Journal of Viral Hepatitis, 4, 199-208. http://dx.doi.org/10.1046/j.1365-2893.1997.00141.x

[16] Berzigotti, A., Gilabert, R., Abraldes, J.G., et al. (2008) Noninvasive Prediction of Clinically Significant Portal Hypertension and Esophageal Varices in Patients with Compensated Liver Cirrhosis. American Journal of Gastroenterology, 103, 1159-1167. http://dx.doi.org/10.1111/j.1572-0241.2008.01826.x

[17] Garcia-Tsao, G., Sanyal, A.J., Grace, N.D. and Carey, W. (2007) Prevention and Management of Gastroesophageal Varices and Variceal Hemorrhage in Cirrhosis. Hepatology, 46, 922-938. http://dx.doi.org/10.1002/hep.21907

[18] Kim, B.K., Han, K.-H., Park, J.Y., HoonAhn, S., Kim, J.K., et al. (2010) Prospective Validation of P2/MS Noninvasive Index Using Complete Blood Counts for Detecting Oesophageal Varices in B-Viral Cirrhosis. Liver International, 30, 860-866. http://dx.doi.org/10.1111/j.1478-3231.2010.02260.x

[19] Blackstone, E.H. (2001) Breaking down Barriers, Helpful Breakthrough Statistical Methods You Need to Understand Better. Journal of Thoracic and Cardiovascular Surgery, 122, 430439. http://dx.doi.org/10.1067/mtc.2001.117536

[20] Rockey, D.C. (2008) Noninvasive Assessment of Liver Fibrosis and Portal Hypertension with Transient Elastography. Gastroenterology, 134, 8-14. http://dx.doi.org/10.1053/j.gastro.2007.11.053

18

Efficacy of Inchinkoto for Liver Cirrhosis in an Infant with Down Syndrome Complicated by Transient Myeloproliferative Disorder

Ryuta Washio, Masaya Takahashi, Sohsaku Yamanouchi, Masato Hirabayashi, Kenji Mine, Yukihiro Noda, Eriko Kanda, Atsushi Ohashi, Hirohide Kawasaki*, Kazunari Kaneko

Department of Pediatrics, Kansai Medical University, Osaka, Japan

Email: *kawasaki@hirakata.kmu.ac.jp

Abstract

Several patients with Down syndrome complicated by transient myeloproliferative disorder may develop liver cirrhosis for which no effective therapeutic agent exists. We report the infant with Down syndrome complicated by transient myeloproliferative disorder and liver cirrhosis who was successfully treated by Inchinkoto, the Japanese herbal medicine. In the present case, Inchinkoto appeared to prevent both histological and serological aggravation of liver cirrhosis. To the best of our knowledge, this is the first report of preventive effect of Inchinkoto on liver cirrhosis, and it can be a choice of treatment for infants with Down syndrome complicated by liver cirrhosis.

Keywords

Inchinkoto, Transient Myeloproliferative Disorder, Down Syndrome, Liver Fibrosis, Japanese Herbal Medicine

1. Introduction

Down syndrome (DS) is the most common chromosome abnormality. It is well known that approximately 10% of DS is complicated by transient myeloproliferative disorder (TMD) [1]. Although most cases of TMD regress spontaneously during the first 3 months of life without treatment [2], 20% - 30% develop letahal liver failure and multiple organ failure [3] [4].

Inchinkoto is a mixture of three medical herbs: Artemisia capillaries spica, Gardenia fructus and Rhei rhizome, and has long been used in Japan, mainly for liver disorders and jaundice [5] [6]. Several authors have claimed that Inchinkoto improves liver function and suppresses liver fibrosis in children with postoperative biliary atresia without serious side effects [6] [7]. However, efficacy of Inchinkoto for liver cirrhosis (LC) for which no effective treatment exists is not yet determined.

Here, we report the efficacy of Inchinkoto for LC in an infant with DS complicated by TMD.

2. Case Description

An infant was born to a 34-year-old multigravida mother. Prenatal ultrasound revealed fetal growth restriction and oligohydramnios at 32 weeks of gestation. At 34 weeks of gestation, Cesarean section under general anesthesia was urgently performed because of non-reassuring fetal status. The infant was male and his Apgar score was 3 and 8 at 1 and 5 min, respectively. His birth length was 42 cm (−1.1 SD) and birth weight was 1726 g (−1.7 SD). He had the phenotypic features of DS such as low-set ears, slanted palpebral fissures and saddle nose. The diagnosis of DS (21 trisomy with male karyotype) and GATA-1 mutation (220 + 2T > C) was later confirmed by chromosomal analysis and genetic sequencing. Peripheral white blood cell count was 58,500/μl with 10% of blasts. Aspartate amino transferase (AST) and alanine aminotransferase (ALT) were 656 and 166 IU/L on postnatal day (PD) 0, respectively. Serum level of direct bilirubin (D-bil) was normal on PD 1, but gradually increased to 4.6 mg/dL on PD 10. Hyaluronic acid (HA) and type IV collagen, both of which are known serum biomarkers for liver fibrosis, were extraordinarily elevated on PD 1 (HA 9570 ng/mL, normal <50 ng/mL; type IV collagen 2519 ng/mL, normal <150 ng/mL, respectively). An ultrasound echocardiography detected atrial septal defect and patent ductus arteriosus necessitating no medical interventions. While he did not reveal any abnormalities of the thyroid or gastrointestinal tract, coagulation test disclosed abnormal findings on PD 0 as following: thromboplastin time 93.9 s (normal: 23 - 35 s); activated partial thromboplastin time 7.6% (normal: 75% - 130%); fibrinogen 20 mg/dL (normal: 150 - 350 mg/dL); anti-thrombin III 10% (normal: 80% - 130%).

From these results, he was diagnosed as having DS complicated by disseminated intravascular coagulation (DIC) associated with TMD and the therapeutic strategy for DIC and TMD was determined: repeated platelet transfusion and administration of fresh frozen plasma and anti-thrombin III resulted in improvement of DIC; low-dose cytarabine therapy for TMD started on PD 11 successfully decreased the number of white blood cell count and eradicated the blasts by PD 16. However, serum markers for liver functions, such as D-bil, AST and ALT continued to increase and reached 32 mg/dL, 246 IU/L and 106 IU/L, respectively on PD 98. In terms of serum biomarkers for liver fibrosis, HA was not normalized by PD 98 (2350 ng/mL) while type IV collagen returned to normal (Figure 1).

To elucidate the progressive liver failure, simultaneous biopsies on liver and bone marrow were performed on PD 76: his bone marrow showed normocellularity without blasts; in contrast, liver biopsy revealed the fibrosis in the portal vein area and hepatic lobules in addition to hyperplasia of the collagen fibers surrounding the hepatocytes (Figure 2). In addition, there was infiltration of lymphocytes and neutrophils, and cholestasis in the hepatocytes and bile ducts. Based on these findings, he was diagnosed as having LC induced by TMD.

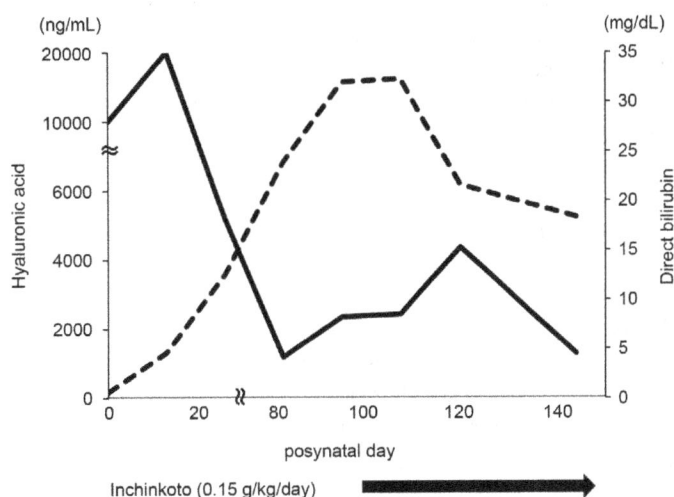

Figure 1. Serial changes in serum levels of direct bilirubin and hyaluronic acid in conjunction with administration of Inchinkoto.

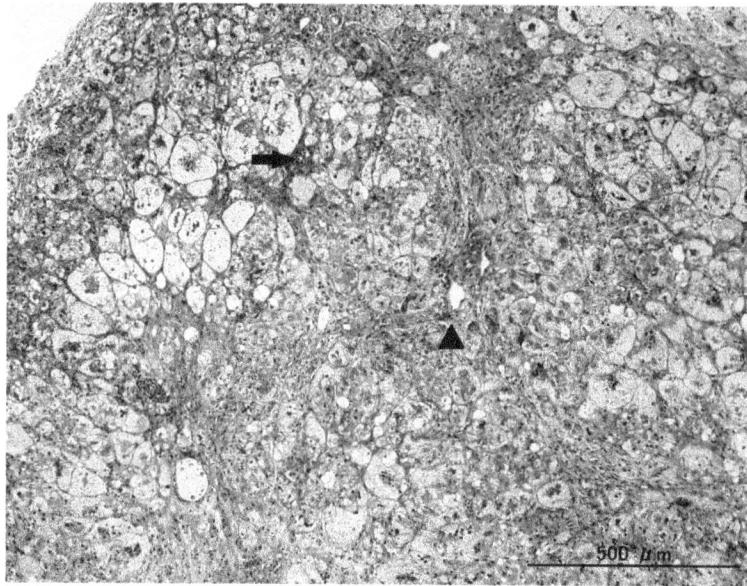

Figure 2. Histopathological findings of liver on postnatal day on 76. Hyperplasia of the collagen fiber surrounding hepatocytes, and fibrosis (▲) in the portal vein areas and the hepatic lobules. The infiltration of lymphocytes and neutrophils (→), and cholestasis in the hepatocytes and bile duct.

Considering its reported efficacy on liver fibrosis in children with postoperative biliary atresia [6] [7], the Japanese herbal medicine, Inchinkoto (Tsumura & Co., Tokyo, Japan) was orally administered at the dose of 0.15 g/kg·per·day) since PD 100. In parallel with the commencement of Inchinkoto, both D-bil and HA gradually decreased to 18.3 mg/dL and 1280 ng/mL, respectively on PD 144 (Figure 1). Improved jaundice made him possible to discharge on PD 144. Unfortunately, however, he developed fatal DIC and multiple organ failure triggered by severe bacterial infection and died on PD 202. It is worthy of special mention that the autopsy findings on liver disclosed no remarkable progressive changes of LC compared to those on PD 76.

3. Discussion

Inchinkoto has long been used to treat various liver disorders in eastern Asia such as Japan and China [5] [6] [7] [8] [9]. Iinuma, et al. reported that Inchinkoto might have a protective and antifibrotic effect for the liver of children with biliary atresia [6] [7]. Though the precise mechanisms of its action on the liver diseases remain unknown, it can be speculated as following: 1) inhibition of hepatocyte apoptosis induced by transforming growth factor-β1 [10]; 2) inhibition of the production of inflammatory cytokines and inducible nitric oxide synthase [11] [12]; and 3) direct suppression of liver fibrosis [13]. Despite the promising effects on diverse liver diseases without serious adverse effects, to the best of our knowledge, Inchinkoto has not previously been given to patients with LC characterized by diffuse nodular regeneration surrounded by fibrous bands [14].

In the present case, it appeared that Inchinkoto prevented the development of fibrosis in LC because D-bil and HA remarkably improved in parallel with commencement of its oral administration. The finding that postmortem liver specimen did not show any progression of fibrotic change compared to biopsy specimen may further support the antifibrotic effect of Inchinkoto on LC. Though measurements of cytokines which may link Inchinkoto with antifibrotic action on liver were not determined, we speculate that Inchinkoto suppressed the production of transforming growth factor-β1 and inflammatory cytokine.

While low-dose cytarabine to treat TMD occasionally not only induces hematological regression but also improves liver fibrosis in some cases [15], these did not fit into our case.

TMD is a well-known hematopoietic disorder that occurs as a complication in approximately 10% of children with DS [5]. Although TMD is a benign disease in most cases, some patients with TMD develop severe liver failure and/or multiple organ failure. Our patient had the GATA-1 mutation (220 + 2T > C), which is known to cause a lack of expression of the full-length GATA-1 protein [16].

The prevalences of GATA-1 mutation in DS have been reported to be 97.3% in patients with TMD and 89.2% in those with acute megakaryoblastic leukemia, respectively [17]. Thus, GATA-1 mutation is thought to play an important role in the pathogenesis of TMD and acute megakaryoblastic leukemia [18]. Interestingly enough, it has been recently demonstrated that GATA-1 expression even enhances the expansion of fetal megakaryocytic precursors, resulting in hepatic fibrosis in a mouse model [19]. Our case presented abnormally high levels of serum HA and type IV collagen even on PD 1. Taken together, we suspect that TMD and liver fibrosis induced by somatic GATA-1 mutation had started in utero.

4. Conclusion

In conclusion, we firstly report the promising efficacy of Inchinkoto for LC for which no effective treatment currently exists except liver transplantation. We therefore believe that Inchinkoto can be a choice of treatment for infants with DS complicated by TMD and LC.

Acknowledgements

The authors thank Dr. Kiminori Terui, Tsutomu Toki and Eetsuro Ito (Department of Pediatrics, Hirosaki Medical University) for GATA-1 mutation analysis. This study was supported by the Mami Mizutani Foundation.

Conflicts of Interest Statement

All authors have declared that they have no conflicts of interest.

References

[1] Zipursky, A. (2003) Transient Leukaemia—A Benign Form of Leukaemia in Newborn Infants with Trisomy 21. British Journal of Haematology, 120, 930-938. https://doi.org/10.1046/j.1365-2141.2003.04229.x

[2] Lange, B. (2000) The Management of Neoplastic Disorders of Haematopoiesis in Children with Down's Syndrome. British Journal of Haematology, 110, 512-524. https://doi.org/10.1046/j.1365-2141.2000.02027.x

[3] Shiozawa, Y., Fujita, H., Fujimura, J., Suzuki, K., Sato, H., Saito, M., Shimizu, T. and Yamashiro, Y. (2004) A Fetal Case of Transient Abnormal Myelopoiesis with Severe Liver Failure in Down Syndrome: Prognostic Value of Serum Markers. Pediatric

Hematology and Oncology, 21, 273-278.

https://doi.org/10.1080/08880010490277088

[4] Hoskote, A., Chessells, J. and Pierce, C. (2002) Transient Abnormal Myelopoiesis (TAM) Causing Multiple Organ Failure. Intensive Care Medicine, 28, 758-762. https://doi.org/10.1007/s00134-002-1305-7

[5] Kiso, Y., Ogasawara, S., Hirota, K., Watanabe, N., Oshima, Y., Konno, C. and Hikiko, H. (1984) Antihepatotoxic Principles of Artemisia Capillaries Buds 1. Planta Medica, 50, 81-85. https://doi.org/10.1055/s-2007-969627

[6] Iinuma, Y., Kubota, M., Yagi, M., Kanada, S., Yamazaki, S. and Kinoshita, Y. (2003) Effects of the Herbal Medicine Inchinkoto on Liver Function in Postoperative Patients with Biliary Atresia—A Pilot Study. Journal of Pediatric Surgery, 38, 1607- 1611. https://doi.org/10.1016/S0022-3468(03)00570-0

[7] Tamura, T., Kobayashi, H., Yamataka, A., Lane, G.J., Koga, H. and Miyano, T. (2007) Inchin-ko-to Prevents Medium-Term Liver Fibrosis in Postoperative Biliary Atresia Patients. Pediatric Surgery International, 23, 343-347. https://doi.org/10.1007/s00383-007-1887-9

[8] Kaiho, T., Tsuchiya, S., Yanagisawa, S., Takeuchi, O., Togawa, A., Okamoto, R., Saigusa, N. and Miyazaki, M. (2008) Effect of the Herbal Medicine Inchin-Ko-To for Serum Bilirubin in Hepatectomized Patients. Hepatogastroenterology, 55, 150- 154.

[9] Takahashi, Y., Soejima, Y., Kumagai, A., Watanabe, M., Uozaki, H. and Fukusato, T. (2014) Japanese Herbal Medicines Shosaikoto, Inchinkoto, and Juzentaihoto Inhibit High-Fat Diet-Induced Nonalcoholic Steatohepatitis in db/db Mice. Pathology International, 64, 490-498. https://doi.org/10.1111/pin.12199

[10] Yamamoto, M., Ogawa, K., Morita, M., Fukuda, K. and Komatsu, Y. (1996) The Herbal Medicine Inchin-ko-to Inhibits Liver Cell Apoptosis Induced by Transforming Growth Factor Beta 1. Hepatology, 23, 552-559.

[11] Yamashiki, M., Mase, A., Arai, I., Huang, X.X., Nobori, T., Nishimura, A., Sakaguchi, S. and Inoue, K. (2000) Effects of the Japanese Herbal Medicine "Inchinko-to" (TJ-135) on Concanavalin A-Induced Hepatitis in Mice. Clinical Science (Lond), 99, 421-431. https://doi.org/10.1042/cs0990421

[12] Matsuura, T., Kaibori, M., Araki, Y., Matsumiya, M., Yamamoto, Y., Ikeya, Y., Nishizawa, M. and Okumura, K.A.H. (2012) Japanese Herbal Medicine, Inchinkoto, Inhibits Inducible Nitric Oxide Synthase Induction in Interleukin-1β-Stimulated Hepatocytes. Hepatology Research, 42, 76-90.

https://doi.org/10.1111/j.1872-034X.2011.00891.x

[13] Inao, M., Mochida, S., Matsui, A., Eguchi, Y., Yulutuz, Y., Wang, Y., Naiki, K., Kakinuma, T., Fujimori, K., Nagoshi, S. and Fujiwara, K. (2004) Japanese Herbal Medicine Inchin-ko-to as a Therapeutic Drug for Liver Fibrosis. Journal of Hepatology, 41, 584-591. https://doi.org/10.1016/j.jhep.2004.06.033

[14] Schuppan, D. and Afdhal, N.H. (2008) Liver Cirrhosis. Lancet, 371, 838-851. https://doi.org/10.1016/S0140-6736(08)60383-9

[15] Kuroiwa, Y., Suzuki, N., Yamamoto, M., Hatakeyama, N., Hori, T. and Mizue, N. (2005) Prognostic Value of Serum Markers for Liver Fibrosis in Transient Abnormal Myelopoiesis (TAM). Rinsho Ketsueki, 46, 1179-1186.

[16] Mansini, A.P., Rubio, P.L., Rossi, J.G., Gallego, M.S., Medina, A., Zubizarreta, P.A., Felice, M.S. and Alonso, C.N. (2013) Mutation Characterization in the GATA-1 Gene in Patients with Down's Syndrome Diagnosed with Transient Abnormal Myelopoiesis or Acute Megakaryoblastic Leukemia. Archivos Argentinos de Pediatria, 111, 532-536.

[17] Roy, A., Roberts, I., Norton, A. and Vyas, P. (2009) Acute Megakaryoblastic Leukemia (AMKL) and Transient Myeloproliferative Disorder (TMD) in Down Syndrome: A Multi-Step Model of Myeloid Leukaemogenesis. British Journal of Haematology, 147, 3-12. https://doi.org/10.1111/j.1365-2141.2009.07789.x

[18] Ahmed, M., Sternberg, A., Hall, G., Thomas, A., Smith, O., O'Marcaigh, A., Wynn, R., Stevens, R., Addison, M., King, D., Stewart, B., Gibson, B., Roberts, I. and Vyas, P. (2004) Natural History of GATA1 Mutations in Down Syndrome. Blood, 103, 2480-2489. https://doi.org/10.1182/blood-2003-10-3383

[19] Birger, Y., Goldberg, L., Chlon, T.M., Goldenson, B., Muler, I., Schiby, G., Jacob-Hirsch, J., Rechavi, G., Crispino, J.D. and Izraeli, S. (2013) Perturbation of Fetal Hematopoiesis in a Mouse Model of Down Syndrome's Transient Myeloproliferative Disorder. Blood, 122, 988-998. https://doi.org/10.1182/blood-2012-10-460998

Abbreviations

TDM	Transient myeloproliferative disorder
DS	Down syndrome
AST	Aspartate amino transferase
ALT	Alanine aminotransferase
D-bil	Direct bilirubin
HA	Hyaluronic acid
DIC	Disseminated intravascular coagulation

19

Nursing Diagnosis in Patients with Liver Cirrhosis in Use of Feeding Tube

Fernanda Raphael Escobar Gimenes[1], Patrícia Costa dos Santos da Silva[2], Andréia Regina Lopes[3], Renata Karina Reis[1], Rebecca Shasanmi[4], Emília Campos de Carvalho[1]

[1]Department of General and Specialized Nursing, University of São Paulo at Ribeirão Preto College of Nursing, Ribeirão Preto, Brazil
[2]Federal University of Uberlândia, Uberlândia, Brazil
[3]Academy of the Brazilian Air Force, Pirassununga, Brazil
[4]Nursing and Public Health Research, Philadelphia, PA, USA
Email: fregimenes@eerp.usp.br, patriciacostaunifenas@hotmail.com, andreiargrigoleto@hotmail.com, rkreis@eerp.usp.br, rshasanmi@gmail.com, ecdcava@eerp.usp.br

Abstract

The objective was to identify the most frequent nursing diagnoses labels in patients with liver cirrhosis in use of feeding tube. A descriptive research was carried out in a Brazilian Hospital with 20 adult patients. Systematic data collection utilized the Conceptual Model of Wanda Horta, the first nurse to introduce the concept of Nursing Process in Brazil. The six phases of the nursing diagnostic reasoning proposed by Risner were used; nursing diagnoses were described according to NANDA-I taxonomy II. Patients were mainly male; half of them were middle age adults; they had an average of 12.8 nursing diagnoses labels; and the most frequent were: risk for aspiration and risk for infection. Nurses needed to develop effective skills to properly diagnose in order to provide safe care and improve patient outcomes.

Keywords

Nursing Diagnosis, Enteral Nutrition, Liver Cirrhosis

1. Introduction

Liver cirrhosis is the leading cause of chronic liver disease in developed countries. In the United States of America, liver cirrhosis results in more than 400,000 hospitalizations and in 27,000 deaths annually. In Taiwan, liver cirrhosis and other chronic liver diseases together are the eighth-leading cause of death overall [1]. This patient group also accounts for 75% of unplanned readmissions to the Gastroenterology and Hepatology unit in an Australian hospital [2]. In Brazil, liver cirrhosis was also the eighth-leading cause of death among men and accounted for almost 9% of hospital admissions in 2010 [3]-[6].

The prevalence of malnutrition in these patients is also a challenge, representing 20% to 90%. Malnutrition is an independent risk factor for morbidity and mortality, because it may result in several complications [7]. Therefore, patients who are unable to meet nutrient needs should be considered candidates for enteral feeding tube in order to ensure daily nutritional requirements [8]-[10].

However, the need for a feeding tube may pose patients at great risk for adverse events due to higher probability of bleeding, especially in the presence of esophageal varices, thrombocytopenia or coagulopathy [11]. Thus, it is a concern for health care team to make an accurate and timely diagnosis of such complications and the delivery of the correct treatment to safe management of patients.

Assisting people with liver cirrhosis can be a challenge for all health care professionals, especially for nurses, because these patients are frequently admitted to hospitals due to the evolution of the disease, and they can deteriorate very quickly requiring constant monitoring and surveillance [4]. Nurses need to apply appropriate clinical judgments and clinical decision-making to reduce the frequency of hospital readmissions and to give safe and qualified care. Clinical reasoning enables nurses to make complex decisions in order to improve patients' outcomes [12]. Therefore, the proper management of patients with liver cirrhosis in use of feeding tube should be the focus of all nursing care plan in order to improve symptom management, to reduce the risks associated with further decompensation, and to enhance patient safety.

Background

The constant changes in clinical status of patients with liver cirrhosis require quick and assertive decision-making. With the aim of providing qualified care, nurses have joined forces to build a body of knowledge focused on evidence-based practices to provide competent and safe care to patients with chronic conditions.

The Nursing Process (NP) is the main methodological framework for the systematic performance of professional practice, or a technological method that nurses use to foster care, and to help in documenting professional practice. Therefore, the deliberate application of NP may contribute to the quality of care, thus improving nurses visibility and professional recognition [13].

Wanda de Aguiar Horta [14], the first nurse to introduce the concept of NP in Brazil, developed a Nursing Conceptual Model based on Maslow's Theory Human Motivation [15]. This theory is based on Basic Human Needs (BHN), which is classified in five levels: physiological needs, safety, love, esteem and self-realization. In addition to Maslow's BHN, Horta adopted the classification proposed by John Mohana. Therefore, the BHN model proposed by Horta is classified into three levels: psychobiological, psychosocial, and psycho spiritual.

The Conceptual Model of Wanda Horta was used in this study because it may help nurses to collect relevant data within the framework of nursing rather than medicine. Thus, the model may assist nurses in critical thinking and give support to practitioners in outcome identification and development of nursing care plans.

Nursing diagnosis is a process of data analysis that uses clinical reasoning to determine whether nursing interventions are indicated, contributing to the quality of care and patient safety through an evidence-based practice[16] [17]. For each nursing diagnosis, nurses select the appropriate interventions suggested by the nursing intervention classification system [17] [18].

Therefore, the present study is justified by the lack of publications addressing the nursing diagnoses in clinical patients with liver cirrhosis in Brazil and worldwide. Several studies identified the most frequent nursing diagnoses in different populations [19]-[21], but none inpatients with liver cirrhosis. Thus the purpose of this study was to identify the most frequent nursing diagnoses labels in patients with liver cirrhosis in use of feeding tube.

2. Method

2.1. Design

A descriptive research design.

2.2. Setting and Sample

The study was carried out in a Brazilian University Hospital, in São Paulo state, from January 2013 to December

2013. Participants consisted of a convenient sample of 20 adult patients with liver cirrhosis in use of feeding tube. Patients in use of percutaneous enteral feeding tubes were excluded. Unconscious patients or patients with cognitive impairment were included in the study after their family's written consent.

The study was approved by the appropriate ethics committee. Patients and/or their families were assured that their identity would remain confidential and they signed a consent form voluntarily.

2.3. Data Collection

A systematic data collection was conducted and it included interaction, observation, and measurement. Data was also collected from other resources, including family and significant others, medical records, results of diagnostic tests, nursing notes, change of shift reports, and health team members. The tool used for data collection was developed by the investigators and it was based on the Conceptual Model of Wanda Horta [22].

2.4. Data Analysis and Rigor

The guidelines proposed by Risner [23] was followed for the diagnostic reasoning, which included six phases for analysis and synthesis of data:
1) Relevant patients' data were categorized. In this study the Conceptual Model of Wanda Horta [14] was used with the aim of revealing relationships among cues, thus making missing data more obvious;
2) Missing information and incongruence were identified to indicate areas for further assessment;
3) Related cues were clustered into patterns to combine patients' elements into a whole, with the aim of constructing patterns containing information about patients' response to an actual or potential health problem, and the factors related to the response;
4) Patients' patterns were then compared with normal ranges, values, expectations, and patient baseline information to identify their health-related responses;
5) Based on patients' responses, inferences were made about their health status, condition, or situation in each of the assessment categories;
6) Finally, etiological relationships were proposed to identify factors influencing or contributing to the patients' responses.

After analysis and synthesis of all patients' relevant data, the nursing diagnosis was described according to NANDA-I taxonomy II [17], that has three levels: domains, classes, and diagnoses. All nursing diagnoses were analyzed and discussed with a panel of three nurses with experience on the Conceptual Model of Wanda Horta and on NANDA-I taxonomy II. The domains and classes of each nursing diagnosis label were also identified.

3. Results

The study sample consisted of 20 hospitalized patients with liver cirrhosis in use of feeding tube. From those, 6 (30%) were females and 14 (70%) were males, with an average age of 56.3 and 59.7, respectively. The age range was from 28 to 81 years, and almost half (n = 11) of the patients were middle age adults (from 41 to 64 years).The most common causes of liver cirrhosis were chronic alcoholism (10, 50%), followed by other etiologies (6, 30%), and viral infection (4, 20%). All patients had other comorbidities, including arterial hypertension, diabetes mellitus, or renal disease.

At hospital admission, patients presented common complications related to liver cirrhosis: hepatic encephalopathy (7, 35%); previous gastrointestinal hemorrhage and spontaneous bacterial peritonitis (6, 30% each); severe weight loss (5, 25%); ascites, esophageal varices, and hepatopulmonary syndrome (3, 15% each); and portal hypertension (2, 10%) (**Table 1**).

From the analysis and synthesis of patients' relevant data, there were 255 nursing diagnoses labels identified. Each patient had an average of 12.8 nursing diagnoses with a minimum of 9 to maximum of 16; 36 different nursing diagnoses labels were identified in the sample and 12 nursing diagnoses labels showed percentage equal to or greater than 50%.

The most frequent domains for these patients were: Domain 4—Activity/Rest (11, 30.6%) and Domain 11—Safety/Protection (10, 27.8%), followed by Domain 3—Elimination and Exchange (5, 13.9%) and Domain 2—Nutrition (4, 11.1%). No diagnoses were identified from Domain 1—Health Promotion, Domain 6—Self Perception, Domain 7—Role Relationships, Domain 8—Sexuality, and Domain 10—Life Principles.

Table 1. Sample and characteristics, according to gender (N = 20).

Variables	Female		Male		Total	
	n	%	n	%	n	%
Age Range						
28 - 40	1	5%	1	5%	2	10%
41 - 64	4	20%	7	35%	11	55%
65 - 74	0	0%	5	25%	5	25%
75 - 81	1	5%	1	5%	2	10%
Causes of cirrhosis						
Chronic alcoholism	1	5%	9	45%	10	50%
Other etiologies	3	15%	3	15%	6	30%
Viral infection	2	10%	2	10%	4	20%
Complications						
Encephalopathy	2	10%	5	25%	7	35%
Hemorrhage	1	5%	5	25%	6	30%
Peritonitis	1	5%	5	25%	6	30%
Severe weight loss	1	5%	4	20%	5	25%
Ascites	0	0%	3	15%	3	15%
Esophageal varices	1	5%	2	10%	3	15%
Hepatopulmonary syndrome	0	0%	3	15%	3	15%
Portal hypertension	0	0%	2	10%	2	10%

Analysis of the class level of the nursing diagnosis identified in the sample, Physical injury (7, 19.4%), Self-care (4, 11.1%), and Gastrointestinal function (4, 11.1%) were the most frequent. **Table 2** shows the domains, classes, and NANDA-I diagnoses labels for patients hospitalized with liver cirrhosis in use of feeding tube.

4. Discussion

Results showed a total of 255 nursing diagnoses labels, with an average of 12.8 nursing diagnoses labels per patient, and 36 different nursing diagnoses labels. The results differ from previous study conducted in an intensive care unit where 1.087 nursing diagnoses were formulated for 44 critical patients, with a mean of 8.5 diagnoses per patient, and 28 different nursing diagnoses labels [24]. Differences may be due to the methodology used in both studies for data collection and analysis.

Risk for aspiration (00039) and Risk for infection (00004) were the most frequent NANDA-I diagnoses labels found in this study. They were presented in 100% of patients. Similar findings were detected by other authors in critical patients [25] [26].

Risk for aspiration (00039) is common in patients in use of feeding tube, especially in those with chronic liver disease and portal hypertension because they have delayed gastric emptying for both the liquid and solid components [27]. In addition, the cirrhosis of the liver does not allow the free passage of blood that accumulates in the gastrointestinal tract and in the spleen, resulting in chronic congestion in this area. Consequently, indigestion due to intra-abdominal pressure and altered bowel function occur [28].

According to Opilla [29], the presence of a feeding tube also increases secretions from tube irritation, impairment of laryngeal function, and disruption of the esophageal sphincters during intubation, thus contributing to the risk for aspiration. It is worth to note that many patients in this study (25%) also had liver encephalopathy. The decreased level of consciousness in these patients and the altered coordination between breathing and swallowing interferes with the patient's ability to protect the airway [30].

Risk for infection (00004) was also presented in all patients in this study. Bacterial infections are a major complication of liver cirrhosis and a serious burden among patients because they may be a triggering factor for the occurrence of gastrointestinal bleeding, hepatic encephalopathy, kidney failure, and further deteriorate liver

Table 2. Domain, classes and NANDA-I diagnosis labels for hospitalized patients in use of feeding tube.

Domain	n	%	Class	n	%	NANDA-I label	n	%
2 - Nutrition	4	11.1	1 - Ingestion	1	2.8	Imbalanced nutrition: less than body requirements (00002)	12	60
			4 - Metabolism	1	2.8	Risk for unstable blood glucose level (00179)	3	15
			5 - Hydration	2	5.5	Excess fluid volume (00026)	12	60
						Risk for imbalanced fluid volume (00025)	1	5
3 - Elimination and exchange	5	13.9	2 - Gastrointestinal function	4	11.1	Risk for constipation (00015)	11	55
						Dysfunctional gastrointestinal motility (00196)	11	55
						Diarrhea (00013)	3	15
						Constipation (00011)	1	5
			4 - Respiratory function	1	2.8	Impaired gas exchange (00030)	2	10
4 - Activity/rest	11	30.6	1 - Sleep/rest	1	2.8	Disturbed sleep pattern (00198)	5	25
						Impaired bed mobility (00091)	8	40
			2 - Activity/exercise	3	8.3	Impaired physical mobility (00085)	3	15
						Impaired walking (00088)	2	10
			3 - Energy balance	1	2.8	Fatigue (00093)	6	30
			4 - Cardiovascular/ pulmonary response	2	5.5	Activity intolerance (00092)	1	5
						Impaired spontaneous ventilation (00033)	1	5
			5 - Self-care	4	11.1	Bathing self-care deficit (00108)	18	90
						Dressing self-care deficit (00109)	17	85.5
						Toileting self-care deficit (00110)	7	35.5
						Feeding self-care deficit (00102)	5	25.5
5 - Perception/cognition	2	5.5	4 - Cognition	2	5.5	Risk for acute confusion (00173)	15	75
						Acute confusion (00128)	5	25
9 - Coping/stress tolerance	2	5.5	2 - Coping responses	2	5.5	Ineffective coping (00069)	2	10
						Anxiety (00146)	2	10
11 - Safety/protection	10	27.8	1 - Infection	1	2.8	Risk for infection (00004)	20	100
			2 - Physical injury	7	19.4	Risk for aspiration (00039)	20	100
						Risk for bleeding (00206)	17	85
						Risk for impaired skin integrity (00047)	16	80
						Risk for falls (00155)	14	70
						Impaired skin integrity (00046)	3	15
						Ineffective airway clearance (00031)	3	15
						Impaired oral mucous membrane (00045)	2	10
			3 - Violence	1	2.8	Risk for suicide (00150)	1	5
			6 - Thermoregulation	1	2.8	Hyperthermia (00007)	1	5
12 - Comfort	2	5.5	1 - Physical comfort	2	5.6	Acute pain (00132)	1	5
						Chronic pain (00133)	1	5

function [31]. Most of the infections in cirrhotic patients are caused by enteric bacteria, accounting for approximately 32% - 34%. This suggests that the defense mechanisms of patients with chronic liver disease fail to prevent the microorganisms present in the intestinal lumen from reaching the systemic circulation, contributing to the risk of spontaneous bacterial peritonitis[4] [28] [32] [33]. Thus, nurses need to act towards the reduction of the negative clinical impact of infections in these patients to reduce repeated hospitalizations and impaired health-related quality of life.

Another nursing diagnoses were frequent in these patients, including: Bathing self-care deficit (00108), Dressing self-care deficit (00109), Risk for bleeding (00206), Risk for impaired skin integrity (00047), Risk for acute confusion (00173), Risk for falls (00155), Imbalanced nutrition: less than body requirements (00002), and Excess fluid volume (00026). Similar results were identified in previous studies involving critical care patients and patients in chronic conditions [25] [34]. In Taiwan, physical symptoms and psychological distress, including abdominal symptoms, fatigue, fluid retention, loss of appetite, systemic symptoms, decreased attention, and bleeding, were common among patients with liver cirrhosis [35]. These NANDA-I diagnoses require from nurses the ability and skills to identify the patients' health status in order to deliver an individualized nursing care plan focused on patient safety.

Self-care deficits related to activities of daily living (eg. Bathing, dressing, feeding, and toileting) were also frequent in patients because they usually suffer from moderate-to-severe fatigue. In this study, 30% of patients presented with fatigue (00093) that can result in decreased motivation, depression, reduced physical activity, and constraints on daily life [1] [36]. Nursing interventions should focus on the maintenance of physical and psychological comfort for these patients to improve their quality of life and to reduce the risks for other injuries, such as falls.

Another potential risk for patients living with liver cirrhosis is the clinical bleeding, most frequently caused by esophageal varices, gastric varices or portal hypertensive gastropathy. In this study, 15% of patients presented esophageal varices and 10% presented portal hypertension at the hospital admission. In chronic liver disease, vitamin K absorption by the liver is decreased. Therefore, the production of coagulation factors (such as II, VII, IX, and X) does not happen, resulting in prolonged time required for a blood sample to clot [4]. Moreover, hospitalized patients with liver cirrhosis require frequent invasive procedures (e.g., peripheral venous catheter for transfusions and the administration of medications and other solutions; venous and arterial puncture for blood samplings; and paracentesis), which may cause complications such as bleeding.

Thrombocytopenia is also a common and persistent problem in cirrhotic patients [7]. In this study, 30% of patients had previous gastrointestinal hemorrhage at the hospital admission, thus the risk for another hemorrhage is real. Variceal bleeding may cause upper gastrointestinal hemorrhage due to portal hypertension. It remains one of the most important complications of chronic liver disease and one of the largest causes of mortality in this group [2] [37]. These patients have a permanent state of hyperdynamic circulation, with pronounced splanchnic vasodilatation. Further bleeding can occur in 60% of patients, with a mortality of up to 33%. Prevention of bleeding is therefore an essential part of the management of these patients [38]. However, the management of patients with acute gastrointestinal bleeding includes not only treatment and control of active bleeding but also the prevention of further bleeding, infections, and renal failure [39].

It is worth to note that 70% of patients had Risk for falls (00155), 35% were elderly, and 25% had Acute confusion (00128). The liver's inability to detoxify the blood results in increased blood circulation of ammonia and other toxic metabolites [4]. The increased concentration of ammonia in the blood causes brain dysfunction and injury, contributing to the hepatic encephalopathy. In the first stage of the hepatic encephalopathy, patients may experience discrete mental changes, as well as motor disorders. As the problem persists, patients may demonstrate mental confusion, mood swings, and sleep pattern changes. All these manifestations, when present, increase the risk for falling. In addition, the incidence of falls in patients over 60 years is almost three times higher than in older adults [40].

Imbalanced nutrition: less than body requirements (00002) was also frequent in these patients (60%). These results differ from those found by Park [41] in patients with heart failure. According to the author, Imbalanced nutrition: less than body requirements (00002) was detected only in 2.7% of patients. Researchers [26] [24] also found different results in intensive care units. In both studies, Imbalanced nutrition: less than body requirements (00002) was detected in 5% of patients. The results differ from the others, perhaps because of patients profile comprising the samples. It is worth to consider that malnutrition is prevalent in patients with liver cirrhosis because the evolution of the disease. According to Tai *et al.* [9] malnutrition was present in 50% of Malaysian patients and the mean caloric intake was low at 15.2 kcal/kg/day. Thus, patients with end stage hepatic failure will present with muscle wasting, decreased fat stores, and overt cachexia [42].

Excess fluid volume (00026) is another common complication in patients with liver cirrhosis because the synthesis of proteins, such as albumin, is impaired. This nursing diagnosis was detected in 60% of all patients. Vargas and França [43] found Excess fluid volume (00026) in a case study conducted in a Brazilian hospital with a patient with liver cirrhosis. Hypoalbuminemia decreases the plasma oncotic pressure deflecting the balance of hemodynamic forces to the accumulation of fluid in interstitial spaces, thus resulting in peripheral edema and ascites. In addition, the liver failure to metabolize aldosterone resulting in increased retention of sodium and water by the kidneys, and in increased potassium excretion [44]. As the problem persists, the retention of sodium and water contributes to the increased blood volume that may cause cardiac overload and hence pulmonary edema.

In relation to the most prevalent NANDA-I domain found in this study, the results were similar to those identified by Park [41] in patients with heart failure. The author also found Physical injury as the most frequently used NANDA-I class in this population. In one study conducted with liver transplant patients, researchers [45] found that the domains mostly affected by patients were activity/rest, safety/protection, elimination and exchange, and comfort. Thus, it is important to identify the specific interventions commonly delivered for specified groups of patients [46] in order to improve patient outcomes and deliver safe care.

Liver cirrhosis is a progressive illness that may culminate in multiple system organ failure and death [47], requiring appropriate health care management, especially from nurses that should act in order to prevent further complications and to improve patient outcomes.

The results of this study sustain that nursing diagnoses should be seen as the basis for independent and collaborative actions because they provide direction for nursing interventions. Thus, nurses need to list nursing diagnoses during the process of care to reflect patients' changing condition and responses in order to individualize patient care.

Limitations

This study presented limitations. Nursing diagnoses labels were identified only in hospitalized patients. In addition, only 20 patients participated in this research, thus future studies should be conducted in multiple healthcare settings, such as ambulatory care, and with larger samples.

5. Conclusion

The most frequent nursing diagnoses labels identified in patients with liver cirrhosis in use of feeding tube were Risk for aspiration (00039) and Risk for infection (00004), requiring from nurses appropriate management of complications. This was the first research conducted in a practice setting with patients with liver cirrhosis. These findings supported that the identification of the most frequent nursing diagnoses in specific population helped nurses to identify the focus of care in patients with complex health problems and to prevent future complications associated with the evolution of disease.

Acknowledgements

F.R.E.G. wish to thank the São Paulo Research Foundation (FAPESP), Brazil, for funding the research (n° 2012/14840-8).

Authors' Contribution

Gimenes F.R.E. contributed to the project design, development of research, data collection, analysis and interpretation of data, writing, critical review of the relevant intellectual content and final approval of the version to be published. Silva P.C.S., Lopes A.R., and Reis R.K. contributed to data analysis, writing, critical review of the relevant intellectual content and final approval of the version to be published. Shasanmi R. made contributions to revisions of article for intellectual content and English language. Campos E.C. made substantial contributions to drafting of the article, and revised the article for important intellectual content.

References

[1]   Wu, L.-J., Wu, M.-S., Lien, G.I.S., Chen, F.-C. and Tsai, J.-C. (2012) Fatigue and Physical Activity Levels in Patients

with Liver Cirrhosis. *Journal of Clinical Nursing*, **21**, 129-138. http://dx.doi.org/10.1111/j.1365-2702.2011.03900.x

[2]   Wigg, A.J., McCormick, R., Wundke, R. and Woodman, R.J. (2013) Efficacy of a Chronic Disease Management Model for Patients with Chronic Liver Failure. *Clinical Gastroenterology and Hepatology*, **11**, 850-858. http://dx.doi.org/10.1016/j.cgh.2013.01.014

[3]   Ministry of Health (2008) [Brazil's Health 2007: An Analysis of the Health Situation]. Ministry of Health, Brasília, 641 p.

[4]   Kelso, L.A. (2008) Cirrhosis: Caring for Patients with End-Stage Liver Failure. *The Nurse Practitioner*, **33**, 24-30. http://dx.doi.org/10.1097/01.NPR.0000325976.85753.dd

[5]   Silva, I.S.S. (2010) [Liver Cirrhosis]. *Revista Brasileira de Medicina*, **67**, 9.

[6]   Barros, M.B.A., Francisco, P.M.S.B., Zanchetta, L.M. and César, C.L.G. (2011) [Trends in Social and Demographic Inequalities in the Prevalence of Chronic Diseases in Brazil. PNAD: 2003-2008]. *Ciência & Saúde Coletiva*, **16**, 3755-3768. http://dx.doi.org/10.1590/S1413-81232011001000012

[7]   Marsano, L.S., Mendez, C., Hill, D., Barve, S. and McClain, C.J. (2003) Diagnosis and Treatment of Alcoholic Liver Disease and Its Complications. *Alcohol Research and Health*, **27**, 247-256.

[8]   Saunders, J., Brian, A., Wright, M. and Stroud, M. (2010) Malnutrition and Nutrition Support in Patients with Liver Disease. *Frontline Gastroenterology*, **1**, 105-111. http://dx.doi.org/10.1136/fg.2009.000414

[9]   Tai, M.-L., Goh, K.-L., Mohd-Taib, S., Rampal, S. and Mahadeva, S. (2010) Anthropometric, Biochemical and Clinical Assessment of Malnutrition in Malaysian Patients with Advanced Cirrhosis. *Nutrition Journal*, **9**, 27. http://dx.doi.org/10.1186/1475-2891-9-27

[10] O'Brien, A. and Williams, R. (2008) Nutrition in End-Stage Liver Disease: Principles and Practice. *Gastroenterology*, **134**, 1729-1740. http://dx.doi.org/10.1053/j.gastro.2008.02.001

[11] Andus, T. (2007) ESPEN Guidelines on Enteral Nutrition: Liver Disease—Tube Feeding (TF) in Patients with Esophageal Varices Is Not Proven to Be Safe. *Clinical Nutrition*, **26**, 272. http://dx.doi.org/10.1016/j.clnu.2006.12.005

[12] Simmons, B. (2010) Clinical Reasoning: Concept Analysis. *Journal of Advanced Nursing*, **66**, 1151-1158. http://dx.doi.org/10.1111/j.1365-2648.2010.05262.x

[13] Garcia, T.R. and Nóbrega, M.M.L. (2009) [Nursing Process: From Theory to the Practice of Care and Research]. *Escola Anna Nery Revista de Enfermagem*, **13**, 188-193.

[14] Horta, W.A. (1979) [Nursing Process]. EPU, São Paulo.

[15] Maslow, A.H. (1943) A Theory of Human Motivation. *Psychological Review*, **50**, 370-396. http://dx.doi.org/10.1037/h0054346

[16] Doenges, M.E. and Moorhouse, M.F. (2012) Application of Nursing Process and Nursing Diagnosis: An Interactive Text for Diagnostic Reasoning. F.A. Davis, Philadelphia.

[17] NANDA International Inc (2012) Nursing Diagnoses: Definitions and Classification. Wiley-Blackwell, Oxford.

[18] Bakken, S., Silveira, D., Gerhardt, L., Dal Sasso, G. and Barbosa, S. (2011) [Systems Decision Support and Patient Safety]. In: Cometto, M., Gómez, P., Dal Sasso, G., Grajales, R., Cassiani, S. and Morales, C., Eds., *Nursing and Patient Safety*, Organização Panamericana de Saúde, Washington, 387-397.

[19] Scherb, C.A., Head, B.J., Maas, M.L., Swanson, E.A., Moorhead, S., Reed, D., Conley, D.M. and Kozel, M. (2011) Frequent Nursing Diagnoses, Nursing Interventions, and Nursing-Sensitive Patient Outcomes of Hospitalized Older Adults with Heart Failure: Part 1. *International Journal of Nursing Terminologies and Classifications*, **22**, 13-22. http://dx.doi.org/10.1111/j.1744-618X.2010.01164.x

[20] Ouslander, J.G., Diaz, S., Hain, D. and Tappen, R. (2011) Frequency and Diagnoses Associated with 7- and 30-Day Readmission of Skilled Nursing Facility Patients to a Nonteaching Community Hospital. *Journal of the American Medical Directors Association*, **12**, 195-203. http://dx.doi.org/10.1016/j.jamda.2010.02.015

[21] Souza, C.C., Mata, L.R.F., Carvalho, E.C. and Chianca, T.C.M. (2013) [Nursing Diagnoses in Patients Classified as Priority Level I and II according to the Manchester Protocol]. *Revista da Escola de Enfermagem da USP*, **47**, 1318-1324. http://dx.doi.org/10.1590/S0080-623420130000600010

[22] Gimenes, F.R.E., Reis, R.K., Silva, P.C.S., Silva, A.E.B.C. and Atila, E. (2015) Nursing Assessment Tool for People with Liver Cirrhosis. *Gastroenterology Nursing*, Publish Ahead of Print.

[23] Risner, P. (1986) Diagnosis: Analysis and Synthesis of Data. In: Griffith-Kenney, J. and Christensen, P., Eds., *Nursing Process Application of Theories, Frameworks, and Models*, 2nd Edition, Mosby, St. Louis, 124-151.

[24] Salgado, P.O. and Chianca, T.C. (2011) Identification and Mapping of the Nursing Diagnoses and Actions in an Intensive Care Unit. *Revista Latino-Americana de Enfermagem*, **19**, 928-935. http://dx.doi.org/10.1590/S0104-11692011000400011

[25] Melo, E., Albuquerque, M. and Aragão, R. (2012) Nursing Diagnosis Prevalence in Patients at an Intensive Care Unit of a Public Hospital. *Journal of Nursing UFPE Online*, **6**, 1361-1368.

[26] Chianca, T.C., Lima, A.P. and Salgado, P.O. (2012) [Nursing Diagnoses Identified in Inpatients of an Adult Intensive Care Unit]. *Revista da Escola de Enfermagem USP*, **46**, 1102-1108. http://dx.doi.org/10.1590/S0080-62342012000500010

[27] Galati, J.S., Holdeman, K.P., Dalrymple, G.V., Harrison, K.A. and Quigley, E.M. (1994) Delayed Gastric Emptying of Both the Liquid and Solid Components of a Meal in Chronic Liver Disease. *The American Journal of Gastroenterology*, **89**, 708-711.

[28] Carrola, P., Militão, I. and Presa, J. (2013) [Bacterial Infections in Patients with Liver Cirrhosis]. *GE Jornal Português de Gastrenterologia*, **20**, 58-65. http://dx.doi.org/10.1016/j.jpg.2012.09.008

[29] Opilla, M. (2003) Aspiration Risk and Enteral Feeding: A Clinical Approach. *Practical Gastroenterology*, **27**, 89-96.

[30] Metheny, N.A. (2002) Risk Factors for Aspiration. *Journal of Parenteral and Enteral Nutrition*, **26**, S26-S33. http://dx.doi.org/10.1177/014860710202600605

[31] Jalan, R., Fernandez, J., Wiest, R., Schnabl, B., Moreau, R., Angeli, P., Stadlbauer, V., Thierry, G.T., Bernardi, M., Canton, R., Agustin Albillos, A., Lammert, F., Wilmer, A., Mookerjee, R., Vila, J., Garcia-Martinez, R., Wendon, J., Such, J., Cordoba, J., Sanyal, A., Garcia-Tsao, G., Arroyo, V., Burroughs, A. and Gines, P. (2014) Bacterial Infections in Cirrhosis. A Position Statement Based on the EASL Special Conference 2013, *Journal of Hepatology*, 60, 1310-1324. http://dx.doi.org/10.1016/j.jhep.2014.01.024

[32] Merli, M., Lucidi, C., Giannelli, V., Giusto, M., Riggio, O., Falcone, M., Ridola, L., Attili, A.F. and Venditti, M. (2010) Cirrhotic Patients Are at Risk for Health Care-Associated Bacterial Infections. *Clinical Gastroenterology and Hepatology*, **8**, 979.e1-985.e1. http://dx.doi.org/10.1016/j.cgh.2010.06.024

[33] Wong, F., Bernardi, M., Balk, R., Christman, B., Moreau, R., Garcia-Tsao, G., Patch, D., Soriano, G., Hoefs, J. and Navasa, M. (2005) Sepsis in Cirrhosis: Report on the 7th Meeting of the International Ascites Club. *Gut*, **54**, 718-725. http://dx.doi.org/10.1136/gut.2004.038679

[34] Almeida, M.A., Aliti, G.B., Franzen, E., Thomé, E.G.R., Unicovsky, M.R., Rabelo, E.R., Ludwig, M.L.M. and Moraes, M.A. (2008) Prevalent Nursing Diagnoses and Interventions in the Hospitalized Elder Care. *Revista Latino-Americana de Enfermagem*, **16**, 707-711. http://dx.doi.org/10.1590/S0104-11692008000400009

[35] Tsai, L.-H., Lin, C.-M., Chiang, S.-C., Chen, C.-L., Lan, S.-J. and See, L.-C. (2014) Symptoms and Distress among Patients with Liver Cirrhosis but without Hepatocellular Carcinoma in Taiwan. *Gastroenterology Nursing*, **37**, 49-59. http://dx.doi.org/10.1097/SGA.0000000000000020

[36] Swain, M.G. (2006) Fatigue in Liver Disease: Pathophysiology and Clinical Management. *Canadian Journal of Gastroenterology*, **20**, 181-188. http://dx.doi.org/10.1155/2006/624832

[37] Hearnshaw, S.A., Logan, R.F., Lowe, D., Travis, S.P., Murphy, M.F. and Palmer, K.R. (2011) Acute Upper Gastrointestinal Bleeding in the UK: Patient Characteristics, Diagnoses and Outcomes in the 2007 UK Audit. *Gut*, **60**, 1327-1335. http://dx.doi.org/10.1136/gut.2010.228437

[38] Bari, K. and Garcia-Tsao, G. (2012) Treatment of Portal Hypertension. *World Journal of Gastroenterology*, **18**, 1166-1175. http://dx.doi.org/10.3748/wjg.v18.i11.1166

[39] Biecker, E. (2013) Gastrointestinal Bleeding in Cirrhotic Patients with Portal Hypertension. *ISRN Hepatology*, **2013**, Article ID: 541836. http://dx.doi.org/10.1155/2013/541836

[40] Sales, M.V.C., Silva, T.J.A., Gil Jr., L.A. and Filho, W.J. (2010) Adverse Events of Hospitalization for the Elderly Patient. *Geriatria & Gerontologia*, **4**, 238-246.

[41] Park, H. (2014) Identifying Core NANDA-I Nursing Diagnoses, NIC Interventions, NOC Outcomes, and NNN Linkages for Heart Failure. *International Journal of Nursing Knowledge*, **25**, 30-38. http://dx.doi.org/10.1111/2047-3095.12010

[42] Krenitsky, J. (2003) Nutrition for Patients with Hepatic. Nutrition Issues in Gastroenterology, *Practical Gastroenterology*, **6**, 23-42.

[43] Vargas, R.S. and França, F.C.V. (2007) [Implementation of the Nursing Process in a Patient with Hepatic Cirrhosis Using the Standardized Terminologies NANDA, NIC and NOC]. *Revista Brasileira de Enfermagem*, **60**, 348-352. http://dx.doi.org/10.1590/S0034-71672007000300020

[44] Brunner, L.S., Smeltzer, S.C.O.C., Bare, B.G., Hinkle, J.L. and Cheever, K.H. (2010) Brunner & Suddarth's Textbook of Medical-surgical Nursing. Wolters Kluwer Health/Lippincott Williams & Wilkins, Philadelphia.

[45] Carvalho, D.V., Salviano, M.E.M., Carneiro, R.A. and Santos, F.M.M. (2007) [Nursing Diagnosis of Post Surgical Patients of Liver Transplantation by Alcoholic and Non Alcoholic Cirrhosis]. *Escola Anna Nery*, **11**, 682-687. http://dx.doi.org/10.1590/S1414-81452007000400020

[46] Park, H. (2010) NANDA-I, NOC, and NIC Linkages in Nursing Care Plans for Hospitalized Patients with Congestive Heart Failure. PhD Dissertation, Univerity of Iowa, Iowa.

[47] Olson, J.C. and Kamath, P.S. (2012) Acute-On-Chronic Liver Failure: What Are the Implications? *Current Gastroenterology Report*, **14**, 63-66. http://dx.doi.org/10.1007/s11894-011-0228-2

20

Noninvasive Fibrosis Scores as Prognostic Markers for Varices Needing Treatment in Advanced Compensated Liver Cirrhosis

Elham Ahmed Hassan1*, Abeer Sharaf El-Din Abd El-Rehim1, Zain El-Abdeen Ahmed Sayed2, Ahmed Mohmmed Ashmawy2, Emad Farah Mohamed Kholef3, Abeer Sabry4, Wael Abd-Elgwad Elsewify5

1Department of Tropical Medicine and Gastroenterology, Faculty of Medicine, Assiut University, Assiut, Egypt 2Department of Internal Medicine, Faculty of Medicine, Assiut University, Assiut, Egypt 3Department of Clinical pathology, Faculty of Medicine, Aswan University, Aswan, Egypt 4Department of Internal Medicine, Faculty of Medicine, Helwan University, Cairo, Egypt 5Department of Internal Medicine, Faculty of Medicine, Aswan University, Aswan, Egypt

Abstract

Background/purpose: Noninvasive assessment of esophageal varices (EVs), their size and bleeding stigmata may reduce endoscopic burden, cost and drawbacks. We aimed to evaluate the diagnostic performance of noninvasive fibrosis scores (AAR, APRI, FIB-4, King and VITRO scores) in predicting the presence of EVs and high risk varices needing treatment (VNT) in HCV-related cirrhosis of Egyptian patients. Methods: This prospective study included 154 HCV-related advanced compensated cirrhotic patients with no history of bleeding who underwent screening endoscopy for EVs. AAR, APRI, FIB-4, King and VITRO scores were assessed. Results: Esophageal varices were found in 120 patients (77.9%) and VNT in 92 patients (59.7%). Apart from AAR, all scores demonstrated statistically significant correlations with the presence and the size of EVs. Using area under receiver operating characteristic curve (AUC), these scores were good predictors for the presence of EVs and VNT, where VITRO score had the highest AUC (0.920 and 0.900) and accuracy (97.1% and 87%), sensitivity (75, 82.6%), specificity (100, 93.5%), PPV (100, 95%) and NPV (53.2, 78.4%) with cutoffs >1.3 and >1.8 respectively. Conclusion: Noninvasive fibrosis scores can predict the presence of EVs and VNT. VITRO score was the best predictor with higher accuracy for clinical applicability than studied scores.

Keywords

Esophageal Varices, Hepatitis C, Liver Cirrhosis, VITRO Score

1. Introduction

Esophageal varices (EVs) contribute to cirrhosis-related morbidity and mortality which are found in 60% 80% of cirrhotic patients and correlated with the severity of liver disease [1] [2] [3]. Mortality from acute variceal bleeding is still very high, about 25% 35% [4] [5]. Moreover, the mortality is up to 3.4 per year in patients with varices who have never bled and 57% per year in patients with variceal bleeding [1] [6]. Thus, endoscopic screening is recommended by all current guidelines at the time of the diagnosis of cirrhosis to identify those at risk of bleed-ing, e.g., large varices (which are found up to 30%), so that prophylactic therapy can be administered [7] [8]. In addition, it should be repeated every 2 3 years in patients who do not have varices and 1 2 years in those with small varices [9].

In order to avoid the endoscopic burden, cost, drawbacks, unpleasant and repeated examinations to the patients, several non-invasive parameters have been investigated for prediction of the presence and the size of EVs [10] [11] [12]. As it was postulated that the progressive fibrotic remodeling of the liver increases the resistance to hepat-ic sinusoidal blood flow and hence, it increases portal venous pressure causing esophageal and gastric varices [3].

In this study, we aimed to investigate the ability of five noninvasive fibrosis scores (AAR, APRI, FIB-4, King and VITRO scores) to predict the presence and the size of EVs in hepatitis C virus (HCV)-related cirrhosis of Egyptian patients in comparison to upper endoscopy.

2.   Materials and Methods

2.1. Study Design

This prospective study was carried out at Al-Rajhi Liver Center, Assiut University Hospital, Assiut Egypt, from May 2016 to February 2017. The study protocol was approved by the local ethics committee of the Assiut Uni-versity Hospital and was in accordance with the previsions of the Declaration of Helsinki. Informed consent was obtained from all the participants before enrollment in the study.

2.2. Patients

This study included 154 adult patients with liver cirrhosis selected consecutively from inpatient wards of the departments of Tropical Medicine and Gastroenterology and Internal medicine, Al-Rajhi Liver Center, Assiut University Hospital.

Cirrhotic patients had diagnostic criteria of liver cirrhosis (LC) by clinical, biochemical and ultrasonographic findings. The cause of liver dysfunction was hepatitis C. The severity of liver cirrhosis was assessed according to Child-Pugh classification [13]. Patients with decompensated cirrhosis (late Child B and Child C), active bleeding, previous endoscopic sclerosis or band ligation of EVs, previous transjugular intrahepatic portosystemic stent shunt or on β-blocker therapy were excluded. Also, patients with severe cardiopulmonary, renal insufficiency, uncontrolled diabetes mellitus, active infections, HIV or HBV co-infections, malignancy, prior antiviral, immu-nosuppressive therapy, recent anticoagulant therapy, alcohol consumption or liver transplantation were excluded.

2.3. Methods

At the study entry, detailed clinical history and examination were taken and abdominal ultrasonography was un-dertaken. Blood samples were collected from stable patients for laboratory investigations included complete blood count, liver, kidney function tests, and serum von-Willebrand factor Antigen (vWF-Ag) levels that were measured by using a fully automated STA analyser and vWF6 Liatest (Diagnostic Stargo, Paris, France) according to the instructions of the manufacturer.

By data collection, non-invasive fibrosis scores were calculated as following:

- AAR = AST (U/L)/ALT (U/L) [14].

- APRI = (AST (U/L)/upper limit of normal)/platelet (109/L) × 100 [15].

- FIB-4 = [age (years) × AST (U/L)]/[platelet (109/L) × ALT (U/L)1/2] [16].

- King score = age (years) × AST (U/L) × INR/platelets (109/L) [17].

- VITRO = vWF-Ag/platelets (109/L) [18].

Upper gastrointestinal endoscopy was done for evaluation of the presence, grade of EVs and stigmata of bleeding by an experienced endoscopist who was blinded to the outcomes of the study. Esophageal varices were graded as following: no varices, small varices without stigmata of bleeding and varices with stigmata of bleeding that need treatment that were large varices and small varices with red signs and known as high risk varices needing treatment (VNT) [6] [19].

2.4. Statistical Analysis

All statistical analyses were conducted using Statistical Package for the Social Sciences (SPSS) for Windows version 16 (SPSS Inc., Chicago, IL, USA) and MedCalc program. The quantitative data were expressed as mean ± standard deviation (SD) and qualitative data were expressed as percentage. Spearman's rank correlation coefficient (r) was used to find correlations. The receiver operating characteristic curves (ROC) were plotted to measure and compare the performance of different noninvasive models for predicting EVs and VNT. Using ROC, The value with the best sensitivity and specificity was chosen as the best cutoff value, in addition, calculation of positive (PPV) and negative (NPV) predictive value, positive and negative likelihood ratio (+LR, −LR) for prediction or exclusion of varices. Logistic regression analysis was used to establish the best model for prediction of high risk esophageal varices needing treatment. All tests were two-tailed and statistical significance was assessed <0.05.

3. Results

3.1. Characteristics of the Studied Patients

This study included 154 patients with HCV-related liver cirrhosis who underwent upper digestive endoscopy; 94 were Child-Pugh class A (61%), and 60 were early Child-Pugh class B (40%). Baseline demographic and clinical characteristics of the studied patients were summarized in Table 1, where, 34 (22.1%) patients had no esophageal varices (EVs), 28 (18.2%) had varices without stigmata of bleeding and 92 (59.7%) patients had high risk varices needing treatment (VNT) that were large and small varices with red signs.

Table 1. Basal characteristics of study patients.

	Total (n = 154)
Age (years, mean ± SD)	50.5 ± 8.5 (40 - 80)
Sex	
Male	92 (59.7%)
Female	62 (40.3%)
Laboratory parameters (mean ± SD)	
S. bilirubin (mg/dl)	1.7 ± 0.6
S. albumin (g/dl)	3.2 ± 0.8
AST (IU L^{-1})	36.4 ± 10.7
ALT (IU L^{-1})	52.2 ± 17.1
INR	1.4 ± 0.2
Platelets (10^9 L^{-1})	83 ± 29
vWF-Ag%	123 ± 27

Child-Pugh score (mean ± SD)	8 ± 3
Non-invasive scores (mean ± SD)	
AAR	0.76 ± 0.4
APRI	1.22 ± 0.54
FIB-4	3.8 ± 1.8
King	37 ± 18.5
VITRO	1.64 ± 0.62
Esophageal varices (%)	
No	34 (22.1%)
Small varices without red signs	28 (18.2%)
Small varices with red signs	30 (19.5%)
Large varices	62 (40.2%)

SD: standard deviation; AST: aspartate aminotransferase; ALT: alanine aminotransferase; INR: international-al normalized ratio; AAR: aspartate aminotransferase-alanine aminotransferase ratio; APRI: AST-platelet ratio index; FIB-4: fibrosis-4 index.

3.2. Noninvasive Fibrosis Scores and Esophageal Varices

Apart from AAR, significant elevations in the mean values of noninvasive fibrosis scores (APRI, FIB-4, King and VITRO) were noted in patients with EVs compared to those without EVs (Table 2), and in patients with high risk VNT than patients without (Table 2).

In addition, these scores (APRI, FIB-4, King and VITRO) were significantly correlated with the grades of esophageal varices, where, VITRO score had the strongest correlation (r = 0.730, P < 0.001). On the other hand, no significant correlation was found between ARR and variceal grades (r = 0.129, P = 0.112) (Table 3).

Table 2. Comparison between noninvasive fibrosis scores as regarding the presence and size of esophageal varices.

Noninvasive score	The presence of esophageal varices		
	Patients without EVs (n = 34)	Patients with EVs (n = 120)	P value
AAR	0.66 ± 0.27	0.79 ± 0.42	0.101
APRI	0.79 ± 0.35	1.35 ± 0.52	<0.001
FIB-4	2.48 ± 1.32	4.15 ± 1.69	<0.001
King	23.27 ± 12.6	40.8 ± 18.1	<0.001
VITRO	0.94 ± 0.2	1.83 ± 0.56	<0.001
	The size of esophageal varices		
	Patients with no or small EVs without red signs (n = 62)	Patients with high risk EVNT (n = 92)	P value
AAR	0.69 ± 0.26	0.80 ± 0.47	0.07
APRI	0.89 ± 0.38	1.45 ± 0.52	<0.001
FIB-4	2.76 ±1.39	4.47 ± 1.64	<0.001
King	26.5 ±13.1	43.9 ± 18.4	<0.001
VITRO	1.13 ± 0.33	1.98 ± 0.54	<0.001

P value < 0.05 = significant. EVs: esophageal varices; EVNT: esophageal varices needing treatment; AAR: aspartate aminotransferase-alanine aminotransferase ratio; APRI: AST-platelet ratio index; FIB-4: fibrosis-4 index.

Table 3. Correlation between noninvasive fibrosis scores and the size of esophageal varices.

	r	P value
AAR	0.129	0.112
APRI	0.546	<0.001
FIB-4	0.511	<0.001
King	0.544	<0.001
VITRO	0.730	<0.001

r: Spearman's rank correlation coefficient; P value < 0.05 = significant. AAR: aspartate aminotransferase-alanine aminotransferase ratio; APRI: AST-platelet ratio index; FIB-4: fibrosis-4 index.

3.3. Diagnostic Performance of Noninvasive Models for Prediction of EVs and VNT

By applying ROC curves, the diagnostic accuracies of AAR, APRI, FIB-4, King and VITRO scores as noninvasive predictors of EVs and VNT were studied to determine which score would have the most clinical utility for prediction (Figure 1 and Table 4). For predicting EVs (Figure 1(a)), the AUC was greatest for VITRO score (0.920) followed by FIB-4 and King scores (0.800 for each) and APRI score (0.795). For predicting VNT (Figure 1(b)), VITRO had the greatest AUC (0.900), followed by FIB-4 score (0.808), APRI score (0.790) and King score (0.783) while the AAR score was <0.70.

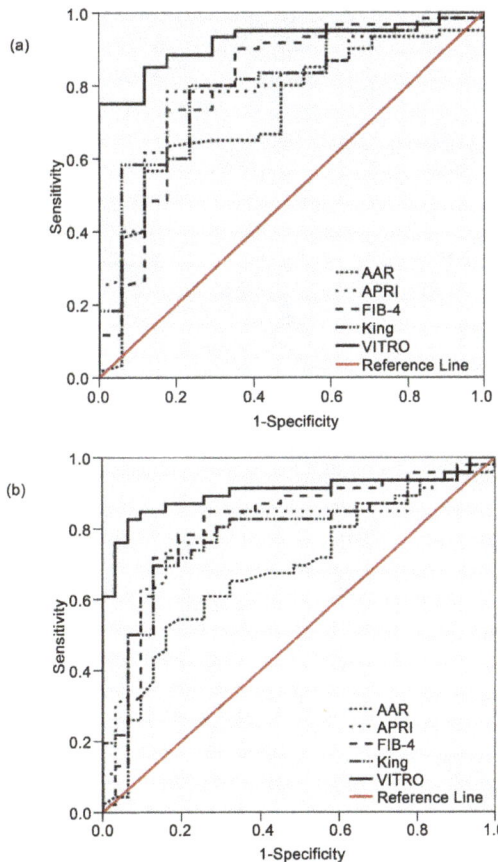

Figure 1. Area under the receiver operating characteristic curve (AUC) of noninvasive fibrosis scores to predict esophageal varices (a) and high risk esophageal varices (b). VITRO score had the highest AUC in predicting esophageal varices (AUC = 0.920) and high risk esophageal varices needing treatment (AUC = 0.900).

The optimum cutoff values of the previously mentioned scores to predict the presence of EVs and VNT were illustrated in Table 4 where; VITRO score had the highest diagnostic indices; with a cutoff value > 1.3, VITRO had 75% sensitivity, 100% specificity, 100% PPV, 53.2% NPV and 97.1% accuracy for the prediction of EVs and at a cutoff value > 1.8, VITRO had 82.6% sensitivity, 93.5% specificity, 95% PPV, 78.4% NPV and 87% accuracy for the prediction of VNT.

By using these scores, we tried to construct a model for predicting the development of EVs and VNT by binary logistic regression analysis (forward: LR) (Table 5). For predicting EVs, the presence or absence of varices was the dependant factor and APRI, FIB-4 King and VITRO scores (significantly associated scores in univariate analysis) were independent variables and the accuracy of this model was 77.9%. After removal of insignificant predictors (i.e., APRI and FIB-4), the accuracy of the model became 85.7%. If only VITRO was used, the accuracy of the model was 87% (odds ratio = 88.03, 95% CI = 18.02 430, P< 0.001) as shown in (Table 5).

Table 4. Diagnostic performance of noninvasive fibrosis scores for prediction of esophageal varices and large esophageal varices.

	Cut-off value	AUC 95% CI	SEN (%)	SPE (%)	PPV (%)	NPV (%)	+LR	−LR	Accuracy (%)
AAR for EV diagnosis	>0.67	0.726 (0.613 - 0.822)	63.3	82.4	92.7	38.9	3.59	0.45	67.5
AAR for EVNT diagnosis	>0.74	0.680 (0.563 - 0.781)	52.2	83.9	82.8	54.2	3.23	0.57	65
APRI for EV diagnosis	>0.85	0.795 (0.687 - 0.878)	78.3	82.4	94	51.9	4.44	0.26	79.2
APRI for EVNT diagnosis	>1.22	0.790 (0.682 - 0.874)	73.9	83.9	87.2	68.4	4.58	0.31	77.9
FIB-4 for EV diagnosis	>2.8	0.800 (0.694 - 0.883)	73.3	82.4	93.6	46.7	4.16	0.32	75.3
FIB-4 for EVNT diagnosis	>3.4	0.808 (0.702 - 0.889)	78.3	74.2	81.8	69.7	3.03	0.29	76.6
King for EV diagnosis	>24.74	0.800 (0.693 - 0.882)	80	76.5	92.3	52	3.4	0.26	79.2
King for EVNT diagnosis	>39.01	0.783 (0.674 - 0.869)	69.6	87.1	88.9	65.9	5.39	0.35	76.7
VITRO for EV diagnosis	>1.3	0.920 (0.835 - 0.969)	75	100	100	53.2	-	0.25	97.1
VITRO for EVNT diagnosis	>1.8	0.900 (0.811 - 0.957)	82.6	93.5	95	78.4	12.8	0.19	87

EV: esophageal varices; EVNT: esophageal varices needing treatment; AAR: aspartate aminotransferase-alanine aminotransferase ratio; APRI: AST-platelet ratio index; FIB-4: fibrosis-4 index; AUC: area under the curve; SEN: sensitivity; SPE: specificity; PPV: positive predictive value; NPV: negative predictive value; +LR: positive likelihood ratio; -LR: negative likelihood ratio.

Table 5. Diagnostic models of esophageal varices and large esophageal varices.

		B	S.E.	Wald	df	Sig.	Exp(B)	95% CI for EXP(B) Lower	95% CI for EXP(B) Upper	Percentage
								Lower	Upper	
Variables in the Equation (for prediction of esophageal varices)										
Step 1	VITRO	4.478	0.809	30.617	1	<0.001	88.026	18.022	429.952	87%
	Constant	−4.536	0.931	23.742	1	<0.001	0.011			
Step 2	King	−0.074	0.034	4.929	1	0.026	0.928	0.869	0.991	85.7%
	VITRO	6.811	1.556	19.161	1	<0.001	907.624	43.001	19,157.1	
	Constant	−5.120	1.103	21.557	1	<0.001	0.006			
Variables in the Equation (for prediction of esophageal varices needing treatment)										
Step 1	VITRO	3.645	0.555	43.101	1	<0.001	38.293	12.897	113.697	81.8%
	Constant	−5.165	0.845	37.362	1	<0.001	0.006			

For prediction of VNT, dependent factors were either small or large varices while the independent factors were APRI, FIB-4 King and VITRO (significantly associated scores in univariate analysis). The accuracy of this model was about 59.7%. After removal of insignificant predictors (i.e., APRI, FIB-4 and King), the accuracy of the model becomes 81.8% where only VITRO was used (odds ratio= 38.3, 95% CI = 12.9 113.7, P < 0.001) (Table 5).

4. Discussion

In this study, we tried to approve the Baveno VI recommendation for prediction of EVs and VNT in compensated HCV-related cirrhosis with non-invasive parameters nearly different from that used in Baveno VI. As in many areas, Transient Elastography was not easily applicable or available. So, simple and easily applicable non invasive fibrosis tests were evaluated. Evaluation of hepatic fibrosis may provide information about the presence and severity of portal hypertension as increased hepatic vascular resistance in cirrhosis is influenced by the presence and the extent of fibrosis [20] [21].

In this study, we demonstrated the ability of noninvasive markers of liver fibrosis to predict the presence of EVs and their size in Egyptian patients with liver cirrhosis and compare them with upper endoscopy. Evaluation of hepatic fibrosis may provide information about the presence and severity of portal hypertension as increased hepatic vascular resistance in cirrhosis is influenced by the presence and the extent of fibrosis [20] [21].

Identification of patients with EVs especially high risk varices by regular screening is fundamental as they candidates for prophylactic therapy [7] [8]. The size of varices has been identified as the principal predictor for variceal bleeding which occurs in up to 30%, and is associated with significant morbidity and mortality [22] [23] [24]. Several non-invasive parameters had been introduced for variceal screening to minimize the usage of endoscopy [10] [11] [12]. We demonstrated that all the studied models (AAR, APRI, FIB-4, King and VITRO scores) had a good performance for the diagnosis of EVs, where, VITRO score is currently the most accurate method for the detection of EVs in patients with LC. We showed a clear correlation between the variceal size and the VITRO score as well as the other noninvasive tests except AAR. In addition, VITRO score has the best performance for the diagnosis of high risk varices needing treatment.

In our work, AAR had the lowest performance in prediction of EVs (AUC = 0.726) and high risk EVs (AUC = 0.648), however these results were much better than that recorded by Deng et al. [24], who showed poor AUCs of AAR for EVs (0.596) and large EVs (0.601).

Previous studies investigating APRI as a predictor for EVs in LC patients showed that a low AUC in predicting EVs (0.62) and Large EVs (0.71) [25] [26]. Deng et al. [24], proposed that at a cutoff value of >0.87, the AUC was 0.539 for the diagnosis of any grade EVs with 68% sensitivity, 46.2% specificity, while at a cutoff value of >0.85, the AUC for predicting Large EVs was 0.506, 68.8% sensitivity, and 41.3% specificity. This study proposed a cutoff value of >0.85 for the diagnosis of EVs with AUC of 0.795. At this cutoff, the sensitivity was 78.3%, specificity was 82.4%, and the overall accuracy was 79.2%. Also, a cutoff of 1.22 for the diagnosis of high risk EVs was proposed at which AUC was 0.790, sensitivity was 73.9%, specificity was 83.9%, and the overall accuracy was 77.9%.

We used FIB-4 cutoff values > 2.8 and 3.4 for which AUCs were 0.8 and 0.808 for diagnosis of EVs and Large EVs with 73.3, 78.3% sensitivity and 82.4, 74.2% specificity respectively. Our findings were compatible with Hassan et al. [12], who reported Fib-4 having AUCs of 76 and 0.76 with 76, 72.9% sensitivity, 80, 66.7% specificity at cutoff > 2.8 and 3.3 for diagnosis of EVs and high risk EVs respectively. However, Fib-4 had been examined in other studies for the prediction of EVs and high risk EVs, having different AUCs and cutoff values; Sebastiani et al. [27], found that AUC was 0.64 for the prediction of EVs at a cutoff value of 3.5, while for the diagnosis of Large EVs, the AUC was 0.63 and the cutoff value was 4.3.

King score had been considered a satisfactory predictor of EVs. In the current study, at a cutoff value of 24.7, the score had an AUC of 0.800, 80% sensitivity, 76.5% specificity, 92.3% PPV, 52% NPV and 79.2% accuracy for the diagnosis of EVs. While for a cut-off value of 39.01, the AUC was 0.783, sensitivity was 69.6%, specificity was 87.1%, PPV was 88.9%, NPV was 65.9% and the accuracy was 76.7% for the prediction of VNT. In the retrospective study of Deng et al. [24], the best cutoff value for the diagnosis of EVs was 17.93, with an AUC of 0.639, 85.3% sensitivity, 44% specificity and 68.7% NPV, and the best cut-off value was 24.80 for diagnosis of high risk EVs, with an AUC of 0.645, 97% sensitivity, 53.6% specificity and 69.8% NPV.

In our study, the VITRO score was significantly higher in patients with EVs than those without. The diagnostic accuracy of VITRO for detecting EVs was significantly better than the other studied tests with an AUC of 0.920 (95%CI 0.835 0.969) with 75% sensitivity, 100% specificity, 100% PPV, 53.2% NPV and the highest accuracy (97.1%) at a cut-off > 1.3. It showed the closest correlation with variceal size, and at cut-off > 1.8 it had AUC, 0.9 (95%CI 0.811 0.957), 82.6% sensitivity, 93.5% specificity, 95% PPV, 78.4% NPV and 87% accuracy in detecting Large EVs suggesting its usefulness in identifying patients with large varices who need endoscopy. Our results supported by Hametner et al. [28], who clearly demonstrated that VITRO score had diagnostic and predictive value in patients with clinically significant portal hypertension (CSPH) assessed by hepatic venous pressure gradient (HVPG) independently of Child-Pugh score and also, it had an impressive correlation with EVs (P < 0.004).

The increased diagnostic accuracy of VITRO score for prediction of EVs and its size may be attributed to incorporation of independent predictors of portal hypertension; platelets and vWF-Ag [29] [30]. Several studies revealed that platelet count was an independent predictor for the presence of esophageal varices [29] [31]. Thrombocytopenia may be partially caused by pooling and sequestration of platelets in an enlarged spleen due to portal hypertension and therefore, it is an indirect marker of portal hypertension [29]. vWF is a marker of endothelial dysfunction that is considered a major determinant of the increased vascular tone of cirrhotic livers and therefore of the development of portal hypertension [30]. Ferlitsch et al. [30], and La Mura et al. [32], declared that circulating levels of vWF had a significant direct correlation with HVPG.

Elevated vWF-Ag levels in liver cirrhosis are partly due to increased synthesis by increased shear stress or bacterial infection associated with endothelial cell damage or reduced clearance by increased activity of ADAMTS13 (vWF cleaving protease) [28]. Thus, VITRO score is significantly superior to AAR, APRI, FIB-4, and King for predicting EVs and high risk VNT. One reason for this superior predictive ability of VITRO is the inclusion of platelets (unlike AAR) and vWF-Ag (unlike all the studied scores), which are well-known predictors of portal hypertension and EVs in cirrhosis as shown in previous studies [29] [30] [31] [32]. In addition, it is simply calculated and its items are easily obtained and measured.

The limitations of this work are a single-centre study and lack of comparison between noninvasive fibrosis scores and measurement of HVPG; an accurate measurement of portal hypertension, as measuring HVPG is not routinely available in our area. These findings are needed to be confirmed by further multicentre prospective studies to validate the usefulness of VITRO score in clinical practice.

5. Conclusion

In conclusion, VITRO score had the best diagnostic performance to predict varices in liver cirrhosis in comparison to the other studied models that may aid in further improvement of the quality of noninvasive screening of EVs and high risk VNT and in further reduction of endoscopic requirement. Hence, it could offer a useful strategy to stratify high-risk patients who would benefit by intensive screening, and to recommend the prophylactic treatment.

Acknowledgements

Authors declared that no financial support from any agency.

Conflicts of Interest

The authors declare that they have no conflict of interest.

References

[1] Jensen, D.M. (2002) Endoscopic Screening for Varices in Cirrhosis: Findings, Implications, and Outcomes. Gastroenterology, 122, 1620-1630. https://doi.org/10.1053/gast.2002.33419

[2] Bosch, J., Berzigotti, A., Garcia-Pagan, J.C. and Abraldes, J.G. (2008) The Management of Portal Hypertension: Rational Basis, Available Treatments and Future Options. Journal of Hepatology, 48, S68-S92. https://doi.org/10.1016/j.jhep.2008.01.021

[3] Park, D.K., Um, S.H., Lee, J.W., et al. (2004) Clinical Significance of Variceal Hemorrhage in Recent Years in Patients with Liver Cirrhosis and Esophageal Varices. Journal of Gastroenterology and Hepatology, 19, 1042-1051. https://doi.org/10.1111/j.1440-1746.2004.03383.x

[4] Pagliaro, L., D'Amico, G., Pasta, L., Politi, F., Vizzini, G., Traina, M., Madonia, S., Luca, A., Guerrera, D., Puleo, A. and D'Antoni, A. (1994) Portal Hypertension in Cirrhosis: Natural History. In: Bosch, J. and Groszmann, R.J., Eds., Portal Hypertension: Pathophysiology and Treatment, Blackwell Scientific, Oxford, 72-92.

[5] Colombo, E., Casiraghi, M.A., Minoli, G., Prada, A., Terruzzi, V., Bortoli, A., Carnovali, M., Gullotta, R., Imperiali, G., Comin, U., et al. (1995) First Bleeding Episode from Oesophageal Varices in Cirrhotic Patients: A Prospective Study of Endoscopic Predictive Factors. Italian Journal of Gastroenterology and Hepatology, 27, 345-348.

[6] de Franchis, R. (2010) Revising Consensus in Portal Hypertension: Report of the Baveno V Consensus Workshop on Methodology of Diagnosis and Therapy in Portal Hypertension. Journal of Hepatology, 53, 762-768. https://doi.org/10.1016/j.jhep.2010.06.004

[7] Garcia-Tsao, G., Bosch, J. and Groszmann, R. (2008) Portal Hypertension and Variceal Bleeding Unresolved Issues. Summary of an American Association for the Study of Liver Diseases and European Association for the Study of the liver. Single-Topic Conference. Hepatology, 47, 1764-1772. https://doi.org/10.1002/hep.22273

[8] Alempijevic, T., Bulat, V., Djuranovic, S., Kovacevic, N., Jesic, R., Tomic, D., Krstic, S. and Krstic, M. (2007) Right Liver Lobe/Albumin Ratio: Contribution to Non-Invasive Assessment of Portal Hypertension. World Journal of Gastroenterology, 13, 5331-5335. https://doi.org/10.3748/wjg.v13.i40.5331

[9] Barnhart, H.X., Haber, M. and Song, J. (2002) Overall Concordance Correlation Coefficient for Evaluating Agreement among Multiple Observers. Biometrics, 58, 1020-1027. https://doi.org/10.1111/j.0006-341X.2002.01020.x

[10] Chalasani, N., Imperial, T.F., Ismail, A., Sood, G., Carey, M. and Wileox, C.M. (1999) Predictors of Large Esophageal Varices in Patients with Cirrhosis. American Journal of Gastroenterology, 94, 3286-3291.

https://doi.org/10.1111/j.1572-0241.1999.1539_a.x

[11] Elalfy, H., Elsherbiny, W., Abdel Rahman, A., Elhammady, D., Shaltout, S.W., Elsamanoudy, A.Z. and El Deek, B. (2016) Diagnostic Non-Invasive Model of Large Risky Esophageal Varices in Cirrhotic Hepatitis C Virus Patients. World Journal of Hepatology, 8, 1028-1037. https://doi.org/10.4254/wjh.v8.i24.1028

[12] Hassan, E.M., Omran, D.A., El Beshlawey, M.L., Abdo, M. and El Askary, A. (2014) Can Transient Elastography, Fib-4, Forns Index, and Lok Score Predict Esophageal Varices in HCV-Related Cirrhotic Patients? Gastroenterology & Hepatolog, 37, 5865. https://doi.org/10.1016/j.gastrohep.2013.09.008

[13] Pugh, R.N., Murray-Lyon, I.M., Dawson, J.L., Pietroni, M.C. and Williams, R. (1973) Transection of the Oesophagus for Bleeding Oesophageal Varices. British Journal of Surgery, 60, 646-649. https://doi.org/10.1002/bjs.1800600817

[14] Sheth, S.G., Flam, S.L., Gordon, F.D. and Chopra, S. (1998) AST/ALT Ratio Predicts Cirrhosis in Patients with Chronic Hepatitis C Virus Infection. The American Journal of Gastroenterology, 93, 44-48.

https://doi.org/10.1111/j.1572-0241.1998.044_c.x

[15] Wai, C.T., Greenson, J.K., Fontana, R.J., Kalbfleisch, J.D., Marrero, J.A., Conjeevaram, H.S. and Lok, A.S. (2003) A simple Noninvasive Index Can Predict Both Significant Fibrosis and Cirrhosis in Patients with Chronic Hepatitis C. Hepatology, 38, 518-526. https://doi.org/10.1053/jhep.2003.50346

[16] Vallet-Pichard, A., Mallet, V., Nalpas, B., Verkarre, V., Nalpas, A., Dhalluin-Venier, V., Fontaine, H. and Pol, S. (2007) FIB-4: An Inexpensive and Accurate Marker of Fibrosis in HCV Infection. Comparison with Liver Biopsy and Fibrotest. Hepatology, 46, 32-36. https://doi.org/10.1002/hep.21669

[17] Cross, T.J.S., Rizzi, P., Berry, P.A., Portmann, B. and Harrison, P.M. (2009) King's Score: An Accurate Marker of Cirrhosis in Chronic Hepatitis C. European Journal of Gastroenterology & Hepatology, 21, 730-738. https://doi.org/10.1097/MEG.0b013e32830dfcb3

[18] Maieron, A., Salzl, P., Peck-Radosavljevic, M., Trauner, M., Hametner, S., Schöfl, R., Ferenci, P. and Ferlitsch, M. (2014) Von Willebrand Factor as a New Marker for Non-Invasive Assessment of Liver Fibrosis and Cirrhosis in Patients with Chronic Hepatitis C. Alimentary Pharmacology & Therapeutics, 39, 331-338. https://doi.org/10.1111/apt.12564

[19] De Franchis, R., Pascal, J.P., Burroughs, A.K., Henderson, J.M., Fleig, W., Groszmann, R.J., Bosch, J., Sauerbruch, T. and Soederlund, C. (1992) Definitions, Methodology and Therapeutic Strategies in Portal Hypertension. A Consensus Development Workshop, Baveno, Lake Maggiore, Italy, April 5 and 6, 1990. Journal of Hepatology, 15, 256-261. https://doi.org/10.1016/0168-8278(92)90044-P

[20] Calvaruso, V., Burroughs, A.K., Standish, R., Manousou, P., Grillo, F., Leandro, G., Maimone, S., Pleguezuelo, M., Xirouchakis, I., Guerrini, G.P., Patch, D., Yu, D., O'Beirne, J. and Dhillon, A.P. (2009) Computer-Assisted Image Analysis of Liver Collagen: Relationship to Ishakscoring and Hepatic Venous Pressure Gradient. Hepatology., 49, 1236-1244. https://doi.org/10.1002/hep.22745

[21] Nagula, S., Jain, D., Groszmann, R.J. and Tsao, G.G. (2006) Histological-Hemodynamic Correlation in Cirrhosis. A Histological Classification of the Severity of Cirrhosis. Journal of Hepatology, 44, 111-117. https://doi.org/10.1016/j.jhep.2005.07.036

[22] The North Italian Endoscopic Club for the Study and Treatment of Esophageal Varices. (1988) Prediction of the First Variceal Hemorrhage in Patients with Cirrhosis of the Liver and Esophageal Varices. New England Journal of Medicine, 319, 983989. https://doi.org/10.1056/NEJM198810133191505

[23] Burroughs, A.K. (1993) The Natural History of Varices. Journal of Hepatology, 17, S10-S13. https://doi.org/10.1016/S0168-8278(05)80448-9

[24] Deng, H., Qi, X., Peng, Y., Li, J., Li, H., Zhang, Y., Liu, X., Sun, X. and Guo, X. (2015) Diagnostic Accuracy of APRI, AAR, FIB-4, FI, and King Scores for Diagnosis of Esophageal Varices in Liver Cirrhosis: A Retrospective Study. Medical Science Monitor, 21, 3961-3977. https://doi.org/10.12659/MSM.895005

[25] Castéra, L., Le Bail, B., Roudot-Thoraval, F., Bernard, P.H., Foucher, J., Merrouche, W., Couzigou, P. and de Lédinghen, V. (2009) Early Detection in Routine Clinical Practice of Cirrhosis and Oesophageal Varices in Chronic Hepatitis C: Comparison of Transient Elastography (FibroScan) with Standard Laboratory Tests and Noninvasive Scores. Journal of Hepatology, 50, 59-68.

https://doi.org/10.1016/j.jhep.2008.08.018

[26] Castéra, L., Sebastiani, G., Le Bail, B., de Lédinghen, V., Couzigou, P. and Alberti, A. (2010) Prospective Comparison of Two Algorithms Combining Non-Invasive Methods for Staging Liver Fibrosis in Chronic Hepatitis C. Journal of Hepatology, 52, 191-198. https://doi.org/10.1016/j.jhep.2009.11.008

[27] Sebastiani, G., Tempesta, D., Fattovich, G., Castera, L., Halfon, P., Bourliere, M., Noventa, F., Angeli, P., Saggioro, A. and Alberti, A. (2010) Prediction of Oesophageal Varices in Hepatic Cirrhosis by Simple Serum Noninvasive Markers: Results of a Multicenter, Large-Scale Study. Journal of Hepatology, 53, 630-638. https://doi.org/10.1016/j.jhep.2010.04.019

[28] Hametner, S., Ferlitsch, A., Ferlitsch, M., Etschmaier, A., Schöfl, R., Ziachehabi, A. and Maieron, A. (2016) The VITRO Score (Von Willebrand Factor Antigen/ Thrombocyte Ratio) as a New Marker for Clinically Significant Portal Hypertension in Comparison to Other Non-Invasive Parameters of Fibrosis Including ELF Test. PLOS One, 17 p.

[29] Thomopoulos, K.C., Labropouou-Karatza, C., M-Timidis, K.P., Katsakoulis, E.C., Icoomou, G. and Nikolopoulou, V.N. (2003) Non-Invasive Predictors of the Presence of Large Oesophagesal Varices in Patients with Cirrhosis. Digestive and Liver Disease, 35, 473-478. https://doi.org/10.1016/S1590-8658(03)00219-6

[30] Ferlitsch, M., Reiberger, T., Hoke, M., Salzl, P., Schwengerer, B., Ulbrich, G., Payer, B.A., Trauner, M., Peck-Radosavljevic, M. and Ferlitsch, A. (2012) von Willebrand Factor as New Noninvasive Predictor of Portal Hypertension, Decompensation and Mortality in Patients with Liver Cirrhosis. Hepatology, 56, 1439-1447. https://doi.org/10.1002/hep.25806

[31] Zaman A., Hapke R., Flora, K., Rosen, H.R. and Bennet, K. (1999) Factors Predicting the Presence of Esophageal or Gastric Varices in Patients with Advanced Liver Disease. The American Journal of Gastroenterology, 94, 3292-3296. https://doi.org/10.1111/j.1572-0241.1999.01540.x

[32] La Mura, V., Reverter, J.C., Flores-Arroyo, A., Raffa, S., Reverter, E., Seijo, S., Abraldes, J.G., Bosch, J. and García-Pagán, J.C. (2011) Von Willebrand Factor Levels Predict Clinical Outcome in Patients with Cirrhosis and Portal Hypertension. Gut. 60, 1133-1138. https://doi.org/10.1136/gut.2010.235689

21

Post Hepatitis C Cirrhosis in Sub-Sahara Kidney Transplant, Treated with Sofosbuvir/Ledipasvir

Tsevi Yawovi Mawufemo1,2*, Lagou Amélie Delphine2, Tia Weu Mélanie2, Coulibaly Pessa Nawo Albert2, Cherif Ibrahima2, Guei Monlet Cyr2, Ackoundou-N'Guessan Kan Clément2

1Nephrology and Dialysis Department of Teaching Hospital Sylvanus Olympio, University of Lomé, Lomé, Togo
2Néphrology, Hémodialysis and Organ Transplant Departement of Yopougon Teaching Hospital, University Felix Houphouet Boigny, Abidjan, Cote d'Ivoire

Email: *seviclaude@gmail.com

Abstract

Background: Viral hepatitis C is a major public health problem in the world. The advent of direct-acting antivirals has revolutionized the taking in charge and prognosis of patients infected with the hepatitis C virus. The interest of this presentation is to draw attention to the problem of therapeutic care posed by viral hepatitis C in kidney transplant patients in Côte d'Ivoire, a country with limited resources where all direct-acting antivirals are not yet available. Patient observation: We report the case of a kidney transplant of 52 years old, chronic bearer of viral hepatitis C virus who after his kidney transplant presented, decompensated active cirrhosis. A treatment based on Sofosbuvir 400 mg/Ledipasvir 90 mg in this patient with genotype 2 for 12 weeks was initiated. A sustained virological response was observed 12 weeks after the end of treatment. Conclusion: Direct-acting antivirals offer the possibility of antiviral C treatment without interferon or ribavirin in cirrhotic renal transplant.

Keywords

Côte D'Ivoire, Direct-Acting Antivirals, Kidney Transplantation, Viral Hepatitis C

1. Introduction

Hepatitis C virus (HCV) infection affects more than 200 million people worldwide and its prevalence is high in patients with end-stage renal disease thus increasing challenges in renal transplant patients infected with HCV. In developed countries around 1.8% to 8% of renal transplant patients are infected with HCV [1]. Despite recent therapeutic advances, the treatment of HCV infection remains a challenge. The arrival of new direct-acting antivirals (DAAs) is a revolution for the treatment of HCV infection and particularly in kidney transplant [2]. In Côte d'Ivoire, a country with limited resources, only a few therapeutic classes of DAAs are available. We report the case of a kidney transplant patient, chronic HCV bearer who presented after his kidney transplant, an array of decompensated cirrhosis in a context of deterioration in general condition and high viral load. The provision of Sofosbuvir/Ledipasvir in Côte d'Ivoire has made it possible to initiate treatment in this patient. All information gathered in this study was processed in accordance with the Code of Ethics. The patient has given his consent regarding this case report.

2. Patient and Observation

A 52-year-old married man residing in Abidjan was hospitalized for general impairment in December 2014. Previously, he has had a kidney transplant since June 2008 from a living donor in New Delhi, India. We do not have a record of his hospitalization for the kidney transplant. His immunosuppressive therapy included Tacrolimus, Mycophenolate Mofetil and steroids. His basic creatinine was between 12 -14 mg/l. He developed post-transplant diabetes treated with insulin since 2013. The glycated hemoglobin was 13.6%. There were no complications related to diabetes. He had viral hepatitis C since the beginning of hemodialysis. He had never received treatment for viral hepatitis C. We do not have information on viral load before kidney transplant, nor on the evaluation of liver fibrosis at the time of hemodialysis. In the history of the disease, he was hospitalized for an alteration of the general condition, digestive disorders (diarrhea made of pasty stools) and impaired renal function. On physical examination, the patient weighed 62 kg for a height of 173 cm, a body mass index of 20.7 kg/m2; the temperature was at 37°C and the blood pressure measured at 140/90 mmHg. The patient was asthenic. There was jaundice, ascites of moderate abundance and discreet edema of the lower limbs. There was no hepatic encephalopathy, hepatomegaly, or splenomegaly. He had benefited from symptomatic treatment that had improved his clinical condition with persisting impaired renal function. At a complete blood count, there were a normochromic normocytic anemia with a hemoglobin rate at 10.2 g/dl, white blood cells to 4100/mm3, platelets at 92,000/mm3. ALT was 69 IU/L, AST at 196 IU/l, total bilirubin 25.8 mg/l, uremia at 0.38 g/l, creatinine 13.7 mg/l, the rate of prothrombin at 42%. Proteinuria of 24 hours was 0.03 g/24h. The total proteins were measured at 95 g/l; albuminemia at 25.3 g/l, gamma globulins at 17.5 g/l with the presence of a beta-gamma block. The inflammater was at 0.82 (A2-A3 activity), the fibro meter at 0.99, and the cirrhometer at 0.97 (F4 fibrosis). Quantitative HCV-RNA was at 5,242,045 UI/ml (6.72 log). The patient had a genotype HCV 2a/2c. Alpha fetoprotein was normal (6.39 ng/ml). Retroviral Human Immunodeficiency Virus (HIV) serology was negative. Abdominal ultrasound had found a liver cirrhosis, a para-umbilical right kidney transplant and left kidney failure Stage 3. The liver stiffness by Fibroscan was 61.5 kPa, IQR/med 14% and 100% success rate. The upper gastrointestinal endoscopy had not shown endoscopic signs of portal hypertension. The liver biopsy puncture was not done. The diagnosis of active decom-pensated viral cirrhosis C in oedemato-ascitic mode was retained with a Child Plugh Turcott B9 score (Ascites: 1; Hepatic encephalopathy: 1; prothrombin: 2; Albumin: 3; Bilirubin: 2). The patient was treated with sofosbuvir 400 mg/lisispavir 90 mg (1 tablet daily) for 12 weeks. Tolerance to anti-HCV therapy was excellent and no side effects were observed. Evolution under treatment of biological, biochemical and virological parameters is shown in Table 1. The patient received treatment based on Sofosbuvir 400 mg/Ledipasvir 90 mg (1 tablet daily) for 12 weeks.

Two years after the end of the treatment, the patient presents a good general condition; the viral load of HCV is always negative (HCV RNA <15 UI/ML), the liver tests are normal (AST at 42 UI/L, ALT at 40 UI/L Rate of prothrombin at 72%.

3. Discussion

3.1. Epidemiology

Chronic infection with HVC is the most common cause of liver diseases after kidney transplantation [3]. The main risk factors for HCV infection in kidney transplant recipient are: dialysis duration and dialysis mode; this is the case of our patient who had contracted HCV probably at his extra-renal cleansing. The risk is higher in hemodialysis than peritoneal dialysis and center hemodialysis than that of home hemodialysis [3].

Table 1. Evolution under treatment of clinical and biological parameters.

Parameters	Week 0	Week 4	Week 8	Week 12	SVR* 12
General condition					
asthenia	Yes	Yes	No	No	No
nausea	Yes	No	No	No	No
ascite	Yes	Yes	No	No	No
headache	No	No	No	No	No
HCV RNA (UI/ml)	5,242,045	<15	<15	<15	<15
HCV RNA (Log)	6.72	<1.2	<1.2	<1.2	<1.2
Platelets (/mm^3)	92,000	80,000	82,000	96,000	120,000
White Blood Cell (/mm^3)	4100	5000	48,000	4800	4400
Neutrophil (/mm^3)	767	2575	1709	1800	1399
Hemoglobin (g/dl)	10.2	10.9	9.5	10.2	11
AST (UI/l)	196	46	43	44	45
ALT (UI/l)	69	28	29	28	44
Creatinin (mg/l)	13.7	15.6	14	13.9	15.6
Rate of prothrombin (%)	42	67	53	54	70

*Sustained virologic response 12 weeks off therapy.

3.2. HCV Viral Load and Liver Fibrosis

We do not have values of HCV viral load and information on liver fibrosis before kidney transplant, but probably he would have had relatively lower hepatic fibrosis and viral load for transplant, compared to viral load and fibrosis after kidney transplant. In the literature, it has been shown a significant increase in HCV viremia a year after the introduction of immunosuppressive treatment with Mycophenolate mofetil [4]. Corticosteroids are also responsible for an increase in viral replication and more rapid progression to cirrhosis in immunocompetent subject [5]. After kidney transplant, there is a rise in serum concentration of HCV RNA which is probably due to the decrease of the immune response under immunosuppressive therapy, and for the establishment of a new balance between the production of virions and their clearance [3] [6] [7] [8].

3.3. Influence of HCV on the Renal Graft Survival

In our case, we observed normal kidney function and an absence of proteinuria six years after kidney transplant. Hestin et al. [9] showed that the presence of anti-HCV antibodies before kidney transplantation was predictive of the appearance of proteinuria after transplantation and graft survival of patients infected with HCV and proteinuric is less than that of non-proteinuria patients. Cosio et al. [10] reported that HCV-infected kidney transplants have a high prevalence of acute or chronic vascular rejection in the first six months after transplantation, resulting in decreased graft survival. Conversely, other authors have reported a lower rate of acute rejection in kidney transplant patients with HCV compared to kidney transplant patients not infected with HCV [11]. This reduction in graft survival seems to be due to the appearance of glomerular diseases attributable to the virus [11]. Recently, it has been shown that in HCV + kidney transplant patients with de novo glomerular diseases, the occurrence of these glomerular was not related to either an increase in the serum HCV RNA concentration after transplantation, nor to an effect describes the virus on kidney cells, but rather to a change in the immune response [7].

3.4. Combination Sofosbuvir/Ledipasvir and HCV Genotype 2

There are no major studies evaluating the combination Sofosbuvir/Ledipasvir in HCV infected patients with HCV genotype 2 and Ledipasvir has a low activity on HCV genotype [12]. The combination Sofosbuvir/Ledipasvir was approved by the European Association for the Study of Liver (EASL) for the treatment of HCV genotype 1, 4, 5, and 6 [13]. Therapeutic options for treating patients infected with HCV genotype 2 are based on the combination Sofosbuvir/Ribavirin or Sofosbuvir/Daclatasvir [13]. Given the recognized side effects of ribavirin such as anemia [14] and the unavailability of DAAs such as Daclat-asvir in Côte d'Ivoire at the time of treatment of our patient, the only treatment option based on DAAs without interferon or ribavirin in this infected HCV kidney transplant recipient genotype 2 was Sofosbuvir/Ledipasvir; which justified its use in our patient. The rapid virological response was observed at initiation of treatment as well as the sustained virological response at 12 weeks after the end of treatment. We did not find in the literature a study evaluating the combination sofosbuvir/ledipasvir in kidney transplant patients infected with HCV genotype 2. However a rapid virologic response was observed in 22 of 25 patients (88%) in the series Kamar et al. [15]; they also found that at the end of treatment, HCV RNA was undetectable in all patients. 4 to 12 weeks after therapy with DAAs, all patients had a sustained virologic response and normalization of transaminases. Tolerance to anti-HCV therapy was excellent and no adverse effects were observed. They therefore concluded that the DAAs are effective and safe for treating HCV infection after kidney transplantation.

4. Conclusion

Hepatitis C in the kidney transplant is common. Here we report the efficacy of the Sofosbuvir/Ledipasvir combination in a genotype 2 positive HCV patient. This is the first case diagnosed in black francophone Africa. The evolution was favorable with a sustained viral response 12 weeks after the end of treatment. The hope of accessible treatment and affordable is therefore possible in black Africa for HCV positive hemodialysis patients as well as HCV positive kidney transplant patients.

References

[1]   Scott, D.R., Wong, J.K., Spicer, T.S., Dent, H., Mensah, F.K., et al. (2010) Adverse Impact of Hepatitis C Virus Infection on Renal Replacement Therapy and Renal Transplant Patients in Australia and New Zealand. Transplantation, 90, 1165-1171. https:// doi.or g/ 10.1097/ TP.0b013e3181f92548

[2]   Baid-Agrawal, S., et al. (2014) Hepatitis C Virus Infection and Kidney Transplantation in 2014: What's New? American Journal of Transplantation, 14, 2206-2220. https:// doi.or g/ 10.1111/ ajt.12835

[3]   Morales, J.M. and Campistol, J.M. (2000) Transplantation in the Patient with Hepatitis C. Journal of the American Society of Nephrology, 11, 1343-1353.

[4]   Rostaing, L., Izopet, J., Sandres, K., Cisterne, J.M., Puel, J. and Durand, D. (2000) Changes in Hepatitis C Virus RNA Viremia Concentrations in Long-Term Renal Transplant Patients after Introduction of Mycophenolate Mofetil. Transplantation, 69, 991-994. https:// doi.or g/ 10.1097/00007890-200003150-00055

[5]   Magrin, S., Craxi, A., Fabiano, C., et al. (1994) Hepatitis C Viremia in Chronic Liver Disease: Relationship to Interferon-Alfa or Corticosteroid Treatment. Hepatology, 19, 273-279. https:// doi.or g/ 10.1002/ hep.1840190203

[6]   Izopet, J., Rostaing, L., Sandres, K., et al. (2000) Longitudinal Analysis of Hepatitis C Virus Replication and Liver Fibrosis Progression in Renal Transplant Recipients. The Journal of Infectious Diseases, 181, 852-858. https:// doi.or g/ 10.1086/ 315355

[7]   Kamar, N., Rostaing, L., Boulestin, A., et al. (2003) Evolution of Hepatitis C Virus Quasispecies in Renal Tranplant Patients with De Novo Glomerolonephritis. Journal of Medical Virology, 69, 482-488. https:// doi.or g/ 10.1002/ jm v.10335

[8]   Alric, L. Di-Martino, V., Selves, J., Cacoub, P., et al. (2002) Long-Term Impact of Renal Transplantation on Liver Fibrosis during Hepatitis C Virus Infection. Gastroenterology, 123, 1494-1499. https:/ / doi.or g/ 10.1053/ gast.2002.36610

[9]   Hestin, D., Guillemin, F., Castin, N., Le Faou, A., Champigneulles, J. and Kessler, M. (1998) Pretransplant Hepatitis C Virus Infection: A Predicator of Proteinuria after Renal Transplantation. Transplantation, 65, 741-744.

      https:/ / doi.or g/ 10.1097/ 00007890-199803150-00024

[10] Cosio, F.G., Sedmak, D.D., Henry, M.L., et al. (1996) The High Prevalence of Severe Early Post-Transplant Renal Allograft Pathology in Hepatitis C Positive Recipients. Transplantation, 62, 1054-1059. https:/ / doi.or g/ 10.1097/ 00007890-199610270-00004

[11] Pascual, J., Crespo, M., Mateos, M.L., et al. (1998) Reduced Severity of Acute Rejection in Hepatitis C Virus Positive Renal Allograft Recipients: Are Milder Immunosuppressive Regimens Advisable? Transplantation Proceedings, 30, 1329-1330. https:/ / doi.or g/ 10.1016/ S0041-1345(98)00263-2

[12] Nkuize, M., Sersté, T., Buset, M. and Mulkay, J.P. (2016) Combination Ledipasvir-Sofosbuvir for the Treatment of Chronic Hepatitis C Virus Infection: A Review and Clinical Perspective. Therapeutics and Clinical Risk Management, 12, 861-872. https:/ / doi.or g/ 10.2147/ TCRM.S77788

[13] European Association for the Study of the Liver (2015) EASL Recommendations on Treatment of Hepatitis C 2015. Journal of Hepatology, 63, 199-236.

      https:/ / doi.or g/ 10.1016/ j.jhep.2015.03.025

[14] Garnier, J.L., Chevallier, P., Dubernard, J.M., Trepo, C., Touraine, J.L. and Chossegros, P. (1997) Treatment of Hepatitis C Virus Infection with Ribavirin in Kidney Transplant Patients. Transplantation Proceedings, 29, 783.

      https:/ / doi.or g/ 10.1016/ S0041-1345(96)00100-5

[15] Kamar, N., Marion, O., Rostaing, L., et al. (2016) Efficacy and Safety of Sofosbuvir-Based Antiviral Therapy to Treat Hepatitis C Virus Infection after Kidney Transplantation. American Journal of Transplantation, 16, 1474-1479.

      https:/ / doi.or g/ 10.1111/ ajt.13518

22

Doppler Ultrasound of Hepatic Vessels in the Diagnosis of Cirrhosis of the Liver in Togo

Lantam Sonhaye1, Abarchi Habibou Boube1, Abdoulatif Amadou1, Bérésa Kolou1, Mohaman Djibril2, Bidamin Ntimon3, Mazamaesso Tchaou1, Nouhou Mamoudou Garba1, Lama Kègdigoma Agoda-Koussema1, Komlavi Adjenou1

1Department of Radiology, University Teaching Hospital of Lomé, Lomé, Togo
2Department of Internal Medicine, University Teaching Hospital of Lomé, Lomé, Togo
3Department of Radiology, University Teaching Hospital of Kara, Kara, Togo

Email: habibouabarchi@gmai.com

Abstract

The aim of this work is to evaluate the role of Ultrasound-Doppler in the hemodynamic study of hepatic vessels during the liver cirrhosis in Togo. Method: This was an analytic cross-sectional study that measured the velocimetric parameters of hepatic vessels in cirrhotic patients and in non-cirrhotic patients. Results: The velocimetric parameters of the hepatic artery, the portal vein, and the hepatic veins were measured in 50 cirrhotic patients and 50 non-cirrhotic The caliber of the portal vein was significantly increased in cirrhotic patients compared to non-cirrhotic patients with 13.11 ± 2.16 mm versus 11.45 ± 1.02 (p < 0.00006). The systolic velocity and the hepatic artery resistance index were significantly raised in the cirrhotic patients compared to the non-cirrhotic with 67.32 ± 22.77 versus 49.97 ± 17.24 (p-value < 0.00004) respectively, and 0.78 ± 0.07 against 0.72 ± 0.08 (p < 0.00006). The caliber of the hepatic veins was significantly decreased in the cirrhotic patients compared to the non-cirrhotic patients (p < 0.0003). There was no correlation between the gender of the patients and the change in the hemodynamics of the hepatic vessels. Conclusion: The hemodynamic study of the hepatic vessels can and must rightly be a diagnostic argument for liver cirrhosis.

Keywords

Doppler, Hepatic Vessels, Diagnosis, Cirrhosis, Togo

1. Introduction

Cirrhosis is defined as a diffuse disorganization of the liver architecture, with annular fibrosis delineating hepatocyte nodules in clusters that are called regeneration nodules [1].

The standard diagnosis of cirrhosis is still histology [2]. However, this technique suffers from a fake negative rate of up to 24% [2]. Also, because of the lack of technical means and the limitations in the histology method within our regions, this diagnosis is based on a bundle of clinical, biological, and radiological arguments [3]. The main radiological method is the Doppler ultrasound.

Several studies have been conducted around the world to determine the contribution of ultrasound coupled or not to Doppler in the diagnosis of cirrhosis [4] [5] [6]. However, in Togo, no study has been conducted on the use of Doppler in the diagnosis of hepatic cirrhosis, despite the frequency of the pathology [3].

We conducted this study to evaluate the Ultrasound-Doppler's hemodynamic variations of the hepatic vessels during the liver cirrhosis in Togo.

2. Method

This was an analytic cross-sectional study that measured the velocimetric parameters of hepatic vessels of the patients treated in the Hepato-gastroenterology department of the Lomé University Hospital Center for the liver's cirrhosis and of volunteers, non-cirrhotic, not having any known liver pathology and having a normal liver function.

We excluded the patients who had a high blood pressure or who got a liver mass, hepatocellular carcinoma, a heart failure, a kidney failure or an abundant ascites.

We utilized the ESAOTE My lab mark of ultrasound provided with multi-frequency probes, a color Doppler module, a pulsed Doppler with its different modes: duplex, triplex, and alternate, with the patients on least six hours of fasting.

All examinations were performed by the same operator, a radiologist with at least five years of experience in the visceral Doppler exam.

The examination began with a study of the hepatic morphology per subcostal, intercostal transverse and sagittal sections.

The filter was set between 50 and 100 Hz. The pulse repetition frequency (PRF) was manually set to operate according to the quality of the resulting plot and the observed velocities. The firing angle was between 30 and 60. The Doppler gate was as possible as possible in the center of the vessel to be surveyed.

All the recordings were made in the subject being in apnea in the middle of breathing. The exploration was done for each of the hepatic vessels.

We identified the portal vein, including the segment between the spleno-portal junction and the intrahepatic bifurcation that was shaped as an anechoic tube within which we measured the largest diameter, as a caliber.

The color Doppler enabled us to detect the blood flow.

The design obtained using pulsed Doppler was secondarily analyzed to determine the speeds and the different indexes.

We identified the hepatic artery by using a costal or intercostal approach. This artery visualization was difficult for some patients, but in most cases, the hepatic artery was anterior and slightly medial to the portal vein.

The pulsed Doppler mode was used to obtain the plot of at least three consecutive pulses that we examined and on which we measured the velocimetry parameters (velocities and indices).

We identified the hepatic veins by using intercostal and recurrent subcostal cuts. We preferably chose the right hepatic vein. The Doppler window was placed in the center of the vein, two centimeters from its meeting with the inferior vena cava.

The size of the portal vein, the direction of its flow, its spectrum, its average velocity (Vm), its maximum systolic velocity (Vsmax), and its diastolic velocity (Vdias) were studied. The vein pulse index of the portal vein (VPI) corresponds to the ratio of the difference between the maximum systolic speed and the diastolic rate on the systolic maximum speed.

The maximum systolic velocity, the average velocity, the diastolic velocity, the resistance index (RI), and the pulse index (PI) were the variables analyzed on the hepatic artery.

The studied parameters of the hepatic veins were the caliber, the spectrum, and the flow.

The hepatic vascular index (HVI) is the ratio of the average velocity of the portal vein to the pulse index (PI).

The arterio-portal ratio (APR) is the ratio of the systolic velocity of the hepatic artery to the maximum systolic velocity of the portal vein.

The Chi-2 statistical test was applied after crossing to determine the interdependence of the qualitative variables with values greater than 2.5. For the variables with values less than 2.5, the Fisher test was used. For the numerical averages, the Student's test was used. The relationship between the continuous values was analyzed using the Pearson correlation test. These tests were significant when the p-value (random part) was less than 0.05.

3. Results

The caliber of the portal vein was abnormal in 52% of cirrhotic patients and in 6% of the case of non-cirrhotic patients. The difference was statistically significant (p-value = 4.75×10^{-7}).

The averages of the velocimetry variables of cirrhotic and non-cirrhotic patients are shown in Table 1.

The Vsmax, Vm, RI, hepatic artery PI, as well as the portal vein caliber, had a statistically significant increase in values in the cirrhotic patients versus noncirrhotic patients (Figure 1 & Figure 2).

The sensitivity and the specificity of the resistance index for a value of 0.77 were respectively 54% and 68%.

Hepatic veins were normal in all the non-cirrhotic patients compared to 36% in the cirrhotic patients. (P-value = 3.01×10^{-11}). The Sensitivity and specificity were 64% and 100%, respectively.

A demodulation of the hepatic vein spectrum was found in 52% of the cirrhotic patients (Figure 3) whereas there was no change in the spectrum in the non-cirrhotic patients (p-value = 1.202×10^{-8}).

The blood flow in the hepatic veins was hepatofugal in both cirrhotic and non-cirrhotic patients.

The relationship between age and the speed parameters assessed by the Pearson correlation test noted a linear and positive correlation between age and RI (p-value = 0.02) and age and PI (p-value = 0.021).

There was no correlation between sex and the statistically significant variations observed between the speed parameters in cirrhotic and non-cirrhotic patients.

4. Discussion

4.1. Portal Vein

We found a statistically significant (p-value = 0.00006) increase in the portal vein average caliber in cirrhotic versus non-cirrhotic patients.

The increase of the caliber is secondary to the increase of the pressure in the portal system. Thus, an increased caliber of the portal vein should be given special attention in the search for other signs of cirrhosis. However, the diameter of the portal vein decrease and become normal again in case of portosystemic diversion. And when these diversion paths are important, the flow reverses and the caliber generally becomes normal or inferior to normal [7].

Table 1. Comparison of the Doppler averages for Cirrhotic patients and Non-Cirrhotic Patients.

	Cirrhotic patients	Non cirrhotic patients	P value
Portal vein			
Caliber (mm)	13.11 ± 2.16	11.45 ± 1.02	0.000066
Average speed (cm/s)	15.50 ± 5.46	15.85 ± 4.18	0.71
Maximum systolic speed (cm/s)	19.30 ± 6.68	19.58 ± 5.03	0.81
Diastolic speed (cm/s)	13.11 ± 6.64	13.88 ± 8.34	0.4
VPI	0.30 ± 0.12	0.27 ± 0.11	0.2
Hepatic artery			
Average speed (cm/s)	30.20 ± 10.94	25.75 ± 13.16	0.06
Maximum systolic speed (cm/s)	67.32 ± 22.77	49.97 ± 17.24	4.09×10^{-5}
Diastolic speed (cm/s)	13.99 ± 6.64	14.28 ± 8.34	0.84
RI	0.78 ± 0.07	0.72 ± 0.08	6.54×10^{-5}
PI	1.82 ± 0.51	1.49 ± 0.40	0.00056
HVI	09.09 ± 4.05	11.52 ± 4.55	0.005
APR	3.80 ± 1.55	2.71 ± 1.16	0.00015

VPI: venous pulse index, RI: resistance index, PI: pulse index, HVI: hepatic vascular index, APR: arterio-portal ratio.

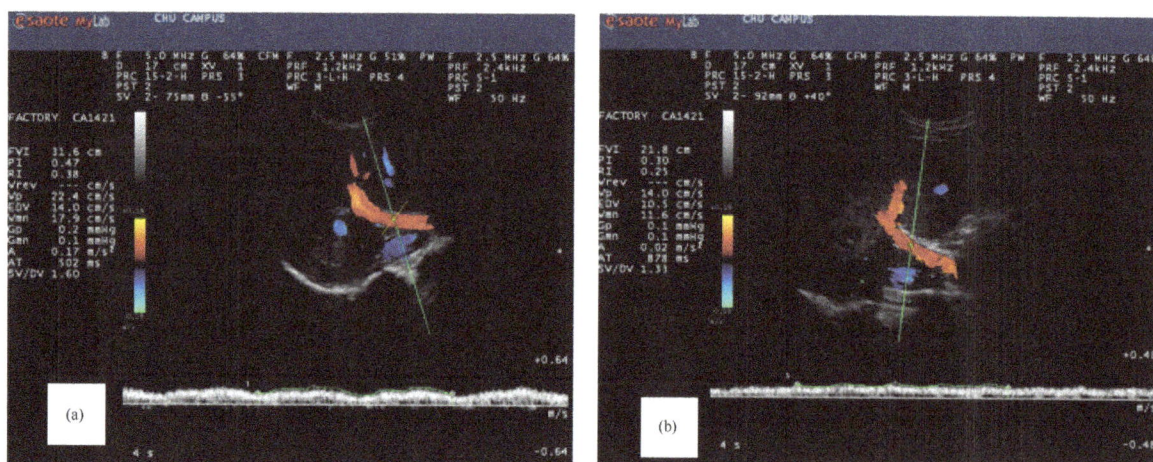

Figure 1. Flow and spectrum of portal vein. (a) In a non cirrhotic patient, hepatopetal flow and normal mean velocity (17, 9 cm/s); (b) in a cirrhotic patient, hepatopetal flow and decreased mean velocity (11.6 cm/s).

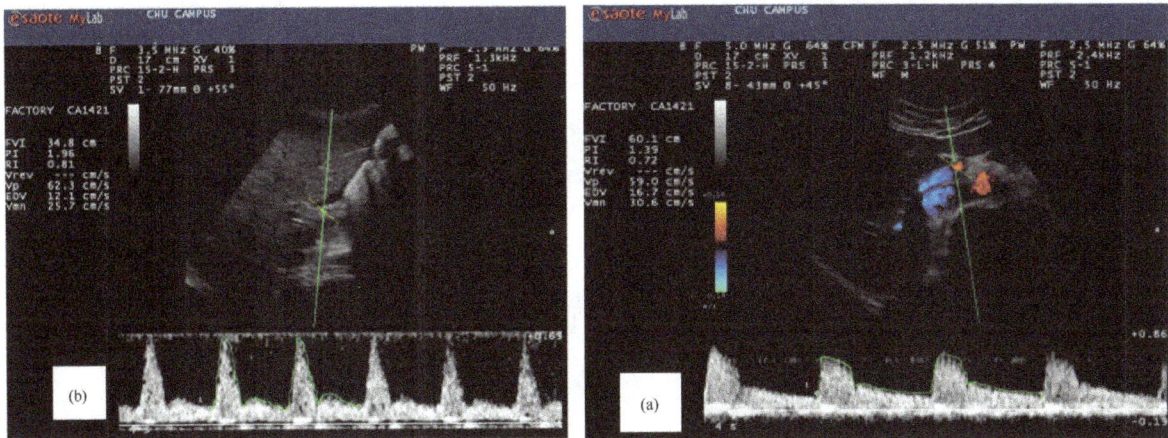

Figure 2. Hepatic artery spectrum. (a) in a non cirrhotic patient, the RI is 0.72; (b) in a cirrhotic patient, the RI is 0.81.

We noticed in this study that a non-statistically significant decrease in the portal average velocity (p-value = 0.71) of systolic and diastolic velocity in cirrhotic versus non-cirrhotic patients. The decrease in these speeds is explained by the loss of the flexibility of the hepatic parenchyma which becomes more difficult to irrigate and behaves as an obstacle to the normal flow of the blood. As a result, the blood flow slows down. This decrease in the velocity of the portal vein has been found by several authors. Iwao et al. [8] and Zironi al [7] respectively found an average of 11.0 ± 2.4 cm/s speed in cirrhotic patients versus 15.9 ± 2.8 cm/s in controls and (13.0 ± 3.2 cm/s against 19.6 ± 2.6 cm/s, p < 0.001) with however a statistically significant difference.

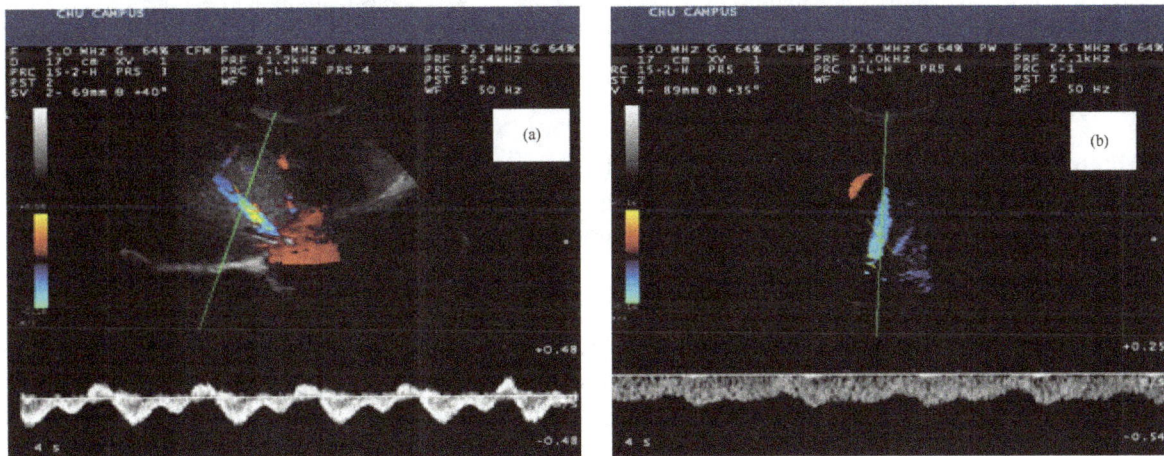

Figure 3. Hepatic vein spectrum. (a) normal waveform in a non cirrhotic patient characterized by a triphasic flow; (b) abnormal hepatic flow pattern in a cirrhotic patient characterized by a flat flow pattern.

For a portal vein average velocity of 15 cm/s, Zironi [7] also found a sensitivity and a specificity of 88% and 96% respectively, not for the diagnosis of cirrhosis but for the detection of portal high blood pressure. This is still important because portal high blood pressure is a major complication of the cirrhosis of the liver.

Then, the lowering of the average speed can be an indirect sign of the diagnosis of hepatic cirrhosis.

Most studies [7] [8] found a decrease in portal vein velocimetry values in cirrhotic patients compared with non-cirrhotic subjects, but with different values.

This variation in values may be related to the difference in the material used, the test conditions and the measurement methods as reported in the literature [7] [9] [10] [11]. It can also be related to the variations in the position of the cursor and the Doppler window [12].

4.2. Hepatic Artery

The maximum systolic rate, as well as the average speed, had a statistically significant increase in cirrhotic patients compared to non-cirrhotic subjects.

The diastolic rate, meanwhile, experienced a non-statistically significant decrease in cirrhotic patients compared to controls.

The increase in the systolic velocity and the decrease in the diastolic velocity might be explained by the increase in the flow of the hepatic artery, which might attempt to curb the decrease the flow of the portal vein. This increased flow leads to the increase of the resistance in the arterial lair and, consequently, lead to the rise of the resistance and the pulse indices [13].

We also found in our study a statistically significant increase of the resistance index (IR) and the pulse index (PI).

This rise of the indices was found by Iwao [8] (PI 1.28 ± 0.18 against 0.95 ± 0.17) Sacerdoti [14] (PI 1.3 ± 0.29 against 0.89 ± 0.09 and RI 0.71 ± 0.07 vs. 0.59 ± 0.04).

Taking into account the normal averages of the RI (0.55 - 0.81) proposed by the literature [15] [16] [17], we found in our study 46% pathological IR and 86% pathological IP. This reflects the sensitivity of these indices to cirrhosis corresponding to the data proposed by Pierce [18] who noted a sensitivity and specificity of 68% and 70% from an IR value of 0.77. However, it should be noted that the increase in the indices is not specific to cirrhosis. Indeed, increased indices are also noted in the chronic diseases of the liver [19], after the meal and with age [20] [21]. It should also be noted that there are significant interand intraobserver variations that might affect the measurement of these indices [22].

4.3. Hepatic Veins

The decreased size of the hepatic veins and the demodulation (decrease in pulse) of the hepatic vein spectrum in cirrhotic patients can be explained by the decrease of the hepatic parenchyma compliance due to fibrosis and the decrease of hepatic vein diameter. resulting from cirrhosis. [23] [24] [25].

Bolondi & al [26] and Colli & al [23] found respectively 50% and 75% abnormal spectrum in the cirrhotic patients.

The decrease of the calibers and the demodulation of the spectrum of the hepatic veins is a sensitive sign of the cirrhosis but nonspecific because the Budd Chiari syndrome, the diffuse hepatic metastases as well as the deep inspiration, obesity, and ascites are the factors that can influence the spectrum of hepatic veins [26] [23] [27] [28] [29] [30].

4.4. Limitations

None of our patients received a liver biopsy puncture which is the gold standard for the diagnosis of cirrhosis [2]. The diagnosis of cirrhosis in the hepato-gastroenterology department was based on a combination of clinical, biological and radiological arguments [3].

5. Conclusions

We found a statistically significant change in hemodynamic parameters of liver vessels in the cirrhotic patient compared with the non-cirrhotic subject. The size of the portal vein; the hepatic artery average and systolic velocities, the resistance and the pulse index had increased in the cirrhotic. The systolic and diastolic average velocities of the portal vein and the caliber of the hepatic veins, on the other hand, had decreased in cirrhotic patients.

The measurement of hemodynamic parameters of hepatic vessels is a sensitive and specific argument for the diagnosis of cirrhosis. It can and must rightly be part of the diagnosis arsenal of cirrhosis.

References

[1] Anthony, P.P., Ishak, K.G., Nayak, N.C., Poulsen, H.E., Scheuer, P.J. and Sobin, L.H. (1977) The Morphology of Cirrhosis Definition, Nomenclature, and Classification. Bull WHO, 55, 521-540.

[2] Regev, A., Berho, M., Jeffers, L.J., Milikoswski, C., Molina, E.G., Pyrsopoulos, N.T., et al. (2002) Sampling Error and Intraobserver Variation in Liver Biopsy in Patients with Chronic HCV Infection. The American Journal of Gastroenterology, 97, 2614-2618. https://doi.or g/ 10.1111/ j.1572-0241.2002.06038.x

[3] Bouglouga, O., Bagny, A., Djibril, M.A., Lawson-Ananissoh, L.M., Kaaga, L., Redah, D., et al. (2012) Aspects épidémiologiques, diagnostiques et évolutifs de la cirrhose hépatique dans le service d'hépato-gastroentérologie du CHU Campus de Lomé. Journal de la Recherche Scientifique de l'Université de Lomé(Togo), 14, 1-7.

[4] Pavlov, C.S., Casazza, G., Semenistaia, M., Nikolova, D., Tsochatzis, E., Liusina, E., et al. (2016) Ultrasonography for Diagnosis of Alcoholic Cirrhosis in People with Alcoholic Liver Disease. Cochrane Database of Systematic Reviews, No. 3. Art. No.: CD011602.

[5] Kok, T.H., van der Jagt, E.J., Haagsma, B.C.M.A., Jansen, P.L.M. and Boeve, W.J. (1999) The Value of Doppler Ultrasound in Cirrhosis and Portal Hypertension. Scandinavian Journal of Gastroenterology, 34, 82-88.

[6] Martinez-Noguera, A., Montserrat, E., Torrubia, S. and Villalba, J. (2002) Doppler in héPatic Cirrhosis and Chronic Hépatitis. Seminars in Ultrasound, CT, and MRI, 23, 19-36. https:// doi.or g/ 10.1016/ S0887-2171(02)90027-2

[7] Zironi, G., Gaiani, S., Fenyves, D., Rigamonti, A., Bolondi, L. and Barbara, L. (1992) Value of Measurement of Mean Portal Flow Velocity by Doppler Flowmetry in the Diagnosis of Portal Hypertension. Journal of Hepatology, 16, 298-303.
https:// doi.or g/ 10.1016/ S0168-8278(05)80660-9

[8] Iwao, T., Toyonaga, A., Kazuhiko, O, Tayama, C., Masumoto, H., Sakai, T., et al. (1997) Value of Doppler Ultrasound Parameters of Portal Vein and Hepatic Artery in the Diagnosis of Cirrhosis and Portal Hypertension. The American Journal of Gastroenterology, 92, 1012-1017.

[9] de Vries, P.J., van Hattum, J., Hoekstra, J.B.L. and de Hooge, P. (1991) Duplex Doppler Measurements of Portal Venous Flow in Normal Subjects. Interand Intra-Observer Variability. Journal of Hepatology, 13, 358-363.

https:// doi.or g/ 10.1016/ 0168-8278(91)90081-L

[10] Bolondi, L., Gaiani, S. and Barbara, L. (1991) Accuracy and Reproducibility of Portal Flow Measurement by Doppler US. Journal of Hepatology, 13, 269-273.
https:// doi.or g/ 10.1016/ 0168-8278(91)90067-L

[11] Sabba, C., Weltin, G.G., Cicchetti, D.V., Ferraioli, G., Taylor, K.J.W., Nakamura, T., et al. (1990) Observer Variability in Echo-Doppler Measurements of Portal Flow in Cirrhotic Patients and Normal Volunteers. Gastroenterology, 98, 1603-1616.
https:// doi.or g/ 10.1016/ 0016-5085(90)91097-P

[12] Nelson, R.C., Lovett, K.E., Chezmar, J.L., Moyers, J.H., Torres, W.E., Murphy, F.B., et al. (1987) Comparison of Pulsed Doppler Sonography and Angiography in Patients with Portal Hypertension. AJR, 149, 77-81.
https:// doi.or g/ 10.2214/ ajr.149.1.77

[13] Marder, D.M., DeMarino, G.B., Sumkin, J.H. and Sheahan, D.G. (1989) Liver Transplant Rejection: Value of the Resistive Index in Doppler Ultrasound of Hepatic Arteries. Radiology, 173, 127-129.
https:// doi.or g/ 10.1148/ r adiology.173.1.2675178

[14] Sacerdoti, D., Merkel, C., Bolognesi, M., Amodio, P., Angeli, P. and Gatta, A. (1995) Hepatic Arterial Resistance in Cirrhosis with and without Portal Vein Thrombosis: Relationships with Portal Hemodynamics. Gastroenterology, 108, 1152-1158.
https:// doi.or g/ 10.1016/ 0016-5085(95)90214-7

[15] Joint, L.K., Platt, J.F., Rubin, J.M., Ellis, J.H. and Bude, R.O. (1995) Hepatic Artery Resistance before and after Standard Meal in Subjects with Diseased and Healthy Livers. Radiology, 196, 489-492. https:/ / doi.or g/ 10.1148/ r adiology.196.2.7617865

[16] Dauzat, M., Lafortune, M., Patriquin, H. and Pomier-Layrargues, G. (1994) Meal Induced Changes in Hepatic and Splanchnic Circulation: A Noninvasive Doppler Study in Normal Humans. European Journal of Applied Physiology, 68, 373-380. https:/ / doi.or g/ 10.1007/ BF00843732

[17] Tanaka, K., Mitsui, K., Morimoto, M., Numata, K., Inoue, S., Takamur, Y., et al. (1993) Increased Hepatic Arterial Blood Flow in Acute Viral Hepatitis: Assessment by Color Doppler Sonography. Hepatology, 18, 21-27.

https:/ / doi.or g/ 10.1002/ hep.1840180105

[18] Pierce, M.E. and Sewell, R. (1990) Identification of Hepatic Cirrhosis by Duplex Doppler Ultrasound Value of the Hepatic Artery Resistive Index. Australasian Radiology, 34, 331-333. https:/ / doi.or g/ 10.1111/ j.1440-1673.1990.tb02667.x

[19] Piscaglia, F., Gaiani, S., Zironi, G., Gramantieri, L., Casali, A., Siringo, S., et al. (1997) Intraand Extrahepatic Arterial Resistances in Chronic Hepatitis and Liver Cirrhosis. Ultrasound in Medicine & Biology, 23, 675-682.

https:/ / doi.or g/ 10.1016/ S0301-5629(97)00012-4

[20] Lafortune, M., Dauzat, M., Pomier-Layrargues, G., Gianfelice, D., Lepanto, L., Breton, G., et al. (1993) Hepatic Artery: Effect of a Meal in Healthy Persons and Transplant Recipients. Radiology, 187, 391-394.

https:/ / doi.or g/ 10.1148/ r adiology.187.2.8475279

[21] Fisher, A.J., Paulson, E.K., Kliever, M.A., Delong, D.M. and Nelson, R.C. (1998) Doppler Sonography of the Portal Vein and Hepatic Artery: Measurement of a Prandial Effect in Healthy Subjects. Radiology, 207, 711-715.

https:/ / doi.or g/ 10.1148/ r adiology.207.3.9609894

[22] Colli, A., Cocciolo, M., Mumoli, N., Cattalini, N., Fraquelli, M. and Conte, D. (1998) Hepatic Artery Resistance in Alcoholic Liver Disease. Hepatology, 28, 1182-1186. https:/ / doi.or g/ 10.1002/ hep.510280503

[23] Colli, A., Cocciolo, M., Riva, C., Martinez, E., Prisco, A., Pirola, M., et al. (1994) Abnormalities of Doppler Waveform of the Hepatic Veins in Patients with Chronic Liver Disease: Correlation with Histologic Findings. AJR, 162, 833-837.

https:/ / doi.or g/ 10.2214/ ajr.162.4.8141001

[24] Walsh, K.M., Leen, E., Macsween, R.N. and Morris, A.J. (1998) Hepatic Blood Flow Changes in Chronic Hepatitis C Measured by Duplex-Doppler Color Sonography. Relationship to Histological Features. Digestive Diseases and Sciences, 43, 2584-2590. https:/ / doi.or g/ 10.1023/ A:1026626505517

[25] Arda, K., Ofelli, M., Calikoglu, U., Olcer, T. and Cumhur, T. (1997) Hepatic Vein Doppler Wave-Form Changes in Early Stage (Child-Pugh A) Chronic Parenchymal Liver Disease. Journal of Clinical Ultrasound, 25, 15-19.

https:/ / doi.or g/ 10.1002/ (SICI)1097-0096(199701)25:1<15::AID-JCU3>3.0.CO;2-N

[26] Bolondi, L., Bassi, S.L., Gaiani, S., Zironi, G., Benzi, G., Santi, V., et al. (1991) Liver Cirrhosis: Changes of Doppler Waveform of Hepatic Veins. Radiology, 178, 513-516. https:/ / doi.or g/ 10.1148/ r adiology.178.2.1987617

[27] Ohta, M., Hashizume, M., Tomikawa, M., Ueno, K., Tanoue, K. and Sugimachi, K. (1994) Analysis of Hepatic Vein Waveform by Doppler Ultrasonography in 100 Patients with Portal Hypertension. The American Journal of Gastroenterology, 89, 170-175.

[28] Ohta, M., Hashizume, M., Kawanaka, H., Akazawa, K., Tomikawa, M., Higashi, H., et al. (1995) Prognostic Significance of Hepatic Vein Waveform by Doppler Ultrasonography in Cirrhotic Patients with Portal Hypertension. The American Journal of Gastroenterology, 90, 1853-1857.

[29] Hosoki, T., Kuroda, C., Tokunaga, K., Marukawa, T., Masuike, M. and Kozuka, T. (1989) Hepatic Venous Outflow Obstruction: Evaluation with Pulsed Duplex Sonography. Radiology, 170, 733-737.

https:/ / doi.or g/ 10.1148/ r adiology.170.3.2644659

[30] Bolondi, L., Gaiani, S., Bassi, S.L., Zironi, G., Bonino, F., Brunetto, M., et al. (1991) Diagnosis of Budd-Chiari Syndrome by Pulsed Doppler Ultrasound. Gastroenterology, 100, 1324-1331. https:/ / doi.or g/ 10.1016/ 0016-5085(91)90785-J

23

Low Serum Free Triiodothyronine Is Associated with Increased Risk of Decompensation and Hepatocellular Carcinoma Development in Patients with Liver Cirrhosis

Ula M. Al-Jarhi[1*], Abeer Awad[1], Mona Mohsen[2]

[1]Department of Medicine, Cairo University, Cairo, Egypt
[2]Department of Chemical Pathology, Cairo University, Cairo, Egypt
Email: *ulamabid.aljarhi@gmail.com

Abstract

Background: FT3 levels in plasma may provide a marker for liver status in cirrhosis. Aim: The aim is to correlate thyroid functions with hepatic status in compensated and decompensated cirrhosis, and to study their effect on development of HCC. Settings and Design: Prospective controlled cohort study. A total of 58 patients with liver cirrhosis were recruited from Kasr AlAiny ER and outpatient clinics. Patients were categorised into compensated (11), decompensated (39) and patients with hepatocellular carcinoma (8). The study also included 12 healthy controls. Methods and Material: Liver function tests, TSH, FT4 and FT3 and abdominal ultrasound and triphasic computed tomography abdominal scans were done. Statistical Analysis Used: Chi-square and unpaired t-tests were used for comparison. One way ANOVA and Kruskal Wallis tests were used to compare more than two groups. Spearman Correlation followed by logistic regression analysis of significant variables was used to find predictors of dependent variables. Results: The frequency of patients with low FT3 was significantly higher in patients with liver cirrhosis (48%), and HCC (50%) than control subjects (12%) (p-value < 0.001). Mean serum FT3 was lowest among decompensated patients (2 pg/ml ± 0.7), followed by patients with HCC (2.5 pg/ml ± 0.7) and highest among compensated patients (3.7 pg/ml ± 0.4), p-value < 0.001. Logistic regression analysis showed that low FT3, male gender, ulcer bleeding and encephalopathy were independently associated with the development of HCC (OR, 95% CI: 1.1, 0.3 - 8). Conclusions: Low FT3 is common among patients with decompensated liver cirrhosis and HCC. FT3 shows a significant negative

---

*Corresponding author.

correlation with severity of liver disease and deterioration of liver function. Low FT3 shows a significant independent association with HCC.

Keywords

Liver Cirrhosis, Thyroid Functions, HCC

1. Introduction

Thyroid hormones are general regulators of tissue metabolism; however, a unique relation with hepatocytes exists. The liver in turn metabolises thyroid hormones and regulates their systemic functions [1].

Low FT3 (free triiodothyronine) is the most frequent disruption encountered in routine screening of thyroid functions in patients with liver cirrhosis. This reflects a reduction in type 1 deiodinase activity, reduced T4 (thyroxine) conversion to T3 and compensatory shift to T4 conversion to rT3 (reversed T3) by type 3 deiodinase [2].

This reduction in FT3 levels may reflect a compensatory hypothyroid state in patients with liver cirrhosis that tunes down hepatocytes' metabolism which may help to preserve liver functions and conserve body protein stores [1]. The prognostic value of low FT3 in liver cirrhosis and its link to encephalopathy have been recognized for a long time [3]. Low serum T3 level is found to be an indicator of poor prognosis for hepatic encephalopathy [4]. However, it is seldom used in spite of its simplicity and low cost.

Overt hypothyroidism has been indirectly linked to HCC (hepatocellular carcinoma) on top of non-alcoholic steatohepatitis [5] [6]. In a previous study, a significantly elevated risk association is found between long term history of hypothyroidism and HCC in women [7]. However, a clear association of abovementioned thyroid derangement with HCC development has not been studied in patients with liver cirrhosis.

The aim of this work was to study to correlate thyroid functions with hepatic status in compensated and decopensated cirrhosis and to determine the effect of thyroid dysfunction in development of HCC.

2. Methods

2.1. Subjects

The study presents a prospective cohort. Patients with non-alcoholic liver cirrhosis, aged between 35 and 65 years were recruited from Kasr Alainy outpatient clinics over three months. Patients were consecutively recruited irrespective to the presenting symptom, hepatic status and presence of complications. Patients with history of head and neck irradiation, known (concurrent or past) extra-hepatic primary tumours or any type of primary liver cancer other than HCC, positive family history of cancer, or primary thyroid disorders were all excluded. A total of 58 patients with liver cirrhosis met the criteria and were included. Decompensated patients were admitted to wards of the internal medicine or to the intensive care according to admission protocols. Patients were followed up till discharge or mortality.

Liver function tests, TSH (thyroid stimulating hormone), FT4 and FT3 and abdominal ultrasound and triphasic computed tomography abdominal scans were done. Lipid profile was done to all patients. The patients were accordingly categorised into three groups: 39 patients with decompensated liver cirrhosis, 11 patients with compensated liver cirrhosis, 8 patients with HCC. The study included 12, age and gender matched healthy controls.

2.2. Hormone Assays

Hormone assays of TSH, FT3 and FT4 was carried out using electro-chemiluminescence immunoassay (ECLIA) on Cobas e411 (Roche Diagnostics International Ltd, Switzerland). Thyroid dysfunction was determined according to normal values by the manufacturer: TSH at 0.72 - 4.2 uiu/ml, FT4 at 0.93 - 1.7 ng/dL and FT3 at 2.57 - 4.43 pg/ml. classification of patients according to the thyroid profile was done following the guidelines of the National Academy of Clinical Biochemistry (NACB) for laboratory diagnosis and monitoring of thyroid diseases [8].

2.3. Ethical Consideration

A written informed consent was obtained from all participants or, if patients were unable to provide consent, from designated surrogates. Confidentiality of data, safe data storage and privacy rights are respected by all who handle patient information. Data was coded and patient names or identity was obscure in all data collection forms and during statistical analysis.

2.4. Statistical Methodology

Analysis of data was done by IBM computer using SPSS (statistical program for social science version 12) as follows: TSH, FT3 and FT4 were used both as continuous and categorical variables. Normal laboratory values were used to form categories for abnormal thyroid functions, e.g. "Low FT3", "High TSH". Description of quantitative variables as mean, SD and range. Description of qualitative variables was done as number and percentage. Chi-square test was used to compare qualitative variables between groups. Unpaired t-test was used to compare quantitative variables, in parametric data (SD < 50% mean). Mann Whitney Willcoxon U test was used in non parametric data instead of unpaired t-test. One way ANOVA (analysis of variance) was used to compare more than two groups as regard quantitative variable. Kruskal Wallis test was used instead of ANOVA test in non parametric data SD > 50% mean. Spearman Correlation test was used to rank variables versus each other positively or inversely. Logistic regression analysis was used to find out the significant independent predictors of dependent variable by backward likelihood ratio technique. p-value > 0.05 was considered as insignificant, p < 0.05 as significant and p < 0.01 as highly significant.

3. Results

As regards general data, all groups had similar age and gender distribution, which was also matching with the control group (**Table 1**).

The frequency of various presenting complications of liver cirrhosis was analysed in decompensated patients (**Figure 1**). SBP was the presenting complication in 5 patients, encephalopathy in 14 patients, GIT haemorrhage in 18 patients. Abdominal ultrasound revealed splenomegaly in all decompensated patients, and ascites in all but one.

Table 1. Comparison between the studied groups as regards socio-demographic characteristics.

Variables	Decompensated N = 39	Compensated N = 11	HCC N = 8	Controls N = 12	$X^{2\,a}$	p-value
Gender						
Male	25 (66.7%)	6 (54.5%)	6 (75%)	8 (66.7%)	0.9	>0.05 NS[b]
Female	13 (33.3%)	5 (45.5%)	2 (25%)	4 (33.3%)		
Age	54.2 ± 10	50 ± 11	54 ± 12.4	49.5 ± 11	1.6	>0.05 NS

[A]chi-square test, [B]non-significant.

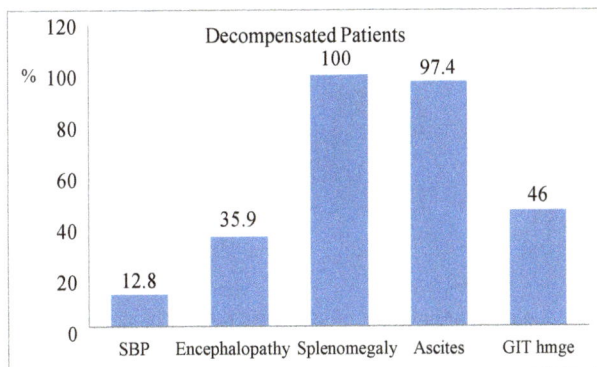

Figure 1. The frequency of various complications in decompensated patients.

Table 2 shows that low FT3 was more frequent among patients with liver cirrhosis as compared to healthy controls with statistically significant difference by using chi-square test. Low FT3 was found in 48% of patients with liver cirrhosis and 50% of patients with HCC, as compared to only 12% in controls (p-value < 0.001).

Figure 2 shows that decompensated patients had the lowest level of FT3; while compensated patients had the highest level with statistically significant difference by using one way ANOVA test. Mean serum FT3 was lowest among decompensated patients (2 pg/ml ± 0.7), followed by patients with HCC (2.5 pg/ml ± 0.7) and highest among compensated patients (3.7 pg/ml ± 0.4). This difference was statistically significant, with p-value < 0.001.

On comparing the relative frequencies of thyroid function categories among various groups, the prevalence of low FT3 was highest in decompensated patients (61.5%), followed by HCC patients (50%) versus none of the compensated patients (**Figure 3**). This difference was statistically significant with a p-value < 0.001. As seen in **Table 3**, other thyroid function abnormalities were matching in frequencies with healthy controls.

On correlation of thyroid functions against other biochemical variables including liver functions and lipid profile, FT3 showed a significant direct correlation with albumin and prothrombin concentration (PC), while an inverse correlation with INR, **Figure 4** and **Figure 5** respectively.

In this study, FT4 level abnormalities were rare and statistically comparable to the control group. Only one patient with liver cirrhosis, hypoalbuminemia and ascites had low FT4. The patient also had low FT3. Elevated FT4 was present in one decompensated patient with low FT3, and one compensated patient with otherwise normal thyroid functions.

Backward likelihood technique of binary logistic regression analysis was done to test the effect of various factors on the incidence of each of the following complications of liver cirrhosis: ascites, SBP (spontaneous bacterial peritonitis), encephalopathy, GIT haemorrhage, HCC.

Table 2. Comparison between total group versus controls as regard thyroid profile.

Variables	Total Cases N = 50	HCC N = 8	Controls N = 12	X^2	P
Low TSH	5 (10%)	2 (25%)	1 (8.3%)	3.9	>0.05
High TSH	4 (8%)	0	1 (8.3%)	1.8	>0.05
Low FT3	24 (48%)	4 (50%)	3 (12%)	25	<0.001
Low FT4	1 (2%)	0	0	0.4	>0.05
High FT4	2 (4%)	0	2 (16.7%)	3.2	>0.05

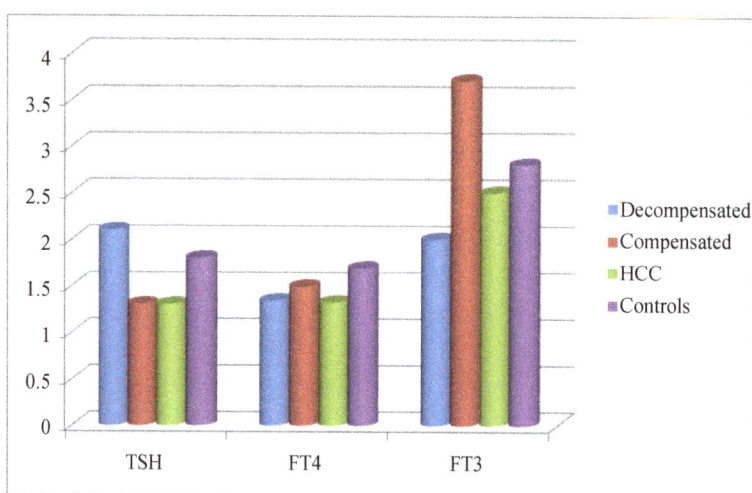

Figure 2. Comparison of the levels of TSH, FT3 and FT4 among various groups. FT3 was significantly higher in the compensated group versus other groups and significantly lower in the decompensated group by one way ANOVA test (p-value < 0.001). No statistically significant differences were noted in TSH and FT4 values among the various groups.

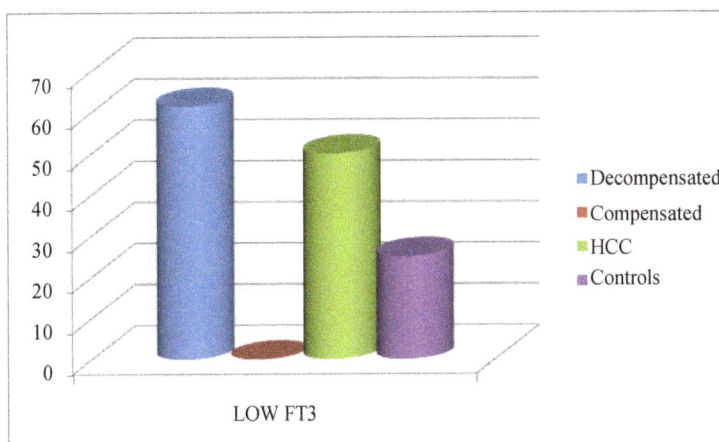

Figure 3. Percentage of subjects with low FT3 in each group. More than 60% of decompensated patients had low FT3, as compared to 50% of patients with HCC, and none of compensated patients. This difference was statistically significant (p-value < 0.001).

Table 3. Comparison between the studied groups as regard thyroid profile categories.

Variables	Decompensated N = 39	Compensated N = 11	HCC N = 8	Controls N = 12	X^2	P
Low TSH						>0.05
Normal	37 (94.4%)	11 (100%)	6 (75%)	11 (91.7%)	4.9	NS[a]
Dysfunction	2 (5.6%)	0	2 (25%)	1 (8.3%)		
High TSH						>0.05
No	35 (89.7%)	11 (100%)	8 (100%)	11 (91.7%)	2	NS
Yes	4 (10.3%)	0	0	1 (8.3%)		
Low FT3						<0.001
No	15 (38.5%)	11 (100%)	4 (50%)	9 (75%)	15	HS[b]
Yes	24 (61.5%)	0	4 (50%)	3 (12%)		
Low FT4						>0.05
No	38 (97.4%)	11 (100%)	8 (100%)	12 (100%)	0.9	NS
Yes	1 (2.6%)	0	0	0		
High FT4						>0.05
No	38 (97.4%)	10 (90.9%)	8 (100%)	10 (83.3%)	4.1	NS
Yes	1 (2.6%)	1 (9.1%)	0	2 (16.7%)		

[a]NS non-significant; [b]HS highly significant.

The presence of ascites, male gender and hypoalbuminemia were found to be independently associated with the development of encephalopathy (p-value < 0.05) (**Table 4**).

Hypoalbuminemia and variceal bleeding were found to be independently associated with the development of ascites (p-value < 0.05) (**Table 5**).

Low FT3, male gender, ulcer bleeding and encephalopathy were independently associated with the development of HCC (**Table 6**). This was evident after adjusting for compounding variables namely splenomegaly, ascites, SBP, encephalopathy, variceal bleeding, liver function tests and lipid profile. Of the 4 independent risk factors for HCC, low FT3 had the greatest contribution to its development, with a beta-coefficient 0.98. This was followed by male gender with a beta-coefficient 0.44, then ulcer bleeding to a lesser degree and lastly encephalopathy.

4. Discussion

The thyroid function derangements found in this study may be attributed either to a true thyroid dysfunction associated with liver disease or the well established entity of "nonthyroidal illness syndrome (NTIS)" formerly known as sick euthyroid syndrome. These findings agree with previous studies that analysed thyroid dysfunction during critical illness. TSH levels are described to be commonly within the normal range in NTIS but may decrease in prolonged illness.

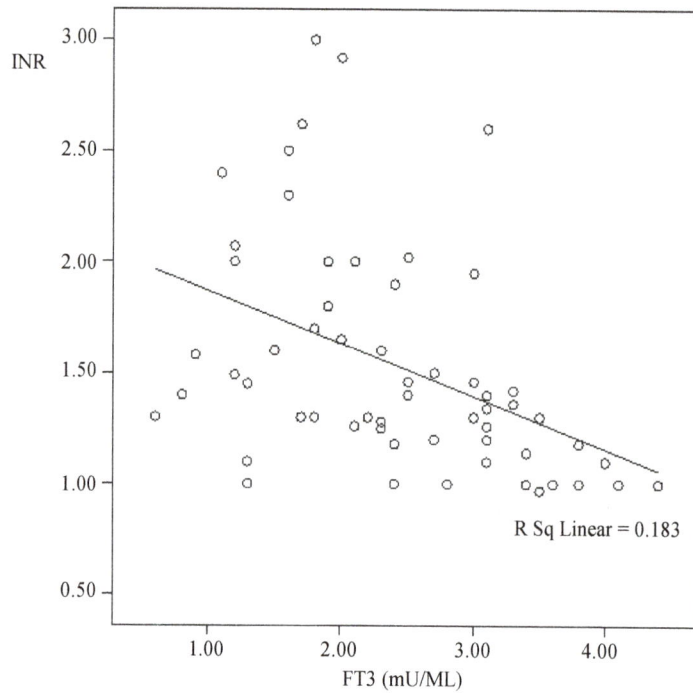

Figure 4. Scatter diagram showing correlation between FT3 and INR among all patients. FT3 shows a significant negative correlation with INR, correlation coefficient −0.30, p-value < 0.05.

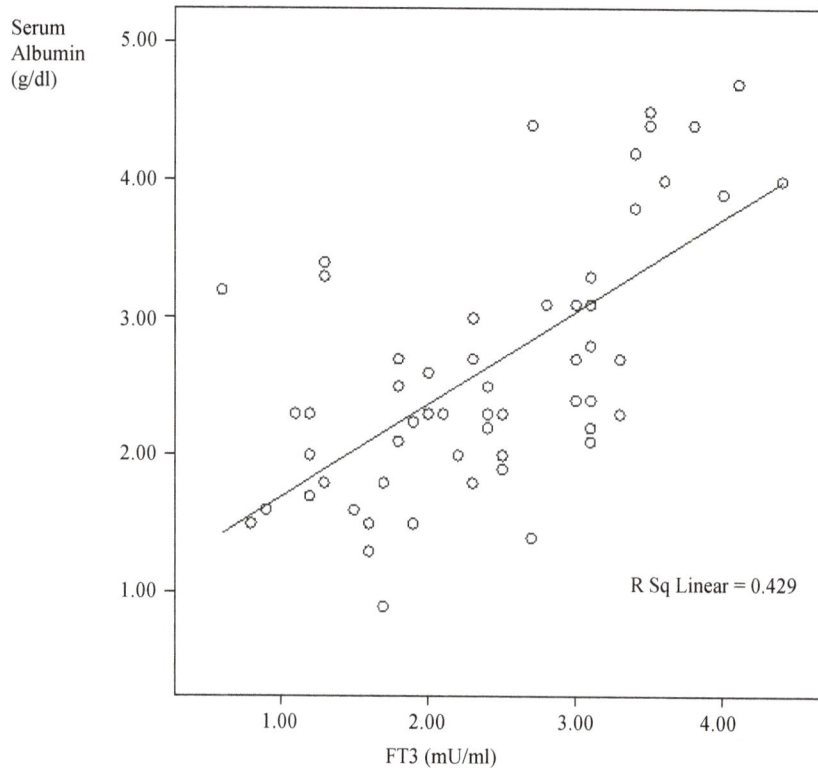

Figure 5. Scatter diagram showing correlation between FT3 and albumin among all patients. FT3 shows a significant positive correlation with albumin, correlation coefficient 0.48, p-value < 0.05.

Table 4. Multivariable analysis between all risk factors versus encephalopathy by logistic regression.

Variables	Beta-coefficient	P	Odd's (95% CI)[a]
Ascites	0.33	<0.05	1.02 (−0.1 - 10.3)
Male gender	0.19	<0.05	1.05 (−0.8 - 12.7)
Low albumin (<3)	0.12	<0.05	1.01 (−0.9 - 22.8)

[a]CI = confidence interval.

Table 5. Multivariable analysis between all risk factors versus ascites by logistic regression.

Variables	Beta-coefficient	P	Odd's (95% CI)[a]
Low albumin (<3)	0.21	<0.05	1.1 (−0.5 - 19.7)
Variceal bleeding	0.20	<0.05	1.08 (−0.3 - 12.8)

[a]CI = confidence interval.

Table 6. Multivariable analysis between all risk factors versus HCC by logistic regression.

Variables	Beta-coefficient	P	Odd's (95% CI)[a]
Low FT3	0.98	<0.05	1.1 (0.3 - 8)
Male gender	0.44	<0.05	1.07 (−0.06 - 11.7)
Ulcer bleeding	0.22	<0.05	1.01 (−0.4 - 13.6)
Encephalopathy	0.13	<0.05	1.02 (−0.9 - 14.3)

[a]CI = confidence interval.

A closer look at these abnormalities is required. In this study, patients with low TSH included 2 patients with decompensated liver cirrhosis. Both had low FT3 and normal FT4. The level of TSH was less than 0.3 but more than 0.05 mU/ml. Only lower TSH levels are associated with true hyperthyroidism in some cases. Such patients fit the definition of NTIS, which is usually found in 60% - 70% of critically ill patients [9].

Also, 2 patients with HCC had low TSH. One had TSH level 0.03 mU/ml, a normal FT4 and low FT3. This can also be attributed to NTIS due to associated low FT3, but later follow up is mandatory due to associated low TSH levels. The other patient had TSH in the range 0.3 - 0.05 (0.2 mU/ml), normal FT3 and FT4. This pattern can also be explained by NTIS, although one control subject had almost the same pattern.

On the other hand, all three patients in this study with high TSH had decompensated liver disease. All three had low FT3, and a TSH level <20 mU/ml. This pattern is quite typical of NTIS. Some hospitalized patients have transient elevations in serum TSH concentrations (up to 20 mU/L) during recovery from nonthyroidal illness. Few of these patients prove to have hypothyroidism when re-evaluated after recovery from their illness. Patients with serum TSH concentrations over 20 mU/L usually have permanent hypothyroidism [10]. Two of our patients had TSH >10 mU/ml should require later follow up.

More importantly, the frequency of patients with FT3 lower than the normal range was significantly higher than control subjects. Low FT3 was found in 48% of patients with liver cirrhosis and 50% of patients with HCC, as compared to only 12% in controls (p-value < 0.001). These figures may reflect NTIS, but again, the difference in distribution of low FT3 among patients with liver cirrhosis is clinically relevant.

By analysis of the relative frequencies of abnormal thyroid function among various groups, the prevalence of low FT3 (below the lower limit) was highest in decompensated patients (61.5%), followed by HCC patients (50%) versus none of the compensated patients. This difference was statistically significant with a p-value < 0.001. Other thyroid function abnormalities were matching with frequencies in healthy controls.

Indeed, samples for thyroid function were collected from decompensated patients who required hospital admissions during an acute illness, in the form of SBP, encephalopathy or GIT haemorrhage. While compensated patients were sampled on an outpatient basis.

These results come in agreement with most previous studies. The majority of critically ill patients have low serum T3 concentrations, as do some outpatients during illness [11].

In addition to lower FT3 levels with clinical decompensation, FT3 also decreases with progression of liver cirrhosis and deterioration of liver function tests.

This association of low FT3 with hepatic decompensation and deterioration of liver functions was demonstrated in previous studies [12]-[14].

It is worthy of note that the single mortality in this study was a patient with the second lowest FT3 level-0.8 pg/ml (normal 2.4 - 4.2). The patient had a PC 60%, albumin 1.5, ascites and presented with hepatic encephalopathy. Other patients with hepatic encephalopathy (a total of 16 patients) survived. Hence, the prognostic value of FT3 level, not only in liver disease severity, but also the outcome in acute complications.

The low total and free T3 levels may be regarded as an adaptive hypothyroid state that serves to reduce the basal metabolic rate within hepatocytes and preserve liver function and total body protein stores. A study in cirrhotic patients showed that the onset of hypothyroidism from intrinsic thyroid disease of various etiologies during cirrhosis resulted in a biochemical improvement in liver function (e.g. coagulation profiles) as compared to cirrhotic controls [15]. Hypothyroidism has also been associated with lesser degrees of decompensation in cirrhosis [16].

The independent association between low FT3 and HCC in the regression model suggests a true association between thyroid dysfunction and HCC, rather than a simple reflection of deterioration in liver functions. Firstly, it had the greatest impact on HCC development. Two complications, ulcer bleeding and encephalopathy had only a minor role. Lastly, all HCC patients were outpatients and not acutely ill. Several previous studies had analysed a similar association with overt hypothyroidism [17] [18].

A possible role for T3 in suppression of HCC cell invasiveness was suggested in a previous study on HCC cell lines [7]. Thus, the independent association of low FT3 with HCC in this study may suggest that the implication of low T3 is more than an index of hepatic disease and HCC. Prolonged depression of FT3 serum levels in patients with liver cirrhosis appears to have a pathogenic role in HCC development.

5. Limitations and Further Investigations

In statistical analysis, the impact of TSH and FT4 derangements as well as risk factors for mortality was impossible due to their rare prevalence among the studied sample. Concerning analysis of HCC risk factors, lipid profile was taken into consideration besides clinical and laboratory measures of liver status. Further studies comprising larger samples should adjust for other well established risk factors for HCC as smoking, obesity and diabetes.

The authors conclude that low FT3 is common among patients with decompensated liver cirrhosis and HCC. FT3 shows a significant negative correlation with severity of liver disease and deterioration of liver function. It may serve also as a prognostic factor in critically ill cirrhotic patients. Also, low FT3 shows a significant independent association with HCC.

The authors suggest that serum FT3 level be an index for liver status, as well as a marker for prognosis in hospitalized liver cirrhosis patients. It may be used as a risk factor and a marker for HCC development.

Acknowledgements

We would like to acknowledge our great Kasr Alainy Hospital, and its workers, nurses and staff members, for all support and help in this study and throughout our careers.

Funding

Authors received no funding for this study.

Declaration of interest

The authors report no conflicts of interest. The authors alone are responsible for the content and writing of the paper.

References

[1]   Malik, R. and Hodgson, H. (2002) The Relationship between the Thyroid Gland and the Liver. *QJM*, **95**, 559-569. http://dx.doi.org/10.1093/qjmed/95.9.559

[2]   Guven, K., Kelestimur, F. and Yucesoy, M.(1993) Thyroid Function Tests in Non-Alcoholic Cirrhotic Patients with Hepatic Encephalopathy. *The European Journal of Medicine*, **2**, 83-85.

[3]   Van Thiel, D.H., Udani, M., Schade, R.R., Sanghvi, A. and Starzl, T.E. (1985) Prognostic Value of Thyroid Hormone Levels in Patients Evaluated for Liver Transplantation. *Hepatology*, **5**, 862-866. http://dx.doi.org/10.1002/hep.1840050526

[4]   Güven, K., Kelestimur, F. and Yücesoy, M. (1993) Thyroid Function Tests in Non-Alcoholic Cirrhotic Patients with Hepatic Encephalopathy. *The European Journal of Medicine*, **2**, 83-85.

[5]   Pucci, E., Chiovato, L. and Pinchera, A. (2000) Thyroid and Lipid Metabolism. *International Journal of Obesity and Related Metabolic Disorders*, **24**, S109-S112. http://dx.doi.org/10.1038/sj.ijo.0801292

[6]   Liangpunsakul, S. and Chalasani, N. (2003) Is Hypothyroidism a Risk Factor for Non-Alcoholic Steatohepatitis? *Journal of Clinical Gastroenterology*, **37**, 340-343. http://dx.doi.org/10.1097/00004836-200310000-00014

[7]   Hassan, M.M., Kaseb, A., Li, D., *et al.* (2009) Association between Hypothyroidism and Hepatocellular Carcinoma: A Case-Control Study in the United States. *Hepatology*, **49**, 1563-1570. http://dx.doi.org/10.1002/hep.22793

[8]   Baloch, Z., Carayon, P., Conte-Devolx, B., *et al.* (2003) Laboratory Medicine Practice Guidelines. Laboratory Support for the Diagnosis and Monitoring of Thyroid Disease. *Thyroid*, **13**, 3-126. http://dx.doi.org/10.1089/105072503321086962

[9]   Economidou, F., Douka, E., Tzanela, M., Nanas, S. and Kotanidou, A. (2011) Thyroid Function during Critical Illness. *Hormones*, **10**, 117-124. http://dx.doi.org/10.14310/horm.2002.1301

[10]   Burman, K.D. and Wartofsky, L. (2001) Endocrine and Metabolic Dysfunction Syndromes in the Critically Ill: Thyroid Function in the Intensive Care Unit Setting. *Critical Care Clinics*, **17**, 43-57. http://dx.doi.org/10.1016/S0749-0704(05)70151-2

[11]   Peeters, R.P., Wouters, P.J., van Toor, H., Kaptein, E., Visser, T.J., Van den Berghe, G., *et al.* (2005) Serum 3,3',5'-Triiodothyronine (rT3) and 3,5,3'-Triiodothyronine/rT3 Are Prognostic Markers in Critically Ill Patients and Are Associated with Postmortem Tissue Deiodinase Activities. *The Journal of Clinical Endocrinology & Metabolism*, **90**, 4559-4565. http://dx.doi.org/10.1210/jc.2005-0535

[12]   Mansour-Ghanaei, F., Mehrdad, M., Mortazavi, S., *et al.* (2012) Decreased Serum Total T3 Level in Hepatitis B and C Related Cirrhosis by Severity of Liver Damage. *Annals of Hepatology*, **11**, 667-671.

[13]   L'age, M., Meinhold, H., Wenzel, K.W. and Schleusener, H. (1980) Relations between Serum Levels of TSH, TBG, T4, T3, rT3 and Various Histologically Classified Chronic Liver Diseases. *Journal of Endocrinological Investigation*, **3**, 379-383. http://dx.doi.org/10.1007/BF03349374

[14]   Faber, J., Thomsen, H.F., Lumholtz, I.B., Kirkegaard, C., Siersbaek-Nielsen, K. and Friis, T. (1981) Kinetic Studies of Thyroxine, 3,5,3'-Triiodothyronine, 3,3,5'-Triiodothyronine, 3',5'-Diiodothyronine, 3,3'-Diiodothyronine, and 3'-Monoiodothyronine in Patients with Liver Cirrhosis. *The Journal of Clinical Endocrinology & Metabolism*, **53**, 978-984. http://dx.doi.org/10.1210/jcem-53-5-978

[15]   Oren, R., Sikuler, E., Wong, F., Blendis, L.M. and Halpern, Z. (2000) The Effects of Hypothyroidism on Liver Status of Cirrhotic Patients. *Journal of Clinical Gastroenterology*, **31**, 162-163. http://dx.doi.org/10.1097/00004836-200009000-00016

[16]   Oren, R., Brill, S., Dotan, I. and Halpern, Z. (1998) Liver Function in Cirrhotic Patients in the Euthyroid versus the Hypothyroid State. *Journal of Clinical Gastroenterology*, **27**, 339-341. http://dx.doi.org/10.1097/00004836-199812000-00012

[17]   Reddy, A., Dash, C., Leerapun, A., Mettler, T.A., Stadheim, L.M., Konstantinos, N., *et al.* (2007) A Possible Risk Factor for Liver Cancer in Patients with No Known Underlying Cause of Liver Disease. *Clinical Gastroenterology and Hepatology*, **5**, 118-123. http://dx.doi.org/10.1016/j.cgh.2006.07.011

[18]   Lin, K.-H., Lin, Y.-W., Lee, H.-F., Liu, W.-L., Chen, S.-T., Chang, K.S.S. and Cheng, S.-Y. (1995) Increased Invasive Activity of Human Hepatocellular Carcinoma Cells Is Associated with an Overexpression of Thyroid Hormone β1 Nuclear Receptor and Low Expression of the Anti-Metastatic nm23 gene. *Cancer Letters*, **98**, 89-95. http://dx.doi.org/10.1016/S0304-3835(06)80015-7 http://dx.doi.org/10.1016/0304-3835(95)04000-T

24

Therapeutic Strategies of Stem Cell Transplantation for Liver Cirrhosis

Fan Chen

Department of Gastroenterology, Fuzhou General Hospital, Fuzhou, China

Email: yangqh4848@sina.com

Abstract

Liver transplantation is widely regarded as the most effective therapy for end-stage liver diseases. However, stem cell-based therapy is being developed as a promising strategy which offers a number of benefits as it is minimally invasive and associated with low immunogenicity and low cost. This paper will review the major clinical issues surrounding the use of stem cell therapy for managing cirrhosis, such as discussing the selection of appropriate subtypes of bone marrow stem cells and the need for pre-differentiation into hepatocyte-like cells prior to transplantation, and providing an overview of the methods to improve cell viability and to prevent the exacerbation of cirrhosis. The role of human umbilical cord blood stem cells and amniotic epithelial cells for the treatment of liver disease will be also introduced.

Keywords

Liver Cirrhosis, Stem Cell Therapy

1. Introduction

Cirrhosis is a common consequence of chronic liver diseases, which leads to the liver failure during the decompensation stage. Liver transplantation is the only therapeutic option for advanced cirrhosis. However, the practicality of liver transplantation is compromised by the severe shortage of donated organs, high medical costs, and long-term requirement for immunosuppression. Sources of hepatocytes for the bio-artificial liver and hepatocyte transplantation are also limited. Recent studies suggest that stem cell transplantation may be a promising new strategy for the treatment of end-stage liver diseases [1] [2]. This treatment has not been widely used currently because of various unknown risks. And the treatment is expected to be gradually mature and standardized through the efforts of scholars.

Here we review therapy-related details on the use of stem cell treatment for liver cirrhosis, and discuss advances in human umbilical cord blood stem cells (UCBSCs) and human amniotic epithelial cells (hAECs) as sources of therapeutic stem cell.

2.   Selection of Bone Marrow Stem Cells for Liver Cirrhosis

2.1. Hematopoietic Stem Cells or Mesenchymal Stem Cells?

2.1.1. Clinical Studies

It is widely accepted that both liver injury and hepatic fibrosis are closely related to the immune response. Hematopoietic stem cells (HSCs) including CD34+, CD133+ and mononuclear cells enhance immune function, whereas, mesenchymal stem cells (MSCs) have an immunosuppressive effect. The type of the stem cells used for managing liver disease, therefore, depends on cellular immune function or the extent of the cellular immune response which varies according to the etiology and stage of liver diseases.

Autoimmune liver diseases are commonly associated with enhanced cellular immune responses, and the use of CD34+ and/ or CD133+ cells exacerbates this condition [3] [4], whereas MSCs provide a safe and effective treatment strategy [5]. While there is no initial evidence of impaired immune response during the early stages of alcoholic liver cirrhosis [6], the immune response may be reduced later as the disease progresses, especially in the advanced stages of liver failure [7]. As a result of this disease profile, HSCs are eliminated by the active immune cells during the treatment of early-stage alcoholic liver cirrhosis, and accompanied by exacerbating the progression of hepatic fibrosis [8]. A previous study shows that two among eight patients receiving HSCs for the treatment of alcoholic liver cirrhosis, there was one case of diabetes (an autoimmune disease) with graft-versus-host disease, and another patient developed deterioration of liver function with the hemorrhage of the upper digestive tract [9]. Jang et al. [10] treated alcoholic cirrhosis with MSCs and found no adverse events.

However, the reduced cellular immune response was usually found in the patients of alcoholic liver cirrhosis complicated with HCV [6], hence the HSCs transplantation did not result in any adverse effect related to the immune response, which has been confirmed by Couto BG [9]. Depressed cellular immune function is often shown in patients with HBVor HCV-induced cirrhosis [11]. Salama et al. [12] treated HCV infection complicated cirrhosis using HSCs (CD34+/CD133+). In these subjects there were no adverse immune responses following HSCs transplantation and HCV titers were reduced in most patients with liver cirrhosis at 6 months post-transplantation. It was therefore, postulated that HSCs treatment may inhibit HBV propagation in patients with HBV-induced cirrhosis [13].

Cellular immune function is generally enhanced during the early stages of cirrhosis complicated by acute-on-chronic liver failure (ACLF), but becomes depressed during the later stages of the disease process [14]. Therefore, appropriate types of stem cells should be selected according to the cellular immune status of patients at the beginning of treatment. However, a recent clinical trail has been undertaken by using allogeneic bone marrow mesenchymal stem cells (BMSCs) through peripheral vein in treatment of patients with different phase of HBV ACLF. The results indicated that transplantation of BMSCs only in plateau phase rather than in advanced phase can improve survival rate of the patients [15].

2.1.2. Summary

Treatment with MSCs may be suitable for patients with early stage alcoholic liver cirrhosis, and for those with autoimmune liver disease, or early stage of liver cirrhosis complicated with ACLF. The use of HSCs is appropriate for late stage alcoholic liver cirrhosis, and for HBV-or HCV-induced cirrhosis, alcoholic liver cirrhosis complicated with HCV or HBV infection, and late stage of liver cirrhosis complicated with ACLF. (As illustrated in Figure 1) [16].

Adipose-derived mesenchymal stem cells (ASCs) should be used instead of BMSCs for early stage of liver cirrhosis complicated with ACLF. This is because BMSCs have to be cultured and propagated in vitro making the procedure too slow to meet the urgent needs of these patients.

It has been suggested that endothelial progenitor cells (EPCs) might be a source of stem cells for cirrhosis therapy [17]. Since EPCs share a range of biological features with HSCs [18], the therapeutic indications for EPCs might also be similar to those for HSCs.

2.2. Predifferentiated or Undifferentiated Stem Cells Prior to Transplantation?

2.2.1. Laboratory Findings

At the laboratory level, predifferentiated stem cells show better therapeutic effects than undifferentiated cells for the treatment of liver fibrosis [19] [20] [21]. There might be multiple reasons listed and discussed as follows: First, induction before transplantation may facilitate stem cell differentiation into hepatocyte, resulting in increased therapeutic effect [22]. Secondly, it is also possible that the pathological hepatic microenvironment which may include inflammation concurrent with fibrosis or cirrhosis may cause stem cells to differentiate into different types of mature cells [23] [24], such as stellate cells and myofibroblasts. This sequence of events would further promote fibrosis. Stem cell induction before transplantation has been used to prevent the differentiation pluripotency. Thirdly, the unpredifferentiated bone marrow stem cells (BMCs) which are undifferentiated juvenile cells, do not enter the hepatic differentiation process and are unstimulated cells. Therefore, they have limited proliferative ability in the microenvironment of the liver and are unable to active the endogenous hepatic stem cells or progenitor cells. By contrast, the pre-differentiated stem cells have entered into the process of differentiation and become stimulated hepatocyte-like cells, thus they have the capacity to propagate rapidly after transplantation, activate endogenous hepatic stem cells and produce a satisfactory therapeutic effect [25].

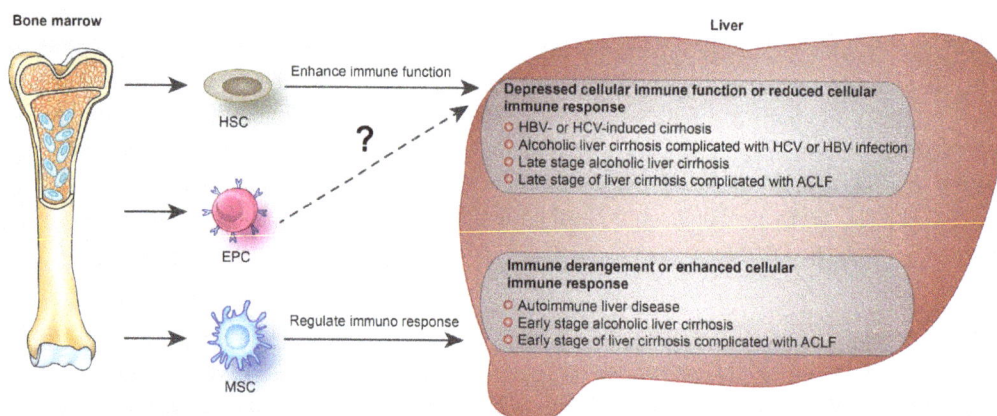

Figure 1. The indications of different bone marrow stem cell types for liver cirrhosis therapy (modified from the figure in reference 16). Abbreviations: HSC: Hematopoietic stem cell; EPC: endothelial progenitor cell; MSC: mesenchymal stem cell; ACLF: acute-on-chronic liver failure.

Other sources of laboratory evidence indicate that undifferentiated stem cells have a better therapeutic effect than predifferentiated stem cells [26] [27] [28]. This is thought to be because the liver regenerating and anti-fibrotic effects of stem cells is the result of their paracrine ability rather than an outcome of directed differentiation. The paracrine capability of predifferentiated stem cells has been shown to be attenuated during in vitro induction [29], though there are different viewpoints [30].

2.2.2. Clinical Studies

Comparison of the effects of transplanting predifferentiated and non-predifferentiated BMSCs via peripheral vein for the treatment of HCV-liver cirrhosis indicates that there is no significant difference with respect to clinical outcome and laboratory examinations at 6 months follow-up [31].

2.2.3. Summary

However, the individual therapeutic merits of predifferentiated and unpredifferentiated stem cell transplantation for the treatment of cirrhosis remain controversial. Based on the published results, thoroughly predifferentiated stem cells appear to be less potent than unpredifferentiated cells [27] [28], and hepatocyte transplantation appears to be less effective than stem cell transplantation [32]. These findings suggest that partial in vitro induction of stem cells, which actives stem cells and preserves their paracrine capability, is the better treatment option [33].

3. Enhancing Stem Cell Survival

The therapeutic effect of stem cell transplantation is seldom satisfactory in late stage cirrhosis, when the microenvironment has been severely damaged. Studies have shown that serum albumin levels (ALB) of cirrhotic patients with Child-Pugh stage A or B hepatic impairment, can be significantly improved by stem cell transplantation, The mean serum ALB value of transplantation group was (33.22 ± 2.85) and (29.65 ± 4.36) g/L for cirrhotic patients with Child-Pugh stage A and B initially. The value rised to (36.74 ± 4.37) g/L and (33.31 ± 5.77) g/L after 24 weeks, when comparing the value before transplantation, $(P < 0.05)$. Whereas, there was no significant improvement in ALB in patients with Child-Pugh stage C cirrhosis. For cirrhotic patients with Child-Pugh C, the mean serum ALB value rised from (27.97 ± 3.04) to (29.24 ± 4.91) after 24 weeks, when comparing the value before transplantation $(P > 0.05)$ [34] [35]. Based on these findings it was postulated that severe damage of the liver function might be closely related to poor stem cell survival post transplantation.

This was further demonstrated by the ex vivo experiments showing that hepatic-directed differentiation was not apparent in BMSCs induced by the serum from patients with 5% liver injury. In the same experiments apoptosis occurred in some of stem cells induced by serum from patients with 10% liver injury and all stem cells died after induction with serum from patients with 15% liver injury [36]. Although, serum from patients with hepatic injury might be able to facilitate the hepatic differentiation of the stem cells in vitro; toxicity from the damaged liver may impair stem cell viability during in vitro induction.

3.1. Methods to Improve the Microenvironment

3.1.1. Hyperbaric Oxygen Treatment

Systematic hypoxemia persists in 30% to 70% of patients with cirrhosis. Long-term hypoxemia results in apoptosis of transplanted stem cells, and affects therapeutic outcomes [37]. Clinical studies have shown that hyperbaric oxygen improves the hypoxemic condition of the hepatic microcirculation, and decreases toxic metabolites in plasma, thereby improving stem cell survival [38]. However, hyperbaric oxygen treatment may not be suitable for patients with esophageal varices. In these subjects normobaric oxygen treatment can be suggested instead.

3.1.2. Medication and (or) Plasma Exchange to Improve the Hepatic Microenvironment

Plasma exchange has been shown to increase serum levels of hepatocyte growth factor (HGF) without changing serum levels of fibroblast growth factors (FGF-4 and bFGF) or epidermal growth factor (EGF). Plasma exchange may, therefore, promote the differentiation of BMCs in a way that allows them to repair and regenerate the impaired hepatocytes in patients with severe hepatitis [39]. Experiments using the pig model of acute liver failure indicate that autologous MSCs transplantation in combination with medication (including 100 ml plasma and 100 mg diammonium glycyrrhizinate intravenously injection) resulted in significantly better outcomes than MSCs transplantation or medication alone [40]. Another clinical trial showed that combined therapy with umbilical cord blood infusion and plasma exchange improved the liver function and immunity to a greater extent than monotherapy in patients with chronic severe hepatitis [41].

Thus, both medication and stem cell transplantation effectively reduce inflammation, improve liver function and prevent further deterioration. These procedures allow toxic metabolites to be quickly removed providing a relatively stable and favorable internal environment for stem cell survival and propagation.

3.1.3. Appropriate Transplantation Site

The spleen is adjacent to liver, and for this reason intrasplenic transplantation is usually preferred when the hepatic environment is not suitable for stem cell transplantation. Transplantation of induced MSCs into the spleen, portal vein, and vena caudalis was shown to improve liver function and the pathohistological profile of patients with chronic liver disease, irrespective of which route was used. However, intrasplenic transplantation was the most effective of the three routes [42].

3.2. Stem Cell Engineering

There are multiple types of stem cell resources for use in treating cirrhosis. These include fetal liver stem cells, embryonic stem cells, UCBSCs, BMCs, amniotic stem cells, and induced pluripotent stem cells [43]. It has long been believed transplantation with a single stem cell type (such as MSCs) might be more effective than using a mixed type (such as a mononuclear cells). However, accumulated evidence suggests that cell-cell interactions may be an important factor for the success of the stem cell therapy, and that cytokines or specific cell types in mixed mononuclear cells, might be associated with improved stem cell survival and therapeutic effect [44]. A previous study reported that co-transplantation of two types stem cell might be an optimal method [45]. Results from another clinical study suggest that UCBSCs transplantation might be superior to BMCs [46]. However, further experimental confirmation is need to determination the types of stem cell that are best adapted to survive the harsh environment of liver.

3.2.1. *In Vitro* Preconditioning

Results from a rat model of liver injury, showed that MSCs co-cultured with hepatocyte growth factor for 2 weeks survived better than control cells, and significantly reduced the degree of liver fibrosis [19]. In another experiment, transplantation of microencapsulated MSCs significantly increase the survival rate of 90% hepatectomized rats. The 2-week survival rate of microencapsulated MSCs group, free MSCs group and 90% hepatectomized group were 91.6%, 25% and 21.4% after transplantation, (P < 0.01). The microencapsulated MSCs exhibited long-term survival, secreted trophic factor and differentiated into hepatocyte-like cells [47]. In another study [48], interleukin 10 gene-modified bone marrow-derived liver stem cells (BDLSCs) were shown to significantly reduce inflammation associated with liver fibrosis. They also promoted liver regeneration, to a greater degree than unmodified BDLSC.

3.2.2. Hepatogenic Differentiation of Stem Cells Prior to Transplantation

A model of rat spinal cord injury [49] showed that predifferentiated MSCs had better motor function and a higher survival rate than undifferentiated MSCs. In another study, MSCs co-cultured with the hepatocyte growth factor and basic fibroblast growth factor were found to a have superior therapeutic effect to control MSCs in a rat model of liver fibrosis [21].

3.2.3. Inhibition Stem Cell Apoptosis

It has been shown that exposure of rat hepatocyte-like cells to ursodeoxycholic acid (UDCA) inhibits deoxycholic acid-induced hepatocyte-like cells apoptosis, by down-regulating the p53/Bax signal pathway [50].

3.3. Summary

Stem cell transplantation has been shown to be associated with an unsatisfactory therapeutic effect in the treatment of liver failure especially for subjects with severe Child-Pugh stage C cirrhosis. One key reason for this is that the transplanted stem cells may not be able to survived the relatively hostile microenvironment of the liver. Promotion of the survival rate of the transplanted stem cells, therefore, appears to be critical for the success of treatment. To achieve this, a suitable transplantation strategy should be considered on basis of the stem cells per se and the hepatic microenvironment (as illustrated in Figure 2).

4.   Prevention of Hepatic Fibrosis

Accumulated studies suggest that fibrosis can be ameliorated by stem cell transplantation [51] [52]. However, there is equally compelling evidence indicating that stem cells have the potential to differentiate into the hepatic stellate cells and myofibroblasts, both of which are involved in the progression of fibrosis [23]. It is therefore, important to address the issue of preventing stem cell induced fibrosis.

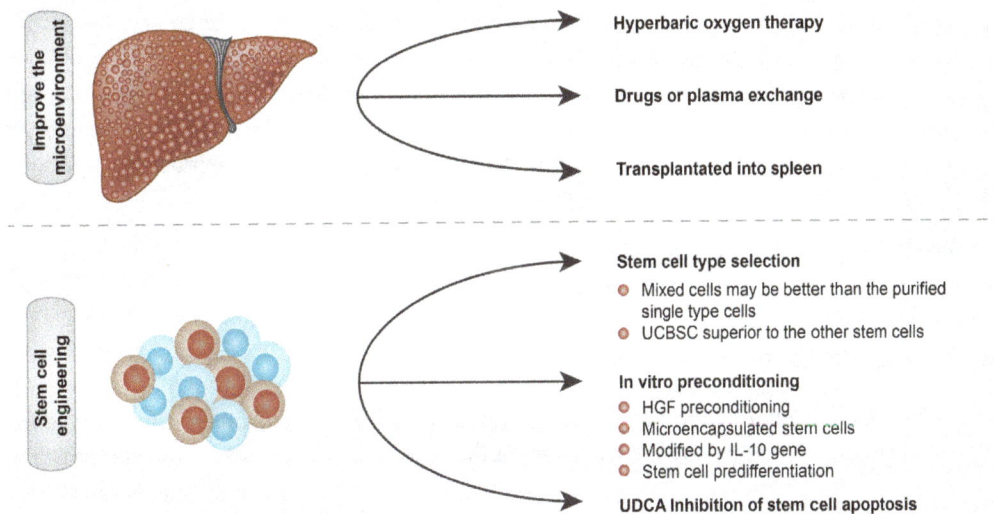

Figure 2. The strategies to improve stem cell survival. Abbreviations: UCBSC: umbilical cord blood stem cells; HGF: hepatocyte growth factor; IL-10: interleukin 10; UDCA: ursodeoxycholic acid.

4.1. Etiological Treatment

Stem cell transplantation is not routinely combined with etiological treatment such as anti-HBV or HCV therapy, and as result liver function is improved, but portal hypertension and upper gastrointestinal hemorrhage are not alleviated [34]. The more recent, combined therapeutic strategy of using stem cell transplantation and etiological treatment has been shown to improve both liver function and upper gastrointestinal hemorrhage [53]. Persisting inflammation in the liver due to lack of etiological treatment not only interferes with the therapeutic effect of transplanted stem cells, but also drives stem cell differentiation towards the production of myofibroblasts and hepatic stellate cells, both of which exacerbate the fibrosis. By contrast, etiological treatment used in conjunction with stem cell transplantation synergistically improves the therapeutic outcome of patients with cirrhosis.

4.2. Anti-Inflammatory Treatment

It is generally believed that under the microenvironment of fibrosis or cirrhosis most transplanted stem cells differentiate into hepatocytes, and hepatic stellate cells by hepatic stellate cells [24] [54]. However, in a mouse model of liver fibrosis or cirrhosis with remarkable inflammation [23], the majority of injected stem cells differentiated into myofibroblasts, resulting in liver fibrosis being aggravated. Under such circumstances an appropriate anti-inflammation therapy should be administered to improve the survival of the transplanted stem cells [40], to avoid stem cells differentiation into myofibroblasts, and thereby improve the therapeutic effect. However the anti-inflammatory effect should not be so strong that it impairs stem cell motility and differentiation [55] [56].

4.3. Selection of an Appropriate Stem Cell Type

The type of stem cells should be selected cautiously and carefully to avoid enhancing the immune response and triggering an immune system disorder [3] [4] [8] [9]. MSCs is preferred over HSCs for the cirrhosis with the severe inflammation. Whole bone marrow cells or bone marrow mononuclear cells (BMNCs) contain a certain proportion of inflammatory cells and a relatively small proportion of HSCs. Consequently, the resulting anti-inflammatory effect is less than that seen with MSCs. Experiments using a rat model of severe cirrhosis showed that liver fibrosis was exacerbated by BMNCs transplantation, but was ameliorated by BMSCs transplantation [57] [58] [59]. Similarly, it was demonstrated by the animal experiments that BMSCs had better therapeutic effect on acute liver injury with the respects to anti-inflammation and anti-fibrosis activity [60].

4.4. Summary

Stem cell transplantation may have either positive or negative outcomes depending on the direction of stem cell differentiation which can be uncertain within the complex microenvironment of the liver. A satisfactory therapeutic effect can be achieved by improving the hepatic microenvironment, by selecting appropriate types of stem cell, and by in vitro manipulation prior to transplantation.

The use of stem cell induction prior to transplantation has been described previously and the therapeutic use of hAECs will be described in a following section.

5.   Umbilical Cord Blood Stem Cells

Umbilical cord blood, has been used for many years in clinical practice as an important source for hematopoietic stem cell transplantation, and has resulted in encouraging therapeutic outcomes. Experimental trials have shown that UCBSCs can be induced and differentiated into hepatocyte-like cells [61]. They stimulate liver regeneration [62], alleviate fibrosis [63], enhance hepatic angiogenesis [64], and inhibit primary hepatocellular carcinoma [65]. They, may, therefore, provide be suitable for use in a wide range of therapeutic applications for end-stage liver disease.

5.1. Experimental Studies with UCBSCs

Liver function and hepatic histological grade were significantly improved in rats with decompensated cirrhotic rats after injection through the tail vein of a nucleated cell suspension of umbilical cord blood containing $> 2 \times 10^6$ CD34+ stem cells [66]. Liver fibrosis and survival rate were also improved relative to control rats although the differences were not statistically significant. In a rabbit model of fibrosis transplantation of 5×10^5 CD34+ UCBSCs via portal vein injection (in the absence of immunosuppression) improved liver function and alleviated fibrosis without causing symptoms of rejection [67]. Based on these findings, it was proposed that following intrahepatic injection into the rats with liver cirrhosis, umbilical cord blood mononuclear cells (UCBMNCs) are differentiated into hepatocytes and inhibit the activation of stellate cells, which is shown by the induction of stellate cells apoptosis [68]. This form of hepatocyte regeneration, coupled with the anti-inflammatory and antioxidant effects of UCBMNCs, is thought to alleviate cirrhosis. The same studies showed that UCBMNCs induced high levels of expression of the APE1 gene which might also be responsible for the prevention and inhibition of hepatic injury.

However, in a similar experimental setting [69], 1×105 transplanted CD34+ UCBMNCs failed to become engrafted in the rat liver, possibly as a result of thioacetamide which was used to induce the cirrhosis. In another study injection of 10×106 UCBMNCs to the liver via the portal vein, resulted in impairment of liver and kidney function, with no positive pathological evidence of improvements in liver tissues. Based on the findings of this study it was concluded that UCBMNCs transplantation exacerbates the liver injury and induces unwanted hepatorenal syndrome [70]. The adverse effects of stem cell therapy on renal and liver function in this study might have been related to an overdose of transplanted stem cells which resulted in a rejection response. The rejection response between the human UCBSCs and human recipients might be less marked than that between human UCBSCs and rat recipients. However, further clinical trials are needed to more precisely elucidate the effect of UCBSCs.

5.2. Clinical Studies with UCBSCs

UCBSCs transplantation via hepatic artery has been undertaken in HBV patients with cirrhosis and Child Pugh stage C liver impairment. These patients were receiving anti-viral agents, HGF, and other symptomatic and supportive treatments. The results indicated significant improvements in symptoms and liver function, and CT examination showed that the liver cross-sectional area was markedly increased compared to pretreatment findings [71]. The satisfactory outcomes reported in this study might in part be explained by the use of associated therapies that addressed the underlying etiology and promoting stem cell survival.

In the another trial [72] UCBSCs were transplanted through the hepatic artery or portal vein in a similar cohort of patients with cirrhosis, HBV and Child Pugh stage C liver impairment. In this study hepatic function and the clinical symptoms significantly improved without obvious adverse effects or complications. Unfortunately, no comparative data exists that compares portal vein and hepatic artery transplantation.

Transplantation of UCBSCs via the hepatic artery of patients with cirrhosis of various etiologies, with Child Pugh stage B or C hepatic impairment, resulted in a slight decrease in serum ALB during the initial stage of treatment, indicating the need to monitor patients closely post transplantation, though clinical symptoms, liver function, and blood coagulation function improved during the subsequent 6 month follow-up period [73].

UCBSCs transplantation has also been used for the treatment of cirrhosis associated with hepatic diabetes. In this study, improvements in cirrhosis were accompanied by normalization of blood glucose levels [74]. In another study UCBSCs transplantation resulted in a general improvement in the overall condition of a patient with Wilson disease, concomitance with increased appetite for food. Serum ceruloplasmin levels were increased and no acute rejection occurred [75].

A study comparing UCBSCs transplantation with the autologous peripheral blood stem cells (PBSCs) transplantation for treatment of severe chronic liver disease with ascitic cirrhosis, indicated that both types of stem cells improved liver function and cirrhosis with equal effect [76]. However, UCBSC transplantation was considered to be more effective than PBSC in patients with the most severe disease. Another study comparing UCBSC with BMCs transplantation showed that higher proportions of CD34+ or CD38+ cells were contained in UCBSCs than in BMCs. Both stem cell types were used to treat HBV and HCV cirrhosis in patients with Child Pugh stage B and C liver impairment. The results showed that the therapeutic effect of UCBSCs transplantation was better than that of BMCs [46].

5.3. Summary

Based on published literature, UCBSCs transplantation for the treatment of cirrhosis appears to be superior to PBSCs or BMCs transplantation. Although the immunogenicity of the UCBSC is very low, it still contains allogeneic cells and close monitoring is required to ensure early detection and prevention of any rejection responses.

6. Amniotic Epithelial Cells and Liver Disease

Recent studies show that human amniotic epithelial cells (hAECs) might possess certain biological characteristics and functions of hepatocytes [77] [78] [79]. Indeed, it has been shown that hAECs can be induced to differentiate into hepatocyte-like cells and cholangiocytes [80] [81]. The potential advantages of using hAECs include the wide range of available sources and the simple collection method. More than 1×108 hAECs can be extracted from a single abandoned placenta amnion [82]. There are no medical ethical issues to address. Undifferentiated cells display low levels of immunogenicity, however, hAECs that have differentiated into hepatocyte-like cells may possess a certain level of immunogenicity [83]. hAECs have also been shown to have antifibrotic properties and anti-inflammatory effects [84], which make them suitable for the treatment of various liver diseases.

6.1. Genetic Liver Diseases

Transplantation of 5×105 hAECs cells every other week via the tail vein prolonged the survival of mice with experimentally induced Niemaoh-Pick disease [85]. Loss of body weight, organ damage, cholesterol deposition, and the relative weight of the liver were all reduced, indicating that hAECs may have a role in the treatment of fatal genetic diseases, including the metabolic liver disease.

6.2. Liver Fibrosis

Amniotic membrane administered as a patch on the surface of the liver has been investigated in a biliary fibrosis model established by bile duct ligation in rats [86]. The amniotic membrane reduced biliary cell inflammation and collagen deposition in the liver. It also inhibited the activity of myofibroblasts, which significantly reduced and delayed the development of liver fibrosis. Based on these findings it was postulated that its therapeutic effect might be due to paracrine activation rather than to the hAECs per se. Other workers have shown that hAECs injection improves liver function, reduces hepatocyte apoptosis, and decreases hepatic inflammation and fibrosis in a CCL4-induced liver fibrosis model in wild-type mice [87]. In this study no obvious rejection response was seen 2 weeks after hAECs transplantation.

These findings suggest that, hAECs may provide an effective treatment for liver fibrosis, as it appears to avoid the exacerbation of fibrosis that is caused by stem cell differentiation. The lack of rejection and exacerbation of fibrosis in the two mouse models also supports the use of hAECs as a potential source of stem cells for the treatment of cirrhosis.

6.3. Summary

The efficacy and safety of human amniotic membrane in the treatment of ophthalmological diseases have been established for a number of years. However, there remains a lack of in-depth studies using hAECs for liver regeneration in liver disease, and therefore requiring rigorous research.

7. Conclusion

Stem cell therapy has been shown to be effective for treating end-stage liver disease. Further investigations using careful study designs should be undertaken to identify the optimal dose of injected stem cells, the times of transplantation, the timing of repeated cell transplantation, and the monitoring on the long-term complications. With the gradual promotion of stem cell therapy, we can focus on the optimization of the treatment methods in the future, especially on the improvement of curative effect and the lasting aspect of therapeutic efficacy. It is believed that the formal clinical application will be realized in the future by the joint efforts of clinical practitioners and basic researchers.

References

[1] Dai, L.J., Li, H.Y., Guan, L.X., Ritchie, G. and Zhou, J.X. (2009) The Therapeutic Potential of Bone Marrow-Derived Mesenchymal Stem Cells on Hepatic Cirrhosis. Stem Cell Research, 2, 16-25.

[2] Eckersley-Maslin, M.A., Warner, F.J., Grzelak, C.A., McCaughan, G.W. and Shackel N.A. (2009) Bone Marrow Stem Cells and the Liver: Are They Relevant? Journal of Gastroenterology and Hepatology, 24, 1608-1616. https://doi.org/10.1111/j.1440-1746.2009.06004.x

[3] Mohamadnejad, M., Namiri, M., Bagheri, M., et al. (2007) Phase 1 Human Trial of Autologous Bone Marrow-Hematopoietic Stem Cell Transplantation in Patients with Decompensated Cirrhosis. World Journal of Gastroenterology, 13, 3359-3363. https://doi.org/10.3748/wjg.v13.i24.3359

[4] Nikeghbalian, S., Pournasr, B., Aghdami, N., et al. (2011) Autologous Transplantation of Bone Marrow-Derived Mononuclear and CD133(+) Cells in Patients with Decompensated Cirrhosis. Archives of Iranian Medicine, 14, 12-17.

[5] Mohamadnejad, M., Alimoghaddam, K., Mohyeddin-Bonab, M., et al. (2007) Phase 1 Trial of Autologous Bone Marrow Mesenchymal Stem Cell Transplantation in Patients with Decompensated Liver Cirrhosis. Archives of Iranian Medicine, 10, 459-466.

[6] Caly, W.R., Strauss, E., Carrilho, F.J., Laudanna, A.A. (2003) Different Degrees of Malnutrition and Immunological Alterations According to the Aetiology of Cirrhosis: A Prospective and Sequential Study. Nutrition Journal, 2, 10. https://doi.org/10.1186/1475-2891-2-10

[7] Liu, Q. (2009) Role of Cytokines in the Pathophysiology of Acute-on-Chronic Liver Failure. Blood Purification, 28, 331-341. https://doi.org/10.1159/000232940

[8] Saito, T., Okumoto, K., Haga, H., et al. (2011) Potential Therapeutic Application of Intravenous Autologous Bone Marrow Infusion in Patients with Alcoholic Liver Cirrhosis. Stem Cells and Development, 20, 1503-1510. https://doi.org/10.1089/scd.2011.0074

[9] Couto, B.G., Goldenberg, R.C., da Fonseca, L.M., et al. (2011) Bone Marrow Mononuclear Cell Therapy for Patients with Cirrhosis: A PHASE 1 Study. Liver International, 31, 391-400. https://doi.org/10.1111/j.1478-3231.2010.02424.x

[10] Jang, Y.O., Kim, Y.J., Baik, S.K., et al. (2014) Histological Improvement Following Administration of Autologous Bone Marrow-Derived Mesenchymal Stem Cells for Alcoholic Cirrhosis: A Pilot Study. Liver International, 34, 33-41. https://doi.org/10.1111/liv.12218

[11] Li, W.Y., Jiang, Y.F., Jin, Q.L., et al. (2010) Immunologic Characterization of Posthepatitis Cirrhosis Caused by HBV and HCV Infection. Journal of Biomedicine and Biotechnology, 2010, Article ID: 138237. https://doi.org/10.1155/2010/138237

[12] Salama, H., Zekri, A.R., Bahnassy, A.A., et al. (2010) Autologous CD34+ and CD133+ Stem Cells Transplantation in Patients with End Stage Liver Disease. World Journal of Gastroenterology, 16, 5297-5305. https://doi.org/10.3748/wjg.v16.i42.5297

[13] Cai, D.C., Li, J., Zeng, Y., Li, Y.G. and Ren, H. (2007) A Study on the Anti-HBV Effect of Dendritic Cell from Human Umbilical Cord Blood. Chinese Journal of Hepatology, 15, 88-91.

[14] Xing, T., Li, L., Cao, H. and Huang, J. (2007) Altered Immune Function of Monocytes in Different Stages of Patients with Acute on Chronic Liver Failure. Clinical & Experimental Immunology, 147, 184-188.

[15] Weng, W.Z., Chen, J.F., Mei, Y.Y., et al. (2013) Treatment Effect of Allogeneic Bone Marrow Mesenchymal Stem Cells Transplantation to Patients with Different Phase of HBV Acute-on-Chronic Liver Failure. Journal of Sun Yat-sen University (Medical Science), 34, 422-428. (In Chinese)

[16] Houlihan, D.D. and Newsome, P.N. (2008) Critical Review of Clinical Trials of Bone Marrow Stem Cells in Liver Disease. Gastroenterology, 135, 438-450. https://doi.org/10.1053/j.gastro.2008.05.040

[17] Liu, F., Liu, Z.D., Wu, N., et al. (2009) Transplanted Endothelial Progenitor Cells Ameliorate Carbon Tetrachloride-Induced Liver Cirrhosis in Rats. Liver Transplantation, 15, 1092-1100. https://doi.org/10.1002/lt.21845

[18] Chen, C., Zeng, L., Ding, S. and Xu, K. (2010) Adult Endothelial Progenitor Cells Retain Hematopoiesis Potential. Transplantation Proceedings, 42, 3745-3749. https://doi.org/10.1016/j.transproceed.2010.07.094

[19] Oyagi, S., Hirose, M., Kojima, M., et al. (2006) Therapeutic Effect of Transplanting HGF-Treated Bone Marrow Mesenchymal Cells into CCl4-Injured Rats. Journal of Hepatology, 44, 742-748. https://doi.org/10.1016/j.jhep.2005.10.026

[20] Mohsin, S., Shams, S., Ali Nasir, G., et al. (2011) Enhanced Hepatic Differentiation of Mesenchymal Stem Cells after Pretreatment with Injured Liver Tissue. Differentiation, 81, 42-48. https://doi.org/10.1016/j.diff.2010.08.005

[21] Li, T.Z., Kim, J.H., Cho, H.H., et al. (2010) Therapeutic Potential of Bone-Marrow-Derived Mesenchymal Stem Cells Differentiated with Growth-Factor-Free Coculture Method in Liver-Injured Rats. Tissue Engineering Part A, 16, 2649-2659. https://doi.org/10.1089/ten.tea.2009.0814

[22] Stock, P., Briickner, S., Ebensing, S., et al. (2010) The Generation of Hepatocytes from Mesenchymal Stem Cells and Engraftment into Murine Liver. Nature Protocols, 5, 617-627. https://doi.org/10.1038/nprot.2010.7

[23] Zhou, W., Chen, P.F., Wu, X.L., Jiang, R. and Xu, Y.H. (2012) Effect of Bone Marrow Mesenchymal Stem Cells on Experimental Liver Fibrosis in Rats and Relevant Mechanism. Chinese Journal of Biologicals, 25, 176-180. (In Chinese)

[24] Zhan, Y., Wang, Y., Wei, L., et al. (2006) Differentiation of Hematopoietic stem Cells into Hepatocytes in Liver Fibrosis in Rats. Transplantation Proceedings, 38, 3082-3085. https://doi.org/10.1016/j.transproceed.2006.08.132

[25] Ma, J.X., Yang, L.P., He, Z.J. and Fang, C.H. (2008) Rat Bone Marrow Mesenchymal Stem Cells Induced Hepatocyte-Like Transplantation for Repairing Acute Hepatic Injury. Journal of Clinical Rehabilitative Tissue Engineering Research, 12, 4026-4030. (In Chinese)

[26] Hwang, S., Hong, H.N., Kim, H.S., et al. (2012) Hepatogenic Differentiation of Mesenchymal Stem Cells in a Rat Model of Thioacetamide-Induced Liver Cirrhosis. Cell Biology International, 36, 279-288. https://doi.org/10.1042/CBI20110325

[27] Hardjo, M., Miyazaki, M., Sakaguchi, M., et al. (2009) Suppression of Carbon Tetrachloride-Induced Liver Fibrosis by Transplantation of a Clonal Mesenchymal Stem Cell Line Derived from Rat Bone Marrow. Cell Transplantation, 18, 89-99. https://doi.org/10.3727/096368909788237140

[28] Piryaei, A., Valojerdi, M.R., Shahsavani, M. and Baharvand, H. (2011) Differentiation of Bone Marrow-Derived Mesenchymal Stem Cells into Hepatocyte-Like Cells on Nanofibers and Their Transplantation into a Carbon Tetrachloride-Induced Liver Fibrosis Model. Stem Cell Reviews and Reports, 7, 103-118. https://doi.org/10.1007/s12015-010-9126-5

[29] Caplan, A.I. and Dennis, J.E. (2006) Mesenchymal Stem Cells as Trophic Mediators. Journal of Cellular Biochemistry, 98, 1076-1084. https://doi.org/10.1002/jcb.20886 [30] Harris, V.K., Faroqui, R., Vyshkina, T. and Sadiq, S.A. (2012) Characterization of Autologous Mesenchymal Stem Cell-Derived Neural Progenitors as a Feasible Source of Stem Cells for Central Nervous System Applications in Multiple Sclerosis. Stem Cells Translational Medicine, 1, 536-547. https://doi.org/10.5966/sctm.2012-0015

[31] Abdel-Aziz, M., Abdel-Hamid, S., Wahdan, O., et al. (2012) Phase II Trial: Undifferentiated versus Differentiated Autologous Mesenchymal Stem Cells Transplantation in Egyptian Patients with HCV Induced Liver Cirrhosis. Stem Cell Reviews and Reports, 8, 972-981. https://doi.org/10.1007/s12015-011-9322-y

[32] Zhang, B., Inagaki, M., Jiang, B., et al. (2009) Effects of Bone Marrow and Hepatocyte Transplantation on Liver Injury. Journal of Surgical Research, 157, 71-80. https://doi.org/10.1016/j.jss.2008.12.013

[33] Zagoura, D.S., Roubelakis, M.G., Bitsika, V., et al. (2012) Therapeutic Potential of a Distinct Population of Human Amniotic Fluid Mesenchymal Stem Cells and Their Secreted Molecules in Mice with Acute Hepatic Failure. Gut, 61, 894-906. https://doi.org/10.1136/gutjnl-2011-300908

[34] Wang, S., Yao, P., Gong, L.J., et al. (2009) Effect of Autologous Bone Marrow Mononuclear Cells Transplantation on Serum Cholinesterase in Patients with Liver Cirrhosis of Different Child-Pugh Scores. Chinese Hepatology, 14, 189-193. (In Chinese)

[35] Wang, S., Yao, P., Gong, L.J., et al. (2010) Effect of Autologous Bone Marrow Mononuclear Cells Transplantation on Serum Albumin in Patients with Liver Cirrhosis. Chinese Journal of Gastroenterology and Hepatology, 19, 509-512.

[36] Luo, L.L., Mu, X.L., Heng, L.N. and Lu, X.H. (2008) Human Bone Marrow Mesenchymal Stem Cells Differentiate into Hepatocyte-Like Cells Induced by Serum of Hepatic Injury in Vitro. China Practical Medicine, 3, 3-4. (In Chinese)

[37] Geng, Y.J. (2003) Molecular Mechanisms for Cardiovascular Stem Cell Apoptosis and Growth in the Hearts with Atherosclerotic Coronary Disease and Ischemic Heart Failure. Annals of the New York Academy of Sciences, 1010, 687-697. https://doi.org/10.1196/annals.1299.126

[38] Yang, P.F., Chen, F.R., Guo, J.B., et al. (2005) Therapeutic Effect of Hyperbaric Oxygen on Cirrhotic Patients with Portal Hypertension. Chinese Journal of Gastroenterology and Hepatology, 14, 293-295. (In Chinese)

[39] Yang, F., Qin, B. and Qin, F. (2011) Influence of Artificial Liver Support System on Bone Marrow Stem Cell Differentiation Factors in Patients with Chronic Severe Hepatitis B. Chinese Journal of Infectious Diseases, 29, 674-678. (In Chinese)

[40] Xu, H.Y., Shi, X.L., Chu, X.H. and Ding, Y.T. (2009) Effects on Swine Acute Liver Failure by Combined Therapy of Autologous Mesenchymal Stem Cell Transplantation and Medical Treatment. World Chinese Journal of Digestology, 17, 962-968. (In Chinese) https://doi.org/10.11569/wcjd.v17.i10.962

[41] Tang, X.P., Zheng, X.H. and Yang, X. (2002) Therapeutic Effect of Umbilical Cord Blood Transfusion Combined with Plasma Exchange on Chronic Severe Hepatitis. Bulletin of Hunan Medical University, 27, 323-325. (In Chinese)

[42] Sun, Y., Chi, B.R., Chen, L., Meng, X.W. and Kong, D.X. (2008) Study on Transplantation of Induced Bone Marrow Mesenchymal Stem Cells via Various Route for the Treatment of Chronic Liver Injury. Chinese Journal of Digestion, 28, 171-174. (In Chinese)

[43] Stutchfield, B.M., Forbes, S.J. and Wigmore, S.J. (2010) Prospects for Stem Cell Transplantation in the Treatment of Hepatic Disease. Liver Transplantation, 16, 827-836. https://doi.org/10.1002/lt.22083

[44] Lu, Q.P., Cao, B.Q. and Wei, W. (2011) Different Component of Bone Marrow Cell Transplantation via the Portal Vein Controlled Study of the Treatment of Hepatic Fibrosis in Rats. Chinese Journal of Clinicians, 5, 4992-4996. (In Chinese)

[45] Wei, Y., Nie, Y., Lai, J., Wan, Y.J. and Li, Y. (2009) Comparison of the Population Capacity of Hematopoietic and Mesenchymal Stem Cells in Experimental Colitis Rat Model. Transplantation, 88, 42-48.

[46] Zhou, H.C., Liu, L., Zhou, J. and Tian, J.L. (2011) Comparison the Percentage of Lymphocytes, Monocytes and CD34, CD38 Isoforms Cells between Bone Marrow and Umbilical Cord Blood Stem Cells in Treatment of Liver Cirrhosis. Chinese Journal of Clinicians, 5, 1731-1733. (In Chinese)

[47] Liu, Z.C. and Chang, T.M. (2009) Preliminary Study on Intrasplenic Implantation of Artificial Cell Bioencapsulated Stem Cells to Increase the Survival of 90% Hepatectomized Rats. Artificial Cells, Blood Substitutes, and Biotechnology, 37, 53-55. https://doi.org/10.1080/10731190802663975

[48] Lan, L., Chen, Y., Sun, C., et al. (2008) Transplantation of Bone Marrow-Derived Hepatocyte Stem Cells Transduced with Adenovirus-Mediated IL-10 Gene Reverses Liver Fibrosis in Rats. Transplant International, 21, 581-592. https://doi.org/10.1111/j.1432-2277.2008.00652.x

[49] Alexanian, A.R., Kwok, W.M., Pravdic, D., Maiman, D.J. and Fehlings, M.G. (2010) Survival of Neurally Induced Mesenchymal Cells May Determine Degree of Motor Recovery in Injured Spinal Cord Rats. Restorative Neurology and Neuroscience, 28, 761-767.

[50] Ji, W.J., Qu, Q., Jin, Y., Zhao, L. and He, X.D. (2009) Ursodeoxycholic Acid Inhibits Hepatocyte-Like Cell Apoptosis by Down-Regulating the Expressions of Bax and Caspase-3. National Medical Journal of China, 89, 2997-3001.

[51] Liao, X., AnCheng, J.Y., Zhou, Q.J. and Liao, C. (2013) Therapeutic Effect of Autologous Bone Marrow-Derived Liver Stem Cells Transplantation in Hepatitis B Virus-Induced Liver Cirrhosis. Hepatogastroenterology, 60, 406-409.

[52] Rabani, V., Shahsavani, M., Gharavi, M., et al. (2010) Mesenchymal Stem Cell Infusion Therapy in a Carbon Tetrachloride-Induced Liver Fibrosis Model Affects Matrix Metalloproteinase Expression. Cell Biology International, 34, 601-605. https://doi.org/10.1042/CBI20090386

[53] Liu, L., Zhou, J., Huang, L.W., He, C.P. and Zhou, H.C. (2011) Clinical Efficacy of Lamivudine, Adefovir Dipivoxil Combined with Autologous Bone Marrow Stem Cell Transplantation in Treatment of Hepatitis B Patients with Decompensated Liver Cirrhosis. Chinese Journal of Gastroenterology and Hepatology, 20, 1092-1094. (In Chinese)

[54] Deng, X., Chen, Y.X., Zhang, X., et al. (2008) Hepatic Stellate Cells Modulate the Differentiation of Bone Marrow Mesenchymal Stem Cells into Hepatocyte-Like Cells. Journal of Cellular Physiology, 217, 138-144. https://doi.org/10.1002/jcp.21481

[55] Di Bonzo, L.V., Ferrero, I., Cravanzola, C., et al. (2008) Human Mesenchymal Stem Cells as a Two-Edged Sword in Hepatic Regenerative Medicine: Engraftment and Hepatocyte Differentiation versus Profibrogenic Potential. Gut, 57, 223-231. https://doi.org/10.1136/gut.2006.111617

[56] Ong, S.Y., Dai, H. and Leong, K.W. (2006) Hepatic Differentiation Potential of Commercially Available Human Mesenchymal Stem Cells. Tissue Engineering, 12, 3477-3485. https://doi.org/10.1089/ten.2006.12.3477

[57] Quintanilha, L.F., Mannheimer, E.G., Carvalho, A.B., et al. (2008) Bone Marrow Cell Transplant Does Not Prevent or Reverse Murine Liver Cirrhosis. Cell Transplantation, 17, 943-953. https://doi.org/10.3727/096368908786576453

[58] Carvalho, A.B., Quintanilha, L.F., Dias, J.V., et al. (2008) Bone Marrow Multipotent Mesenchymal Stromal Cells Do Not Reduce Fibrosis or Improve Function in a Rat Model of Severe Chronic Liver Injury. Stem Cells, 26, 1307-1314. https://doi.org/10.1634/stemcells.2007-0941

[59] Pulavendran, S., Rose, C. and Mandal, A.B. (2011) Hepatocyte Growth Factor Incorporated Chitosan Nanoparticles Augment the Differentiation of Stem Cell into Hepatocytes for the Recovery of Liver Cirrhosis in Mice. Journal of Nanobiotechnology, 9, 15. https://doi.org/10.1186/1477-3155-9-15

[60] Pulavendran, S., Vignesh, J. and Rose, C. (2010) Differential Anti-Inflammatory and Anti-Fibrotic Activity of Transplanted Mesenchymal vs. Hematopoietic Stem Cells in Carbon Tetrachloride-Induced Liver Injury in Mice. International Immunopharmacology, 10, 513-519. https://doi.org/10.1016/j.intimp.2010.01.014

[61] Kakinuma, S., Tanaka, Y., Chinzei, R., et al. (2003) Human Umbilical Cord Blood as a Source of Transplantable Hepatic Progenitor Cells. Stem Cells, 21, 217-227. https://doi.org/10.1634/stemcells.21-2-217

[62] Piscaglia, A.C., Zocco, M.A., Di-Campli, C., et al. (2005) How Does Human Stem Cell Therapy Influence Gene Expression after Liver Injury? Microarray Evaluation on a Rat Model. Digestive and Liver Disease, 37, 952-963. https://doi.org/10.1016/j.dld.2005.06.012

[63] Henning, R.J., Aufman, J., Shariff, M., et al. (2010) Human Umbilical Cord Blood Mononuclear Cells Decrease Fibrosis and Increase Cardiac Function in Cardiomyopathy. Regenerative Medicine, 5, 45-54. https://doi.org/10.2217/rme.09.71

[64] Elkhafif, N., El-Baz, H., Hammam, O., et al. (2011) CD133+ Human Umbilical Cord Blood Stem Cells Enhance Angiogenesis in Experimental Chronic Hepatic Fibrosis. APMIS, 119, 66-75. https://doi.org/10.1111/j.1600-0463.2010.02693.x

[65] Wulf-Goldenberg, A., Eckert, K. and Fichtner, I. (2011) Intrahepatically Transplanted Human Cord Blood Cells Reduce SW480 Tumor Growth in the Presence of Bispecific EpCAM/CD3 Antibody. Cytotherapy, 13, 108-113. https://doi.org/10.3109/14653249.2010.515577

[66] Wang, Z.Y., Zhu, Y.J., Wang, Z. and Du, B. (2012) Therapeutic Effect of Umbilical Cord Blood Stem Cells Transplantation for Rat Model of Decompensated Cirrhosis. National Medical Frontiers of China, 7, 4-6. (In Chinese)

[67] Mehanna, R.A., Habachy, N.M., Sharara, G.M., et al. (2012) Transplantation of Human Umbilical Cord Blood Stem Cells in Rabbits' Fibrotic Liver. Journal of American Science, 8, 83-94.

[68] Bassiouny, A.R., Zaky, A.Z., Abdulmalek, S.A., et al. (2011) Modulation of AP-Endonuclease1 Levels Associated with Hepatic Cirrhosis in Rat Model Treated with Human Umbilical Cord Blood Mononuclear Stem Cells. International Journal of Clinical and Experimental Pathology, 4, 692-707.

[69] Sáez-Lara, M.J., Frecha, C., Martin, F., et al. (2006) Transplantation of Human CD34+ Stem Cells from Umbilical Cord Blood to Rats with Thioacetamide-Induced Liver Cirrhosis. Xenotransplantation, 13, 529-535. https://doi.org/10.1111/j.1399-3089.2006.00344.x

[70] Álvarez-Mercado, A.I., García-Mediavilla, M.V., Sánchez-Campos, S., et al. (2009) Deleterious Effect of Human Umbilical Cord Blood Mononuclear Cell Transplantation on Thioacetamide-Induced Chronic Liver Damage in Rats. Cell Transplantation, 18, 1069-1079. https://doi.org/10.3727/096368909X12483162197088

[71] Yang, H.L., Yu, F.T., Li, G.J., et al. (2010) Umbilical Cord Blood Stem Cell Transplantation for 86 Patients with Decompensated Cirrhosis. Shandong Medical Journal, 50, 16-17. (In Chinese)

[72] Zhang, L.X., Xing, L.H., Zhang, L.L., et al. (2010) Umbilical Cord Blood Stem Cell Transplantation in Treatment of Decompeusated Cirrhosis: A Preliminary Clinical Observation. Chinese General Practice, 13, 2680-2682. (In Chinese)

[73] Hu, X.X., Chen, H.O., Qian, L., et al. (2011) Umbilical Cord Blood Stem Cell Transplantation via Hepatic Artery in Treatment of Liver Cirrhosis. Chinese Journal of General Practitioners, 10, 58-60. (In Chinese)

[74] Xu, S. and Wang F. (2011) Umbilical Cord Blood Stem Cell Transplantation in Treatment of Decompensated Cirrhosis Complicated with Hepatic Diabetes: A Case Report. Journal of Clinical Hepatology, 14, 306-311. (In Chinese)

[75] Li, Q. and Li, C.H. (2010) Perioperative Nursing for a Patient with Wilson Disease in Unrelated Cord Blood Stem Cell Transplantation. Chongqing Medicine, 39, 3448-3449. (In Chinese)

[76] Li, C.Y., Zhao, J.L., Zhang, L., et al. (2010) Efficacy Comparison between Umbilical Cord Blood and Autologous Peripheral Blood Stem Cell Transplantation in the Treatment of Ascitic Cirrhosis. Chinese Journal of Blood Transfusion, 23, 182-184. (In Chinese)

[77] Sakuragawa, N., Enosawa, S., Ishii, T., et al. (2000) Human Amniotic Epithelial Cells Are Promising Transgene Carriers for Allogeneic cell Transplantation into Liver. Journal of Human Genetics, 45, 171-176. https://doi.org/10.1007/s100380050205

[78] Davila, J.C., Cezar, G.G., Thiede, M., et al. (2004) Use and Application of Stem Cells in Toxicology. Toxicological Sciences, 79, 214-223. https://doi.org/10.1093/toxsci/kfh100

[79] Takashima, S., Ise, H., Zhao, P., Akaike, T. and Nikaido, T. (2004) Human Amniotic Epithelial Cells Possess Hepatocyte-Like Characteristics and Functions. Cell Structure and Function, 29, 73-84. https://doi.org/10.1247/csf.29.73

[80] Marongiu, F., Gramignoli, R., Dorko, K., et al. (2011) Hepatic Differentiation of Amniotic Epithelial Cells. Hepatology, 53, 1719-1729. https://doi.org/10.1002/hep.24255

[81] Moritoki, Y., Ueno, Y., Kanno, N., et al. (2007) Amniotic Epithelial Cell-Derived Cholangiocytes in Experimental Cholestatic Ductal Hyperplasia. Hepatology Research, 37, 286-294. https://doi.org/10.1111/j.1872-034X.2007.00049.x

[82] Miki, T. (2011) Amnion-Derived Stem Cells: In Quest of Clinical Applications.

Stem Cell Research & Therapy, 2, 25. https://doi.org/10.1186/scrt66

[83] Tee, J.Y., Vaghjiani, V., Liu, Y.H., et al. (2013) Immunogenicity and Immunomodulatory Properties of Hepatocyte-Like Cells Derived from Human Amniotic Epithelial Cells. Current Stem Cell Research & Therapy, 8, 91-99. https://doi.org/10.2174/1574888X11308010011

[84] Manuelpillai, U., Moodley, Y., Borlongan, C.V. and Parolini, O. (2011) Amniotic Membrane and Amniotic Cells: Potential Therapeutic Tools to Combat Tissue Inflammation and Fibrosis? Placenta, 32, S320-S325. https://doi.org/10.1016/j.placenta.2011.04.010

[85] Hong, S.B., Seo, M.S., Park, S.B., et al. (2012) Therapeutic Effects of Human Amniotic Epithelial Stem Cells in Niemann-Pick Type C1 Mice. Cytotherapy, 14, 630-638. https://doi.org/10.3109/14653249.2012.663485

[86] Sant'Anna, L.B., Cargnoni, A., Ressel, L., Vanosi, G. and Parolini, O. (2011) Amniotic Membrane Application Reduces Liver Fibrosis in a Bile Duct Ligation Rat Model. Cell Transplantation, 20, 441-453. https://doi.org/10.3727/096368910X522252

[87] Manuelpillai, U., Tchongue, J., Lourensz, D., et al. (2010) Transplantation of Human Amnion Epithelial Cells Reduces Hepatic Fibrosis in Immunocompetent CCl4-Treated Mice. Cell Transplantation, 19, 1157-1168. https://doi.org/10.3727/096368910X504496

Abbreviations

UCBSCs: human umbilical cord blood stem cells hAECs: human amniotic epithelial cells

HSCs: Hematopoietic stem cells MSCs: mesenchymal stem cells ACLF: acute-on-chronic liver failure

BMSCs: bone marrow mesenchymal stem cells ASCs: adipose-derived mesenchymal stem cells EPCs: endothelial progenitor cells

ALB: serum albumin levels HGF: hepatocyte growth factor FGF: fibroblast growth factors IL-10: interleukin-10

EGF: epidermal growth factor

BDLSCs: bone marrow-derived liver stem cells UDCA: ursodeoxycholic acid

BMNCs: bone marrow mononuclear cells UCBMNCs: umbilical cord blood mononuclear cells PBSCs: peripheral blood stem cells

Permissions

All chapters in this book were first published in OJGAS, by Scientific Research Publishing; hereby published with permission under the Creative Commons Attribution License or equivalent. Every chapter published in this book has been scrutinized by our experts. Their significance has been extensively debated. The topics covered herein carry significant findings which will fuel the growth of the discipline. They may even be implemented as practical applications or may be referred to as a beginning point for another development.

The contributors of this book come from diverse backgrounds, making this book a truly international effort. This book will bring forth new frontiers with its revolutionizing research information and detailed analysis of the nascent developments around the world.

We would like to thank all the contributing authors for lending their expertise to make the book truly unique. They have played a crucial role in the development of this book. Without their invaluable contributions this book wouldn't have been possible. They have made vital efforts to compile up to date information on the varied aspects of this subject to make this book a valuable addition to the collection of many professionals and students.

This book was conceptualized with the vision of imparting up-to-date information and advanced data in this field. To ensure the same, a matchless editorial board was set up. Every individual on the board went through rigorous rounds of assessment to prove their worth. After which they invested a large part of their time researching and compiling the most relevant data for our readers.

The editorial board has been involved in producing this book since its inception. They have spent rigorous hours researching and exploring the diverse topics which have resulted in the successful publishing of this book. They have passed on their knowledge of decades through this book. To expedite this challenging task, the publisher supported the team at every step. A small team of assistant editors was also appointed to further simplify the editing procedure and attain best results for the readers.

Apart from the editorial board, the designing team has also invested a significant amount of their time in understanding the subject and creating the most relevant covers. They scrutinized every image to scout for the most suitable representation of the subject and create an appropriate cover for the book.

The publishing team has been an ardent support to the editorial, designing and production team. Their endless efforts to recruit the best for this project, has resulted in the accomplishment of this book. They are a veteran in the field of academics and their pool of knowledge is as vast as their experience in printing. Their expertise and guidance has proved useful at every step. Their uncompromising quality standards have made this book an exceptional effort. Their encouragement from time to time has been an inspiration for everyone.

The publisher and the editorial board hope that this book will prove to be a valuable piece of knowledge for researchers, students, practitioners and scholars across the globe.

List of Contributors

Emiddio Barletta, Lucia Cannella, Vincenza Tinessa, Domenico Germano and Bruno Daniele
Department of Medical Oncology, G. Rummo Hospital, Benevento, Italy

Yunfu Lv
Department of General Surgery, People's Hospital of Hainan Province, Haikou, China

Hasan Sedeek Mahmoud and Shamardan Ezz El-Din S. Bazeed
Department of Tropical Medicine and Gastroenterology, Qena Faculty of Medicine, South Valley University, Qena, Egypt

Gilmar Amorim de Sousa
The Integrated Department of Medicine, Federal University of Rio Grande do Norte, UFRN, Natal, Brazil

Iris do Céu Clara Costa, Dyego Leandro Bezerra de Souza and Fabia Barbosa de Andrade
The Post-Graduate Program in Collective Health of UFRN, Natal, Brazil

Lívia Medeiros Soares Celani
Gastroenterology Program of the University Hospital Onofre Lopes/UFRN, Natal, Brazil

Ranna Santos Pessoa, Marlon César de Souza Filho, Daniel Fernandes Mello de Oliveira, Luana Lopes de Medeiros, Lucila Samara Dantas de Oliveira and Maria Flávia Monteiro
The Medical School of UFRN, Natal, Brazil

Patrick P. Basu, Mark M. Aloysius and Robert S. Brown
Columbia University College of Physicians & Surgeons, New York, USA

Patrick P. Basu and Mark M. Aloysius
King's County Hospital, New York, USA

Niraj James Shah
James J. Peters VA Medical Center, New York, USA

Lifen Hu, Xihai Xu, Zhongsong Zhou, Guoshen Chen, Huafa Ying, Ying Ye and Jiabin Li
Department of Infectious Diseases, the First Affiliated Hospital of Anhui Medical University, Hefei, China

Lifen Hu
Department of Center Laboratory, the First Hospital of Anhui Medical University, Hefei, China

Jiabin Li
Department of Infectious Diseases, the Affiliated Chaohu Hospital of Anhui Medical University, Chaohu, China

Iliass Charif, Kaoutar Saada, Ihssane Mellouki, Mounia El Yousfi, Dafrallah Benajah, Mohamed El Abkari, Adil Ibrahimi and Nourdin Aqodad
Department of Gastroenterology and Hepatology, Hassan II University Hospital, Fez, Morocco
Faculty of Medicine and Pharmacy of Fez, Sidi Mohammed Ben Abdellah University of Fez, Fez, Morocco

Rym Ennaifer, Myriam Cheikh, Rania Hefaiedh, Hayfa Romdhane, Houda Ben Nejma and Najet Bel Hadj
Department of Hepato-Gastro-Enterology, Mongi Slim Universitary Hospital, Tunis, Tunisia
Faculty of Medicine, University of Tunis El Manar, Tunis, Tunisia

Naoki Hotta
Department of Internal Medicine, Division of Hepatology, Masuko Memorial Hospital, Aichi, Japan

Helen Ngo and Raymund Gantioque
Patricia A. Chin School of Nursing, California State University, Los Angeles, USA

Doffou Adjeka Stanislas, Kouame Hardryt Dimitri, Bangoura Demba, Kissi Anzouan-Kacou and Attia Koffi Alain
Gastrointestinal Unit, Cocody Teaching Hospital Center, Abidjan, Côte d'Ivoire

Assi Constant, Ndjitoyap Ndam Antonin Wilson, Ouattara Amadou and Lohoues-Kouacou Marie-Jeanne
Gastrointestinal Unit, Yopougon

Ashraf A. Hammam
Internal Medicine Department, Zagazig University, Zagazig, Egypt

Amal A. Jouda
Tropical Medicine Department, Zagazig University, Zagazig, Egypt

Mona E. Hashem
Clinical Pathology Department, Zagazig University, Zagazig, Egypt

Fakhar Ali Qazi Arisar and Syed Hasnain Ali Shah
Section of Gastroenterology, Department of Medicine, The Aga Khan University Hospital, Karachi, Pakistan

Tanveer Ul Haq
Section of Interventional Radiology, Department of Radiology, The Aga Khan University Hospital, Karachi, Pakistan

Comlan Albert Dovonou, Cossi Adébayo Alassani, Kadidjatou Sake, Cossi Angelo Attinsounon and Agossou Romaric Tandjiekpon
Medicine Department and Medical Specialities, Medical Faculty, Parakou University, Parakou, Benin

Comlan Albert Dovonou and Cossi Angelo Attinsounon
Internal Medicine Department, Departmental Hospital Center of Borgou, Parakou, Benin

Angèle Azon-Kouanou, Djimon Marcel Zannou and Fabien Houngbe
Medicine Department and Medical Specialities of Cotonou Health and Science Faculty, Cotonou, Benin

Mamert Fulgence Yao Bathaix, Akelesso Bagny, Kouamé Alassane Mahassadi, Anassé Jean-Baptiste Okon, Ya Henriette Kissi-Anzouan, Stanislas Doffou, Aboubacar Demba Bangoura, Hatrydt Dimitri Kouamé, Kadiatou Diallo, Antonin N'Dam, Aoudi Ousman De, Koffi Alain Attia and Aya Thérèse N'dri Yoman
Department of Hepatology and Gastroenterology of the Centre Hospitalier Universitaire de Yopougon (CHU-Y), Abidjan, Côte d'Ivoire

Andrew Villion, Michael Lishner, Michal Chowers and Sharon Reisfeld
Sackler School of Medicine, Tel Aviv University, Tel Aviv, Israel

Michael Lishners and Sharon Reisfeld
Department of Medicine A, Meir Medical Center, Kfar Saba, Israel

Michal Chowers
Infectious Diseases Unit, Meir Medical Center, Kfar Saba, Israel

Mona A. Amin, Ahmed E. El-Badry, May M. Fawzi and Shorouk M. Moussa
Internal Medicine, Cairo University, Cairo, Egypt

Dalia A. Muhammed
Community Medicine, Cairo University, Cairo, Egypt

Ryuta Washio, Masaya Takahashi, Sohsaku Yamanouchi, Masato Hirabayashi, Kenji Mine, Yukihiro Noda, Eriko Kanda, Atsushi Ohashi, Hirohide Kawasaki and Kazunari Kaneko
Department of Pediatrics, Kansai Medical University, Osaka, Japan

Fernanda Raphael Escobar Gimenes, Renata Karina Reis and Emília Campos de Carvalho
Department of General and Specialized Nursing, University of São Paulo at Ribeirão Preto College of Nursing, Ribeirão Preto, Brazil

Patrícia Costa dos Santos da Silva
Federal University of Uberlândia, Uberlândia, Brazil

Andréia Regina Lopes
Academy of the Brazilian Air Force, Pirassununga, Brazil

Rebecca Shasanmi
Nursing and Public Health Research, Philadelphia, PA, USA

Elham Ahmed Hassan and Abeer Sharaf El-Din Abd El-Rehim
Department of Tropical Medicine and Gastroenterology, Faculty of Medicine, Assiut University, Assiut, Egypt

Zain El-Abdeen Ahmed Sayed and Ahmed Mohmmed Ashmawy
Department of Internal Medicine, Faculty of Medicine, Assiut University, Assiut, Egypt

Emad Farah Mohamed Kholef
Department of Clinical pathology, Faculty of Medicine, Aswan University, Aswan, Egypt

Abeer Sabry
Department of Internal Medicine, Faculty of Medicine, Helwan University, Cairo, Egypt

Wael Abd-Elgwad Elsewify
Department of Internal Medicine, Faculty of Medicine, Aswan University, Aswan, Egypt

Tsevi Yawovi Mawufemo
Nephrology and Dialysis Department of Teaching Hospital Sylvanus Olympio, University of Lomé, Lomé, Togo

Tsevi Yawovi Mawufemo, Lagou Amélie Delphine, Tia Weu Mélanie, Coulibaly Pessa Nawo Albert, Cherif Ibrahima, Guei Monlet Cyr and Ackoundou-N'Guessan Kan Clément
Néphrology, Hémodialysis and Organ Transplant Departement of Yopougon Teaching Hospital, University Felix Houphouet Boigny, Abidjan, Cote d'Ivoire

Lantam Sonhaye, Abarchi Habibou Boube, Abdoulatif Amadou, Bérésa Kolou, Mazamaesso Tchaou, Nouhou Mamoudou Garba, Lama Kègdigoma Agoda-Koussema and Komlavi Adjenou
Department of Radiology, University Teaching Hospital of Lomé, Lomé, Togo

Mohaman Djibril
Department of Internal Medicine, University Teaching Hospital of Lomé, Lomé, Togo

Bidamin Ntimon
Department of Radiology, University Teaching Hospital of Kara, Kara, Togo

Ula M. Al-Jarhi and Abeer Awad
Department of Medicine, Cairo University, Cairo, Egypt

Mona Mohsen
Department of Chemical Pathology, Cairo University, Cairo, Egypt

Fan Chen
Department of Gastroenterology, Fuzhou General Hospital, Fuzhou, China

Index

www.ingramcontent.com/pod-product-compliance
Lightning Source LLC
Chambersburg PA
CBHW082013190326
41458CB00010B/3178